Radar Sensor Engineering

Radar Sensor Engineering

Christian G. Bachman
KRYBOK Associates

LexingtonBooks
D.C. Heath and Company
Lexington, Massachusetts
Toronto

Library of Congress Cataloging in Publication Data

Bachman, Christian G., 1931–
  Radar sensor engineering.

  Includes bibliographical references and index.
  1. Radar. I. Title.
  TK6580.B25   621.3848'1   81-48004
  ISBN 0-669-05233-7   AACR2

Copyright © 1982 by D.C. Heath and Company

All rights reserved. No part of this publication may be reproduced or transmitted in any form or by any means, electronic or mechanical, including photocopy, recording, or any information storage or retrieval system, without permission in writing from the publisher.

Published simultaneously in Canada

Printed in the United States of America

International Standard Book Number: 0-669-05233-7

Library of Congress Catalog Card Number: 81-48004

Contents

	Figures	vii
	Tables	xi
	Acknowledgments	xiii
	Introduction	xv
Part I	Radar-Performance Prediction	1
Chapter 1	**The Radar Range Equation**	3
	Waveform Energy	4
	Antenna Transmit Gain	4
	Antenna Receive Gain	5
	Target Cross-Section	5
	Propagation Factor	6
	Signal-to-Noise Ratio	6
	Noise-Power Density	14
	Radar-Loss Factor	18
Chapter 2	**Radar Resolution and Accuracy**	35
	Angular Accuracy	35
	Angular Resolution	38
	Range-Measurement Accuracy	39
	Range Resolution	40
	Doppler-Measurement Accuracy	41
	Doppler Resolution	42
Chapter 3	**Radar Performance in Clutter**	43
	MTI Operation and Performance Limits	43
	Pulse-Doppler Operation and Performance Limits	82
	Computer Simulation of Performance in Clutter	103
Chapter 4	**Radar Performance in an ECM Environment**	111
	Performance in the Presence of Barrage Jamming	111
	Performance in the Presence of Pulsed Repeater Jammers	112

	References: Part I	119
Part II	*Radar-Target Detection*	121
Chapter 5	**Radar Cross-Section Models in Detection**	123
	Introduction	123
	Discussion	123
	RCS-Averaging Concepts	126
Chapter 6	**Required S/N per Pulse *versus* Detection Probability**	133
	Appendix 6A: Approximate Formula for n'	140
	Appendix 6B: No Target Signal: False-Alarm Probability	141
	Appendix 6C: Nonfluctuating Target Signal	142
	Appendix 6D: Swerling Case 1	143
	Appendix 6E: Swerling Case 2	145
	Appendix 6F: Swerling Case 3	146
	Appendix 6G: Swerling Case 4	148
	Appendix 6H: Chi-Square Family	149
	Appendix 6I: Confluent Hypergeometric Functions	151
	Appendix: Radar Detection Tables	153
	Index	281
	About the Author	285

Figures

1–1	Probability of Detection for a Sine Wave in Noise as a Function of the Signal-to-Noise (Power) Ratio and the Probability of False Alarm	8
1–2	Additional Signal-to-Noise Ratio Required to Achieve a Particular Probability of Detection	10
1–3	Integration-Improvement Factor as a Function of the Number of Pulses Integrated for the Five Cases of Target Fluctuations Considered ($P_{FA} = 10^{-8}$)	11
1–4	Comparison of Operator Integration on a PPI Display with a Perfect Video Integrator	13
1–5	R/R_1 versus Delta	15
1–6	R/R_1 versus Delta	15
1–7	R/R_1 versus Delta (Case I Swerling)	16
1–8	R/R_1 versus Delta (Case I Swerling)	16
1–9	R/R_1 versus Delta (Case II Swerling)	17
1–10	R/R_1 versus Delta (Case III Swerling)	17
1–11	Antenna-Noise Temperature for Typical Conditions of Cosmic, Solar, Atmospheric, and Ground Noise	18
1–12	Scanning Loss versus Scan Speed	19
1–13	Typical Pencil-Beam Rasters	21
1–14	Pattern-Loss Factor for Fan-Beam Radars	22
1–15	Pattern-Loss Factors for Pencil-Beam Radars	23
1–16	Comparison of Power Averaging with Statistical Averaging	24
1–17	Collapsing Loss versus Collapsing Ratio	25
1–18	Effect of Bandwidth and Passband Shape, Calculated for Several Pulse Shapes, Assuming Automatized Detection	26
1–19	Range-Gate Straddling Loss	27
1–20	Filter-Spacing Loss	27

1–21	Radar Atmospheric Attenuation for Different Elevation Angles	28
1–22	Radar Attenuation for Transversal of Entire Troposphere at Various Elevation Angles, Applicable for Targets outside the Troposphere	31
1–23	Theoretical Values of Attenuation in Rain and Fog	32
2–1	σ_{min} as a Function of S/N, N, and β for $f(u) = e^{-u}$	36
2–2	Error Slope and Crossover Loss	39
3–1	Phase-Processing MTI System	44
3–2	Target Plus Clutter-Signal Characteristics	44
3–3	Power Response of Single-Delay-Line Phase Processor	46
3–4	Envelope-Processing MTI System	47
3–5	Vector-Processing MTI	48
3–6	An Example of a Clutter-Locking Circuit	49
3–7	Multiple-Delay-Line Cancelers	50
3–8	Canceler Configuration Useful in Staggered-PRF Systems	51
3–9	Relative-Power Response Curves for Staggered-PRF System	52
3–10	Basic Digital MTI Canceler	52
3–11	Clutter Improvement for a Single-Canceler System as Limited by the Spectrum Width of Clutter	54
3–12	Clutter Improvement for a Double-Canceler System as Limited by the Spectrum Width of Clutter	55
3–13	Loss in Clutter Improvement Due to an Average-Velocity Component of the Clutter Spectrum	56
3–14	Velocity Spread for Coherently Detected Sea-Clutter Signals	57
3–15	Variation of Mean Doppler with Wind Speed	58
3–16	Single-Canceler Clutter Improvement as Limited by Sea Clutter	59

Figures

3-17	Double-Canceler Clutter Improvement as Limited by Sea Clutter	60
3-18	Standard Deviation of Doppler Spectrum for Precipitation as a Function of Azimuth Angle	61
3-19	Model for Wind Speed as a Function of Altitude	62
3-20	Single-Canceler Clutter Improvement as Limited by Precipitation Clutter—Clutter Locked	63
3-21	Double-Canceler Clutter Improvement as Limited by Precipitation Clutter—Clutter Locked	65
3-22	Single-Canceler Clutter Improvement as Limited by Precipitation—No Clutter Locking	66
3-23	Double-Canceler Clutter Improvement as Limited by Precipitation—No Clutter Locking	67
3-24	Frequency Spectra of Various Types of Fixed Targets	68
3-25	Effect of Internal Fluctuations on Clutter Attenuation	68
3-26	Environmental Diagram for Air-Defense Radar	69
3-27	Normalized Standard Deviation of Clutter Spectrum Due to Antenna Scanning	71
3-28	Loss in Clutter Attenuation for Staggered-PRF System Compared with Unstaggered Double Canceler	72
3-29	Limitation on Clutter Improvement Due to Second-Time-Around Clutter for Double Canceler	72
3-30	Area Illumination for Pulse Radar	75
3-31	Volume Illumination for Pulse Radar	76
3-32	Variation of σ^0 for Sea Echo as a Function of Grazing Angle and Frequency	77
3-33	Variation of σ^0 for Sea Echo as a Function of Grazing Angle at Low (Small Angles)	77
3-34	Normalized Sea-Clutter Cross-Section *versus* Grazing Angle for L-, S-, and C-Bands	78
3-35	Radar Reflectivity *versus* Precipitation Rate	79

3–36	Backscatter Coefficients as a Function of Grazing Angle	80
3–37	Probability Distribution of Land Cross-Section per Unit Area, σ_0	81
3–38	Unambiguous Velocity *versus* Unambiguous Range	84
3–39	Ambiguity Diagram for Uniform Pulse Train	85
3–40	Envelope of the Correlation Function for a Nonuniformly Spaced Pulse Train of N Pulses	87
3–41	Pulse-Doppler Point-Clutter Improvement Factor	87
3–42	Detection Efficiency *versus* Relative Clutter Suppression with a 21-Pulse Train	89
3–43	Power Spectra at Target Range for 17 Pulse Transmissions	90
3–44	Pulse-Doppler Waveform and Basic Receiver Block Diagram	91
3–45	Spectrum of Returns from a Coherent Pulse Train	92
3–46	(a) Integration-Improvement Factor, Square-Law Detector, P_d = Probability of Detection, n_f = False-Alarm Number; (b) Integration Loss as a Function of n, the Number of Pulses Integrated P_d, and n_f	93
3–47	Zero *IF* (Homodyne) Pulse-Doppler Receiver	94
3–48	Tapped Delay-Line Receiver for Coherent Pulse Train	94
3–49	Comb-Filter Processors	96
3–50	Clutter Rings	104
3–51	Effective Clutter Spectral Densities Due to One Sideband	104
3–52	Total Spectral Density of Clutter	104
3–53	Clutter Rejection by MTI	105
3–54	Target Visibility—Antenna Average Sidelobes = 100 dB	110

Tables

1–1	Effects of Amplitude Taper on Antenna Parameters	5
3–1	Target Spectra	62
3–2	Characteristics of Clutter Spectra	64
3–3	System Limitations for Example MTI	74
3–4	Ratio of Target Response to Doppler-Shifted Interference	86
3–5	Optimum Number of Uniformly Spaced Pulses for Extensive Clutter	101
3–6	Optimum Improvement Factor I (in dB) for Uniformly Spaced Pulses and Extensive Clutter	102
3–7	System Parameters	107
3–8	Clutter Geometry	108
3–9	Doppler Processor Outputs	109
3–10	Doppler-Processor Output Summary	109
4–1	ECCM against Repeaters	114
4–2	ECCM against Noise Jamming	117
5–1	RCS Sample Data Set (Amplitude Only)	126

Acknowlegments

My wife and family have my gratitude for their understanding during manuscript preparation.

Members of the technical staff of the Ground Systems Division of Hughes Aircraft provided much material extracted from original government reports and from the open engineering literature, which formed a large portion of this book. The radar-detection probability tables were generated by Murray F. Kaye of Raytheon Company and were made available during his graduate state-of-the-art seminar on radar-systems analysis at the Raytheon Institute. His contributions to my awareness and understanding of this complex subject are gratefully acknowledged.

This book is dedicated to the U.S. aerospace and electronics systems engineers, whose experience and expertise constitute a vital national resource, and who should be treated accordingly.

Introduction

This book presents a compilation of analytical techniques most often needed by the engineer who is evaluating the performance and operational capabilities of pulse radars. The intent is to provide a complete and consistent set of equations and definitions that will allow persons who are generally familiar with radar theory to arrive at reasonable and consistent estimates of radar performance in a variety of natural and man-made environmental conditions. No attempt is made to show the derivation. Text is included to explain the meaning of the various parameters and loss factors that make up the performance equations. From this complete set the reader can select those factors that apply to the particular radar being analyzed. Methods for arriving at proper values for the radar parameters and loss factors are described by means of equation or graphs.

Chapter 1 deals with the classical radar range equation applicable to radars operating in a benign environment that includes no interference or clutter. Because this equation is the one most commonly used to characterize radar performance, this book emphasizes complete and accurate descriptions of its parameters. Some of the loss factors shown do not apply to all radars, and certain others will have negligible values in many cases. The analyst should consider each factor carefully to determine whether it applies to the case at hand.

Chapter 2 considers the inherent measurement accuracy and resolution of radars. The intent is to place bounds on radar performance based on the theoretical limits imposed by system noise. Very few radars can achieve these performance levels as a consequence of compromises in their implementation. Because of the wide variety of approaches to measurement and resolution incorporated in current radars, it is not feasible to attempt further detail in this limited study.

Chapter 3 illustrates the effects of land, sea, and weather clutter on radar performance and shows the effectiveness of clutter-discrimination techniques. The two most common approaches to clutter discrimination, moving-target indication (MTI) and pulse-doppler processing, are treated. Equations are included that allow the analyst to compute clutter-rejection performance as limited by clutter fluctuation and by nonideal radar-equipment limitations. The effects of various forms of radar instabilities and incoherencies are illustrated by equations and graphs.

Chapter 4 provides an overview of radar performance in the presence of barrage-type electronic countermeasures (ECM). The effects of more sophisticated ECM techniques must be evaluated for each particular radar implementation; their inclusion here is therefore limited to a general discussion.

The intent of this section is simply to delineate appropriate equations and to describe their application to performance analysis.

Inevitably, the reader will encounter problems in analyses that are not covered herein. In most cases, however, consistent application of these methods will result in performance estimates that are reasonable, repeatable, and supportable.

**Part I
Radar-Performance
Prediction**

1 The Radar Range Equation

A good approximation of the maximum range at which a radar can successfully detect or track a target in a benign environment is provided by the classical radar range equation.

$$R = \left[\frac{EG_t G_r \lambda^2 \sigma_t F^4}{(4\pi)^3 (S/N) k T_{NI} L} \right]^{1/4}, \quad (1.1)$$

where E = transmitted waveform energy (joules);

G_t = antenna transmit power gain;

G_r = antenna receive power gain;

λ = wavelength (meters);

σ_t = target effective cross-section (square meters);

F = propagation factor;

S/N = minimum required signal-to-noise ratio;

K = Boltzmann's constant = 1.38×10^{-23} joules/degree Kelvin;

T_{NI} = effective input-noise temperature (degrees Kelvin);

L = total radar losses;

R = range in meters.

Consistent and enlightened application of equation 1.1 will produce answers that are sufficiently accurate for most purposes. Certain refinements are possible when all radar parameters and target characteristics are accurately known. However, more precise performance calculations usually involve a degree of complexity that virtually necessitates the use of computer simulations. Those factors that are better represented by computer simulations are noted in the detailed discussions that follow.

Waveform Energy

Use of the waveform-energy term (E) automatically accounts for such factors as pulse compression, multiple-pulse waveforms, and nonrectangular pulses. In the general case it is necessary to integrate the transmitter power output over the waveform period. In most cases it is adequate simply to form the product.

$$E = P_p \tau N, \qquad (1.2)$$

where P_p = peak transmitter power;

τ = half-power pulse width;

N = number of pulses in the waveform.

E is measured at the transmitter output port; all losses that occur beyond that point are included in the loss term L. Note that N is the number of pulses in a single cycle of a repetitive waveform. N does not include multiple hits seen by a scanning radar; these are accounted for in computing S/N.

Antenna Transmit Gain

Antenna transmit gain (G_t) is conventionally computed as follows:

$$G = \frac{4\pi A \eta}{\lambda^2}, \qquad (1.3)$$

where A = projected aperture area (square meters);

η = efficiency factor.

The efficiency factor includes the effect of amplitude taper, array blockage, spillover losses, mismatch to free space, mechanical errors, and phase errors. η does not include ohmic losses associated with antenna microwave components such as rotary joints or feed lines, which are included in L.

Table 1–1 illustrates the effect of several possible amplitude tapers on gain and beam shape. Note that the gain factor represents only one plane; some antennas use different tapers in the two planes. The product of two gain factors constitutes the efficiency factor η.

It should be noted that the foregoing relationships are valid for reflector antennas and for array antennas composed of elements having approximately unity gain. If the elements of an array individually provide significant gain,

The Radar Range Equation

Table 1-1
Effects of Amplitude Taper on Antenna Parameters

Aperture Distribution	Aperture Efficiency ($\sqrt{\eta}$)	Half-Power Beamwidth (Degrees)	Highest Sidelobe Level (dB below Peak)
Uniform	1.00	50.4 λ/D	13.2
$\cos \theta$	0.81	68.7 λ/D	23.0
$\cos \theta + 0.1$	0.85	65.2 λ/D	22.8
$\cos \theta + 0.4$	0.92	59.5 λ/D	20.5
$\cos \theta + 0.7$	0.95	57.3 λ/D	17.7
$\cos \theta + 1.0$	0.97	57.1 λ/D	17.7
$\cos^2 \theta$	0.67	83.0 λ/D	32.0
$\cos^2 \theta + 0.08$	0.74	75.1 λ/D	42.0
$\cos^2 \theta + 0.2$	0.79	68.6 λ/D	33.9
$\cos^2 \theta + 0.3$	0.83	68.5 λ/D	31.0
$\cos^2 \theta + 0.4$	0.86	63.2 λ/D	29.3
$\cos^2 \theta + 0.5$	0.89	61.5 λ/D	28.0
$\cos^2 \theta + 0.6$	0.91	60.5 λ/D	26.9
$\cos^3 \theta$	0.575	95.1 λ/D	40.0
$\cos^4 \theta$	0.515	111.6 λ/D	48.0
Taylor 30 dB, $n=4$	0.85	68.2 λ/D	30.0
Taylor 25 dB, $n=4$	0.90	63.2 λ/D	25.0
Taylor 35 dB, $n=5$	0.81	73.2 λ/D	35.9
Taylor 40 dB, $n=6$	0.77	76.8 λ/D	40.0
$1 - X (X < 1)$	0.75	73.4 λ/D	26.4
$1 - (1 - 0.8) X^2 \; (X < 1)$	0.994	52.8 λ/D	15.8
$1 - (1 » 0.5) X^2 (X < 1)$	0.97	55.6 λ/D	17.1
$1 - X^2 \; (X < 1)$	0.83	65.9 λ/D	20.6

as illustrated by an array of Yagi or log-periodic arrays, then the gain of the composite array will be greater than equation (1.3) would indicate. A good approximation is obtained by multiplying the element gain by the number of elements.

Antenna Receive Gain

Antenna receive gain (G_r) is computed in the manner just described for transmit gain. In many cases it is identical.

Target Cross-Section

Target cross-section (σ_t) is defined as a mean value, generally specified as a function of target aspect angle. Scintillation effects are included in the evaluation of minimum S/N.

Propagation Factor

The propagation factor (F) represents the actual electric-field strength existing at the target divided by that which would exist if only direct-path radiation were present. It has a range of values from 0 to about 2, with values greater than 2 possible but rare. A value of 1 represents no excess energy and is normally used when the radar main lobe (free-space pattern) does not intersect the earth. A perfectly reflecting flat earth with both the target and the point of reflection near beam center would produce values oscillating between 0 and 2. This model is often taken for fan-beam search radars. Atmospheric refraction should also be included in F, although it is common to handle refraction by plotting detection contours on special range-altitude charts that include refraction effects [2]. Further details on propagation factor are explained [1]. Atmospheric absorption losses are not included in F; they are accounted for in L.

Signal-to-Noise Ratio

Signal-to-noise ratio (S/N) is referenced to the IF stage, where it is equal to one-half the peak ratio of signal power to rms noise power that would exist at the output of a hypothetical matched filter. It is also defined as an energy ratio, as follows:

$$\frac{S}{N} = \frac{E_r}{N} = \frac{E_r}{kT_{NI}}, \qquad (1.4)$$

where E_r = signal energy at the antenna input terminals;

kT_{NI} = previously defined.

As defined for equation 1.1, S/N is the minimum ratio of waveform energy to noise-power density that provides the specified probability of detection and false alarm for a particular type of target under specified circumstances.

The factors necessary to define S/N are as follows:

1. probability of detection (P_d);
2. probability of false alarm (P_{FA});
3. target-scintillation model;
4. number of waveforms integrated per observation;
5. efficiency of integration;
6. number of independent observations.

The Radar Range Equation

Probability of Detection

Probability of detection is the starting point in defining S/N and is usually specified in advance based on the radar's function. It is often necessary to compute P_{FA}, starting with some maximum allowable false-alarm rate such as one per second (or some other specified value). The number of independent chances for false alarm during a one-second interval is equal to the number of resolution cells generated per second.

For a simple pulse radar,

$$N_{RC} = T_R B \bar{f}_r, \qquad (1.5)$$

where T_R = active radar ranging or display time (sec);

B = radar bandwidth (Hz);

$\bar{f}_r$ = average pulse-repetition frequency (sec^{-1}).

In many instances a radar has very little dead time and one may simply use the approximation,

$$N_{RC} = B. \qquad (1.6)$$

Having established the allowable false-alarm rate (R_{FA}), one computes P_{FA} as follows:

$$P_{FA} = \frac{R_{FA}}{N_{RC}}. \qquad (1.7)$$

The work of Marcum and Swerling [3] substitutes the symbol n, which is 0.69 times the reciprocal of P_{FA}. Figure 1–1, adapted from Marcum's work by Skolnik [4] shows required S/N as a function of P_D and P_{FA} for a nonscintillating target.

Target Scintillation

Target scintillation models are almost universally based on the work of Swerling [3]. The reader is directed to the reference for detailed analysis of the subject. Swerling's models are intended to apply to conventional scanning search radars that achieve several hits on a target during each scan with several seconds elapsing between observations (typical example: 10 hits per

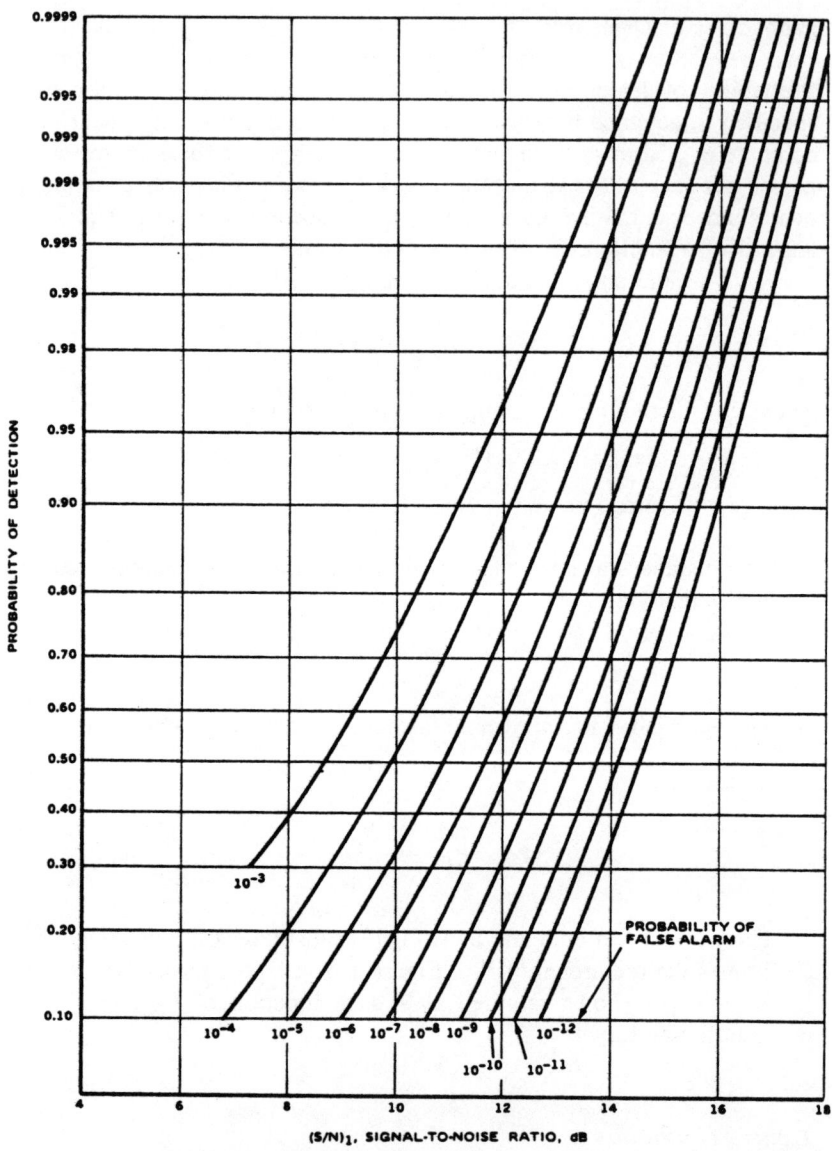

Figure 1-1. Probability of Detection for a Sine Wave in Noise as a Function of the Signal-to-Noise (Power) Ratio and the Probability of False Alarm

The Radar Range Equation

scan, 10 seconds between observations). Four scintillation models are presented, differing in the length of their amplitude-correlation interval and their reflecting characteristics.

Case I assumes the correlation interval is longer than the time on target during a scan but is shorter than the scan period. This is called scan-to-scan scintillation or *decorrelation*. The model is assumed to be composed of several independent reflectors of equal size, characterized by the signal-amplitude distribution

$$W(X,\bar{X}) = \frac{1}{\bar{X}} exp\left[-\frac{X}{\bar{X}}\right], \qquad (1.8)$$

where X = input S/N;

$\bar{X}$ = average of all input S/N.

Case II assumes pulse-to-pulse scintillation or decorrelation with the amplitude distribution of [8].

Case III assumes scan-to-scan scintillation with a target model consisting of one large reflector and several small ones. This produces the amplitude distribution

$$W(X,\bar{X}) = \frac{4X}{\bar{X}^2} exp\left[-\frac{2X}{\bar{X}}\right]. \qquad (1.9)$$

Case IV assumes pulse-to-pulse scintillation with the amplitude distribution of [9]. Nonscintillating targets are occasionally called case V [3,4]. The terms may be considered interchangeable.

Figure 1–2 summarizes the effect of scintillation on S/N requirement for a single pulse. The larger penalty of pulse-to-pulse scintillation at high P_D is obvious. However, if several hits are integrated each scan, the pulse-to-pulse model begins to average toward the nonscintillating case and produces a smaller penalty than in the scan-to-scan cases. Figure 1–3 illustrates this point, based on the assumption that linear integration follows square-law detection. Both figures are from [4], second printing, and are based on Swerling's original work. In order to use these curves, first find the required S/N for a nonscintillating target in Figure 1–1; then add to it the scintillation penalty of figure 1–2; then, if more than one pulse is integrated, subtract the integration improvement factor of figure 1–3 to arrive at a final required single pulse S/N. Note that the last step is valid only if a linear video integrator is used. CRT integration does not necessarily follow the same curves.

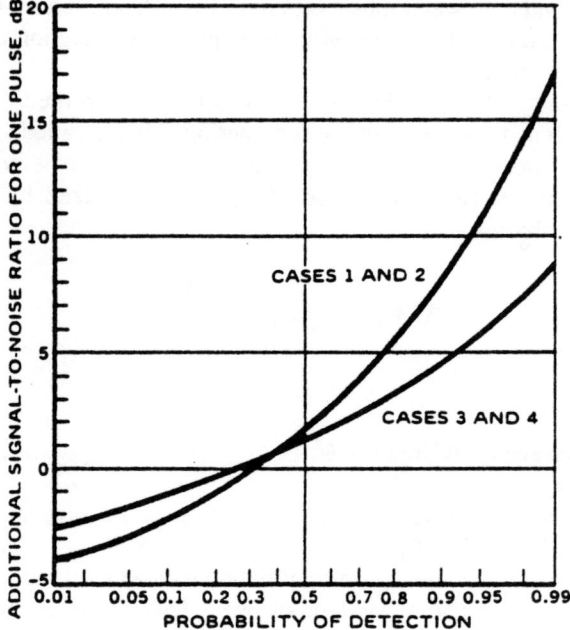

Figure 1-2. Additional Signal-to-Noise Ratio Required to Achieve a Particular Probability of Detection

Pulses or Pulse Bursts

The number of pulses or pulse bursts integrated (N_i) by a scanning radar is a function of beam width, PRF (or burst rate), and scan rate. The shape of the antenna pattern also enters the problem. In this chapter the effect of beam shape is accounted for as a loss factor. In computing the correct value for S/N, the beam is assumed to be rectangular with a width corresponding to the 3-dB beamwidth. Thus,

$$N_i = f_r \frac{\theta}{\omega}, \qquad (1.10)$$

where θ = 3-dB beamwidth in degrees;

ω = scan rate in degrees/sec.

Integration Efficiency

Integration efficiency is a function of the type of integrator and of whether integration occurs before or after the second detector. Predetection integra-

The Radar Range Equation

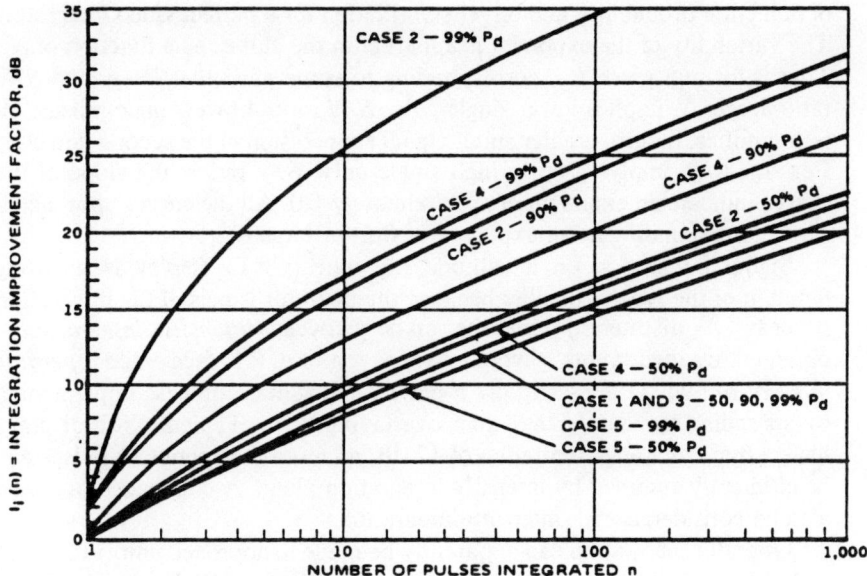

Figure 1-3. Integration-Improvement Factor as a Function of the Number of Pulses Integrated for the Five Cases of Target Fluctuations Considered ($P_{FA} = 10^{-8}$)

tion has unity efficiency with the exception of losses that may be associated with range gating and filtering processes. In this book such losses are considered under the system-loss factor, and predetection integration is taken as 100 percent efficient. Postdetection integration may suffer a loss as a result of poor efficiency in the integrator and another because of small-signal suppression effects in the second detector.

Small-signal suppression in the second detector, sometimes called *noise capture*, causes a variable loss in S/N for ratios smaller than approximately 6 dB. Larger S/N ratios are not affected by the second detector. Suppression becomes progressively worse as S/N decreases below 6 dB, eventually approaching a limit where the effective output integrated S/N is approximated by

$$(S/N)_o = (S/N)_i \, (N_i)^{1/2} \tag{1.11}$$

where $(S/N)_O$ = integrated S/N; (output of second detector)

$(S/N)_i = IF$ S/N ratio. (input to second detector)

thus the exponent of N_i can vary between .05 and 1.0. Figure 1-3 shows curves of integration-improvement factor as a function of N_i for several cases of detection probability and target scintillation for a perfect video integrator. The variability of the exponent is apparent in the curves as a function of N_i. That is, for any given P_D, corresponding to some particular integrated S/N ratio, larger N_i implies lower single pulse S/N ratio. Lower single-pulse S/N ratio implies, in turn, greater small-signal suppression in the second detector. For small N_i, implying fairly high single-pulse S/N ratios, the slope of the curves indicate an exponent of approximately 1.0. All the curves approach a slope corresponding to an exponent of 0.5 for large N_i.

Signal integration on a cathode ray tube (CRT) display is a strong function of the type of display because the operator is part of the integration process. A distinction must be made between *intensity integration*—obtained by overlaying several sweeps on a CRT face—and *operator integration*, which is essentially a pattern-recognition process implemented by spreading the data rather than overlaying them. Typical CRT displays have a dynamic range limitation of 12 dB; no more than about 16 pulses can be efficiently summed by intensity integration alone. A collapsing loss must also be considered with intensity integration.

Operator integration can apparently be made to approach unity efficiency with the proper display format. Figure 1–4, taken from Blake [2], shows the results of a classical experiment originally conducted at the Massachusetts Institute of Technology (MIT) and documented in the Radiation Laboratory Series [5, figure 9.2]. Examination of the original test implementation indicates that variations in N_i were implemented by simulating a wider antenna beamwidth. Pulse repetition frequency (PRF) and antenna scan rate were held constant. Thus, as N_i increased, the width of the target trace on the plan position indicator (PPI) display increased. In the limit, at about $N_i = 6,000$, the target formed a complete circle on the PPI. This procedure fits the criteria for operator integration in that the data are spread on the display so that the operator can simultaneously view all of the data. The lower curve on figure 1–4 represents a perfect video integrator. Both curves represent a detection probability of 0.5. The probability of false alarm is 10^{-8} for the perfect integrator and is unknown for the experimental data. However, the close match for N_i less than a few hundred pulses indicates a similar false-alarm probability. It is interesting to note that the curves diverge at a point where the number of pulses integrated exceeds about 300. Since 6,000 pulses corresponds to a full circle on the display, 300 pulses corresponds to an arc of about 18° wide; the loss of operator efficiency for larger N_i could be attributed to the difficulty of correlating data that are spread over an excessively large area.

Signal integration on a signal amplitude versus time display (A-scope) is less efficient because less persistence is used and the operator does not have the opportunity to view all the data simultaneously. Lawson and Uhlenbeck [5; figure 8.7] indicate an efficiency factor approximately equal to the square root of the number of hits per scan.

The Radar Range Equation

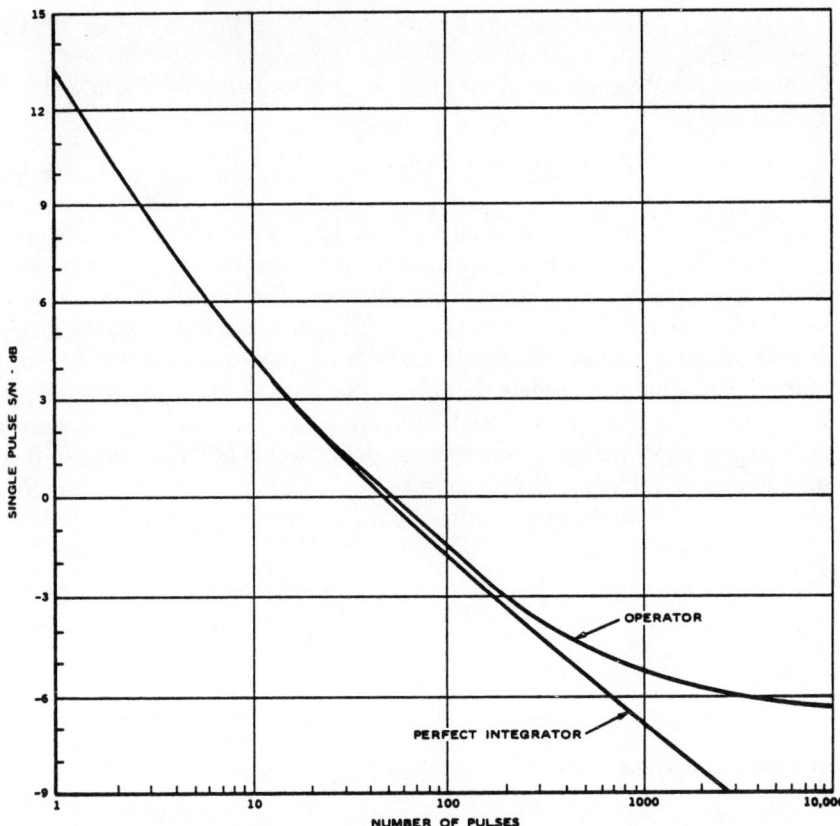

Figure 1-4. Comparison of Operator Integration on a PPI Display with a Perfect Video Integrator

Number of Independent Observations

Target motion has a strong effect on the required S/N for any P_D and P_{fa}. In many cases it will be possible to view the target on several scans before the ground rules of the problem require that a detection be made. If *detection* is defined to mean the recognition of a target on any one of several observations (scans), then one may compute a cumulative probability of detection (P_C) that exceeds P_D for single observations. Two cases are considered here: targets that are stationary in range and targets that have a finite closing-velocity component. The first case might be typical of a missile in its early launch phase, passing vertically through a detection fence. The second is typical of aircraft targets or missiles in their terminal phase.

The first case is computed from consideration of elementary statistics. For M observations,

$$P_C = 1 - (1 - P_D)^M. \quad (1.12)$$

Conversely, if P_C is specified and it is desired to determine P_D in order to specify S/N,

$$P_D = 1 - (1 - P_C)^{1/M}. \quad (1.13)$$

A convenient means of evaluating P_C for targets having a closing velocity has been supplied in a paper by Mallett and Brennan [6]. Their work is based on the assumption that a target is periodically observed (at the radar scan rate) as it closes from some very great range to the required detection range. In practice, observations that begin at about 1.5 times the required detection range are adequate to validate the results. Because of its length, Mallett and Brennan's work is not reproduced here. However, a minor manipulation of their equations follows as an aid to using their results. In the notation of their revised paper, published in *Proceedings of the IEEE* (June 1964), the S/N that must exist with the target at the detection range is expressed as follows:

$$\left(\frac{S}{N}\right)_R = \left(\frac{R_0}{R}\right)^4, \quad (1.14)$$

where R_0 = range for unity S/N;

R = detection range.

From the equations derived in the chapter,

$$\left(\frac{R_0}{R}\right)^4 = \frac{R_1^3 \Delta R}{R^4} = \frac{\Delta R/R}{(R/R_1)^3} = \left(\frac{S}{N}\right)_R, \quad (1.15)$$

where ΔR = target motion between observations;

$R_1 = R_0/\Delta R$.

Because the curves of the revision paper are plotted as functions of $\Delta R/R$ and R/R_1, equation 1.15 provides a convenient way to evaluate S/N at the detection range. If multiple hits per scan are integrated, then this is the effective S/N after integration. Figures 1-5 through 1-10 show the relationships of $\Delta R/R$ and R/R_1 for several classes of targets.

Noise-Power Density

The effective input-noise temperature (T_{NI}) is identical to that defined by Blake [2]. It includes receiver-noise figure, ohmic losses in the microwave-receive network, and antenna temperature.

The Radar Range Equation

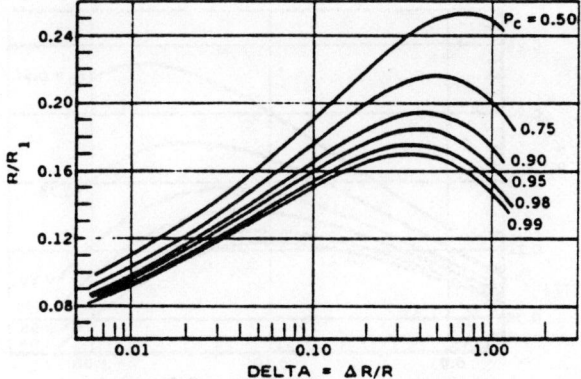

Note: P_c is the cumulative probability of detection by range R for a nonfluctuating target model.

Figure 1–5. R/R_1 *versus* Delta

$$T_{NI} = T_a + T_0 (L_r NF - 1), \qquad (1.16)$$

where T_a = antenna temperature from figure 1–11;

T_0 = ambient temperature = 290°K normally;

L_r = microwave receive losses;

NF = receiver noise figure.

Noise-power density is simply kT_{NI}, where k is Boltzman's constant.

If for any reason the reader desires to reference noise temperature to

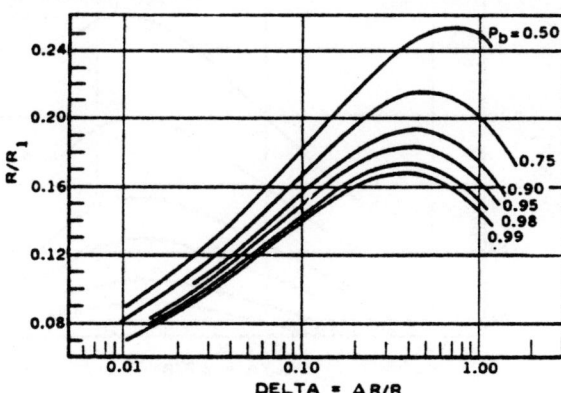

Note: P_b is the probability of detection by range R on a single-scan nonfluctuating target.

Figure 1–6. R/R_1 *versus* Delta

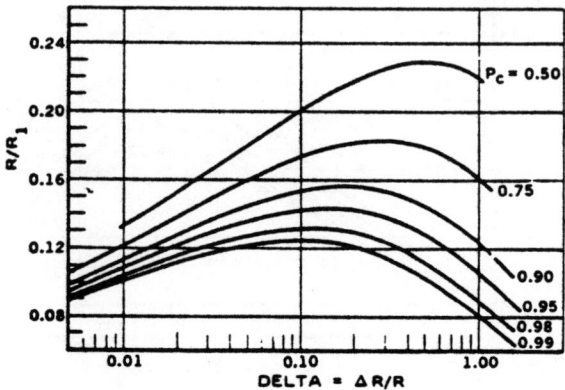

Note: P_c is the cumulative probability of detection by range R.
Figure 1–7. R/R_1 *versus* Delta (Case I Swerling)

some point in the system other than the antenna input terminals, this is very simply done. Total receive-microwave losses are divided into two groups: those that precede the reference point (that is, those located between the antenna input from free space and the point of reference), which we will call L_{r1}, and those located between the reference point and the receiver stage where sufficient amplification is incorporated to fix the noise power, called L_{r2}. Now, $L_{r1} + L_{r2} = L_r$. The new reference temperature (figure 1–11) is computed as

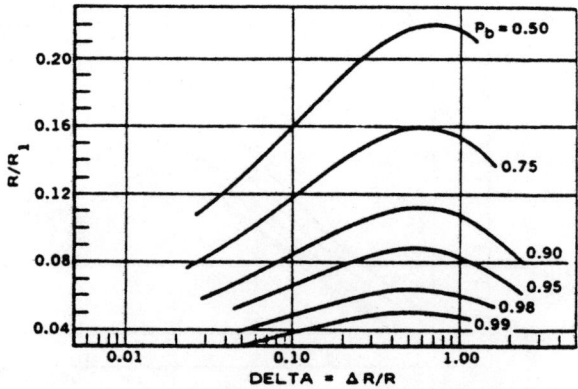

Note: P_b is the probability of detection by range R on a single scan.
Figure 1–8. R/R_1 *versus* Delta (Case I Swerling)

The Radar Range Equation

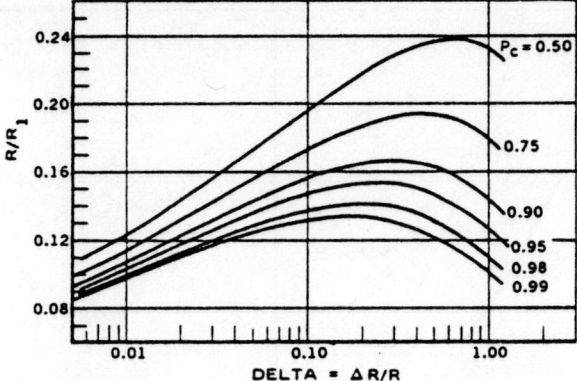

Note: P_c is the cumulative probability of detection by range R.

Figure 1–9. R/R_1 *versus* Delta (Case II Swerling)

$$T_{NX} = \frac{T_a}{L_{r1}} + T_0 \left[(L_{r2} - 1) + L_{r2}(NF - 1) \right].$$

Note that L_{r1} must now be added to total system-loss factor L. In general, T_{NI} is a simpler approach because all of the receive loss is accounted for in one location.

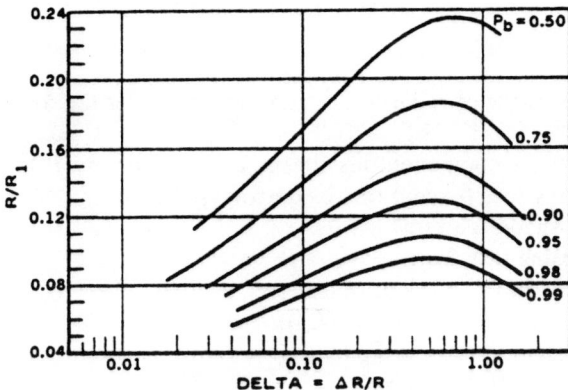

Note: P_b is the probability of detection by range R on a single scan.

Figure 1–10. R/R_1 *versus* Delta (Case III Swerling)

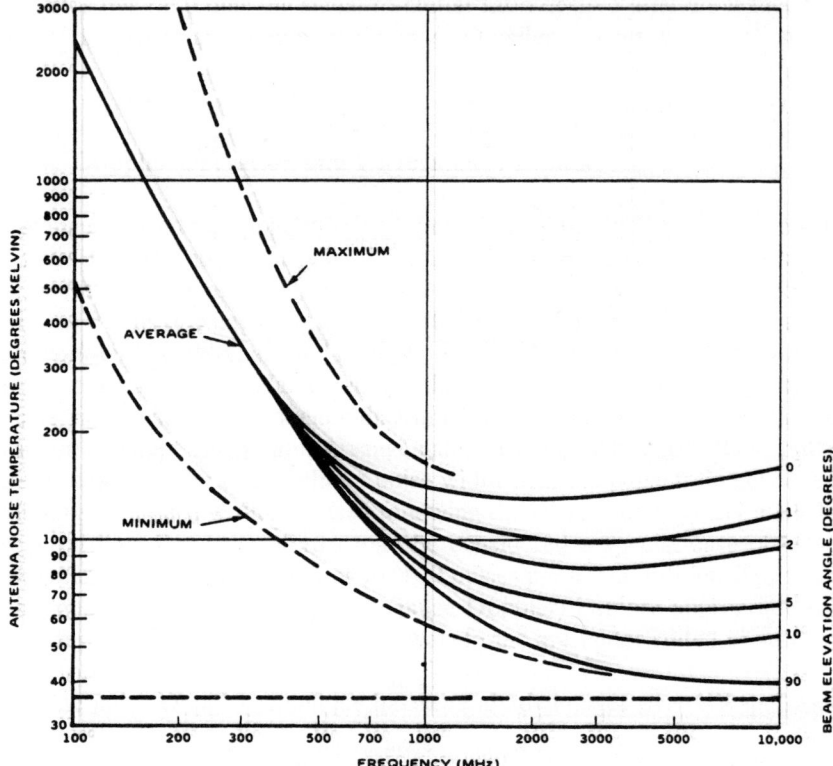

Figure 1-11. Antenna-Noise Temperature for Typical Conditions of Cosmic, Solar, Atmospheric, and Ground Noise

Radar-Loss Factor

A complete listing of radar losses (L) must consider the following:

1. microwave component loss in the transit network (L_t);
2. scanning loss (except for fixed arrays) (L_S);
3. beam shape (pattern) loss (L_p)
4. collapsing loss (L_C);
5. receiver-mismatch loss (L_M);
6. range-gate straddling loss (L_G);
7. Doppler-filter straddling loss (L_f)
8. atmospheric loss (L_a);
9. precipitation loss (L_w);
10. operator loss (L_o);
11. off-broadside loss (L_b).

The Radar Range Equation

Microwave-Component Loss

All ohmic losses between the point at which transmitter power is rated and the point at which *RF* energy is radiated are included in the microwave-component loss (L_t) tabulation. This factor should include radome losses when applicable but should not include feed mismatch to free space, which is considered part of the antenna-gain computation. Only the one-way loss in the transmission network is included.

Scanning Loss

Scanning loss (L_S) is defined as the reduction in signal power due to motion of the antenna between transmission and reception [7]. It is thus directly proportional to target range and scan rate and inversely proportional to beamwidth. Effectively, antenna scanning causes a reduction in effective two-way beamwidth and in peak two-way gain. The latter is due to the fact that transmission and reception cannot both occur at beam center. The narrower two-way beamwidth reduces the number of pulses that are integrated per scan. Figure 1–12 shows the effect of peak gain reduction and the

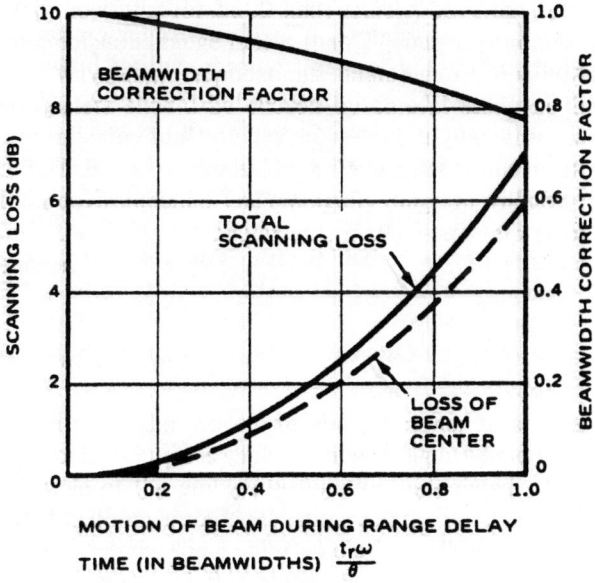

Figure 1–12. Scanning Loss *versus* Scan Speed

total scanning loss due to both effects. It is normalized with respect to range, scan rate, and beamwidth so that 1.0 equals an antenna motion of one 3-dB beamwidth between transmission and reception.

Beam Shape (Pattern) Loss

Beam shape loss (L_p) is caused by the fact that targets generally are not located at beam center and, therefore, do not benefit from peak antenna gain. Pattern loss is simply the average antenna gain (two-way) with respect to beam center over all possible target locations.

With fan-beam scanning radars gain is usually averaged only in one plane because normal aircraft targets are nearly always near peak gain in elevation plane. Gain is averaged in the azimuth plane to simulate the beam scanning past a target. It is standard practice to average only between 3-dB points. Certain target classes, such as missiles traveling in a near-vertical trajectory, traverse the entire elevation beamwidth. Beam shape loss in both planes should be considered for these targets.

Pencil-beam search radars form a raster of beam positions in space and usually obtain far fewer hits per scan than do fan-beam radars—often only one. Gain is averaged in both the elevation and azimuth planes over an area bounded by the lines of overlap with adjacent beams. Figure 1-13 shows two typical pencil-beam rasters and indicates the area for gain averaging.

Bradley [8] shows an analysis of pattern-loss factors for both fan- and pencil-beam radars. Bradley does not consider multiple-hit integration for the fan-beam case, but plots L_p as a function of beam motion between transmissions. That is, if the beam moves 0.1 beamwidth between transmissions, he averages gain over a ± 0.05-beamwidth sector centered on the beam nose. Figure 1-14 provides the loss factor normally associated with scanning fan-beam radars by reading it at the 3-dB beam-crossover point. Figure 1-15 shows similar results for pencil beams. The results of figure 1-15 are somewhat limited in utility because they assume equal beam-crossover points in azimuth and elevation.

It should be noted here that the concept of beam shape loss causes an error in performance predictions as a consequence of the nonlinear relationship between average S/N and average P_D, which is the parameter we seek. Additionally, the small signal-suppression effect that occurs in the second detector causes a nonlinear relationship between IF and video S/N. Accurate manual computations of average P_D are tedious, particularly if a range of target parameters or performance levels is to be considered. Thus, pattern-loss factors are widely used for convenience even though the concept is somewhat erroneous. If accurate computations are required, however, the

The Radar Range Equation

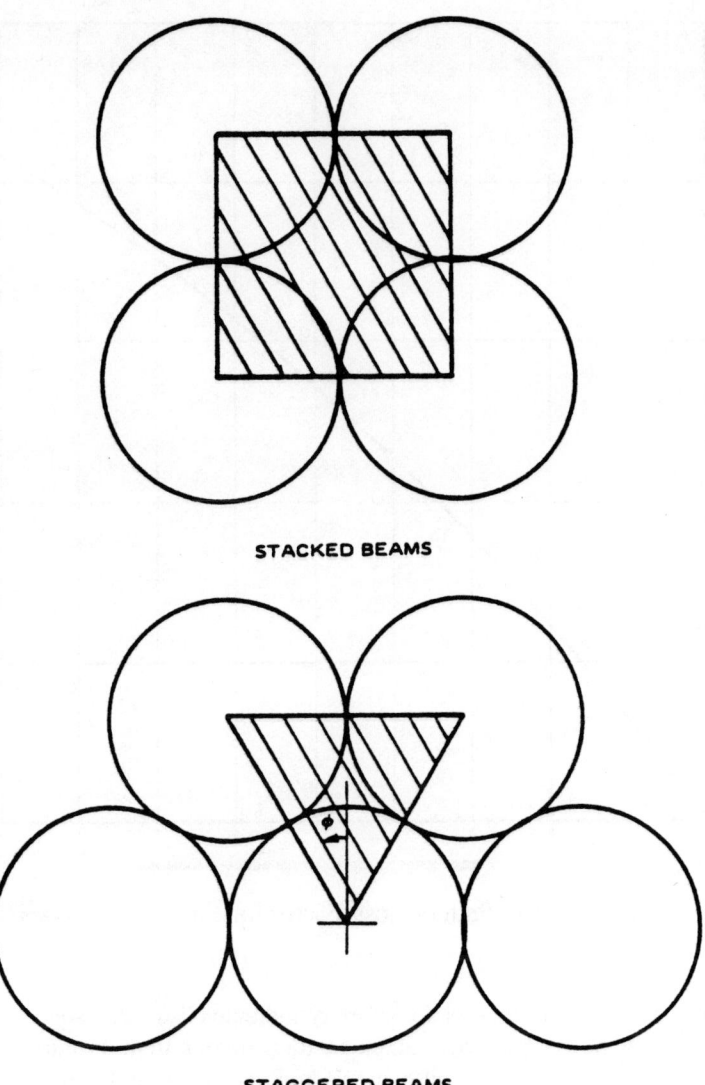

Figure 1–13. Typical Pencil-Beam Rasters

use of digital-computer detection simulations to find average P_D is almost mandatory.

Figure 1–16 shows a comparison of the results that are obtained by averaging gain and by averaging P_D for three different stack-factor products (SFP). *Stack factor* is defined as the ratio of beamwidth to beam spacing in

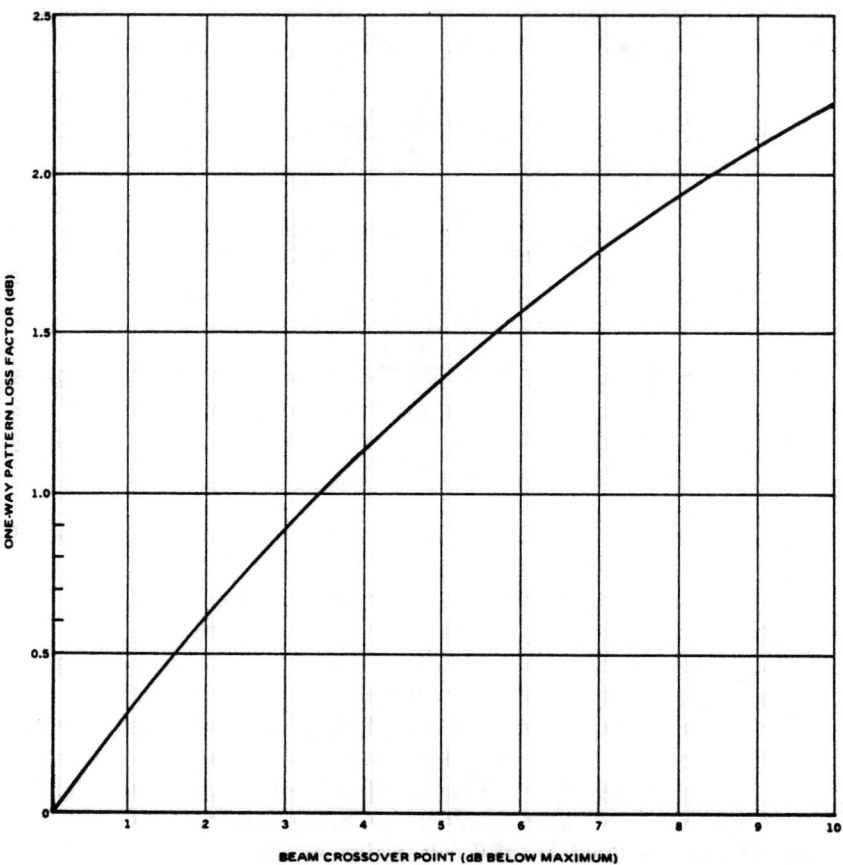

Figure 1-14. Pattern-Loss Factor for Fan-Beam Radars

the search raster. A stack factor of unity indicates 3-dB crossover points of the one-way power patterns. Stack factors greater than 1 indicate closer spacing. SFP applies to pencil-beam radars and is simply the product of the azimuth-plane and elevation-plane stack factors. Note that the curves cross at $P_D = 0.5$, indicating that the conventional pattern-loss factor is valid at the P_D. The conventional approach gives optimistic results for higher P_D. Because average detection probability is a complicated function of beam shape, stack factor in each plane, false-alarm probability, and the detection criteria of a particular radar, it is necessary to perform a computer simulation for each radar. If this is done, L_p is absorbed in the S/N term and is removed from the system-loss factor.

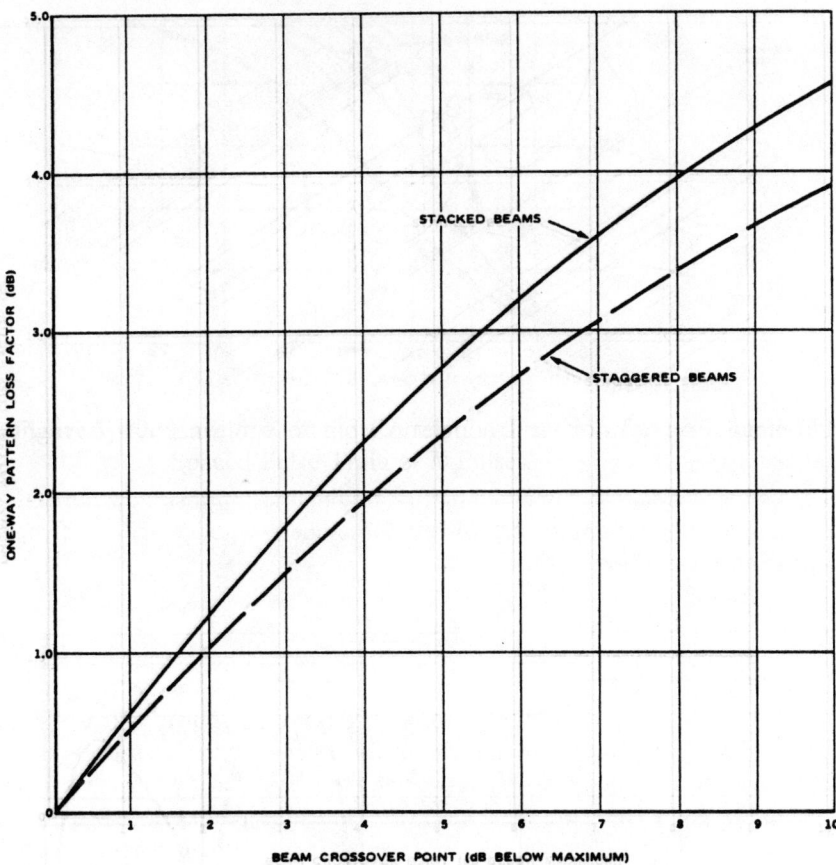

Figure 1-15. Pattern-Loss Factors for Pencil-Beam Radars

Collapsing Loss

Collapsing loss (L_c) occurs when a signal sample is forced to compete with more than one noise sample. This can be caused by such factors as receiver-bandwidth mismatch, excessive range-gate width, summation of video from more than one receiver channel, or excessively slow writing speeds on CRT displays. In all these cases several noise samples are "collapsed" onto a single signal sample, thereby requiring a higher original S/N ratio in order to maintain the same probability of detection and false-alarm rate. A good analytical treatment of collapsing loss is presented by Marcum [3] and by Barton [7].

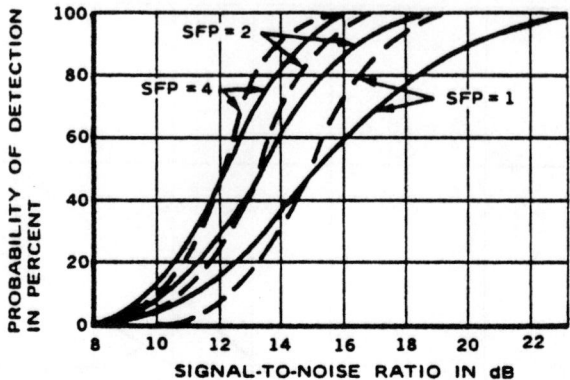

Figure 1–16. Comparison of Power Averaging with Statistical Averaging

Collapsing loss is computed by first finding the *collapsing ratio* (ρ). If n signal-plus-noise samples are forced to compete with m additional noise samples, the collapsing ratio is:

$$\rho = \frac{m+n}{n}. \qquad (1.17)$$

Collapsing loss is related to collapsing ratio, as shown in figure 1–17, derived by Marcum. Additional curves for other P_D and P_{FA} are shown [3]. The two curves illustrate the two principle ways that collapsing loss can occur. In going from the ideal case ($\rho = 1$) to the real case ($\rho > 1$), if the number of independent chances for false alarm remains constant, figure 1–17(a) is used. For example, if several receiver channels are summed together, the number of chances for false alarm remains constant as the number of channels increases from one to several (from $\rho = 1$ to $\rho > 1$) and figure 1–17(a) applies. Another example for figure 1–17(a) is three-dimensional radar data that is collapsed to a two-dimensional display. If the number of chances decreases in inverse ratio to the increase in ρ, figure 1–17(b) is used. An example of this type of collapsing loss is illustrated by excessively narrow video bandwidths. As video bandwidth decreases, for constant pulse width and ranging time, the number of chances for false alarm is reduced, making figure 1–17(b) the appropriate graph.

Receiver-Mismatch Loss

Radar receivers generally do not represent matched filters and, therefore, cause a reduction in the theoretical S/N ratio. Figure 1–18 provides a means

The Radar Range Equation

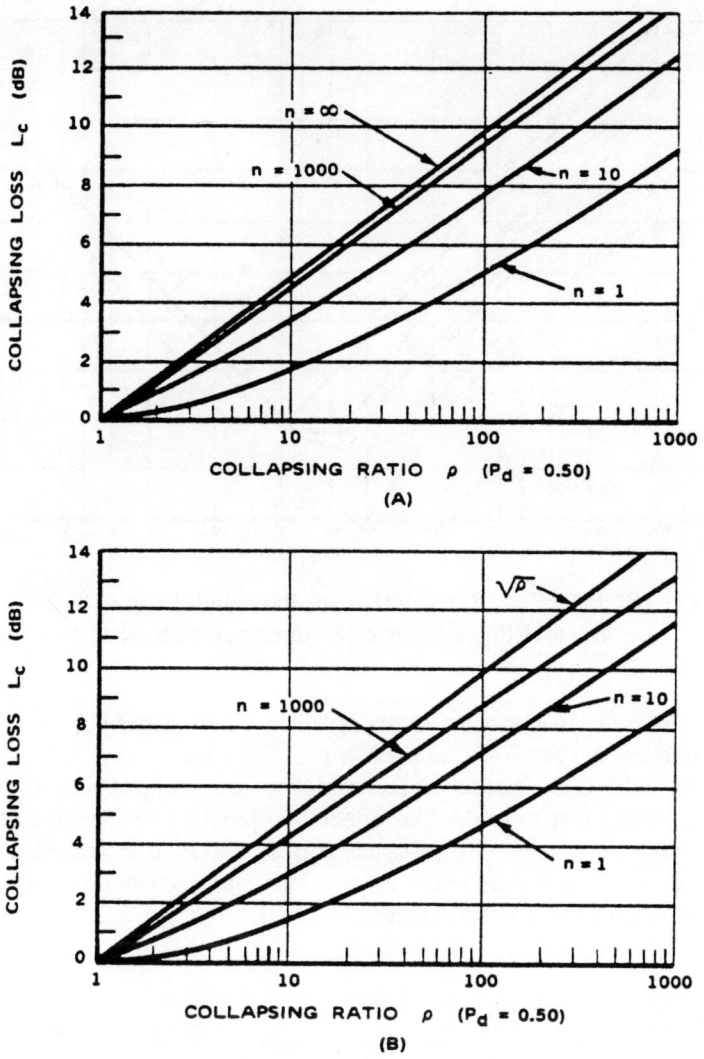

Note: (a) Collapsing loss for constant $(n + n)$; (b) collapsing loss for constant $p/(m + n)$.

Figure 1-17. Collapsing Loss *versus* Collapsing Ratio

to extimate receiver-mismatch loss (l_m) if pulse and receiver-passband shape are known. The Gaussian pulse–Gaussian passband case represents a matched filter and is used as a reference point. It is seen that the square-pulse–square-passband case is about 1 dB worse than the matched filter at their respective optimum bandwidths. The other two curves, probably more

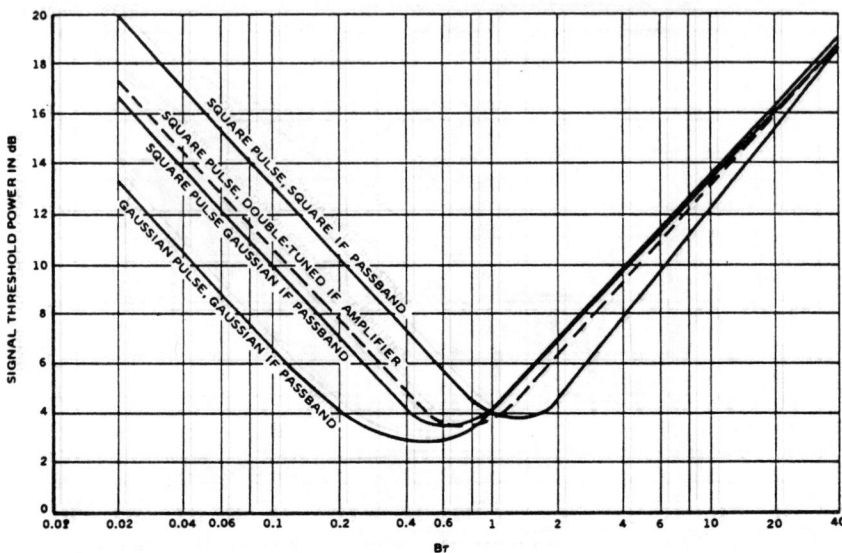

Figure 1-18. Effect of Bandwidth and Passband Shape, Calculated for Several Pulse Shapes, Assuming Automatized Detection

nearly representative of actual radar receivers, show about 0.5 dB loss. For the derivation of these results, see [5], pp. 204–210.

If the radar is designed to detect rapidly moving targets such as ballistic missiles, one must consider the effect of Doppler shift on the receiver bandwidth. Under some circumstances it is necessary to increase bandwidth significantly above the optimum point in order to accommodate a range of possible target Doppler frequencies.

Range-Gate Straddling Loss

Systems that use range gating in the detection process are subject to a loss due to the fact that the target signal never exactly coincides in time with the range gate. Consequently, the gate will be only partly filled with signal power and completely filled with noise, causing a reduction in effective S/N ratio. This loss (L_G) can be reduced by using more range gates than range bins, producing an overlap of range gates. In the limit, with very small center-to-center gate spacing, L_G approaches 0 dB because the signal is very nearly centered in one of the gates. The actual loss is sensitive to gate shape and pulse shape as well as to gate spacing. Figure 1–19 shows typical results that are based on rectangular gates and pulses.

The Radar Range Equation

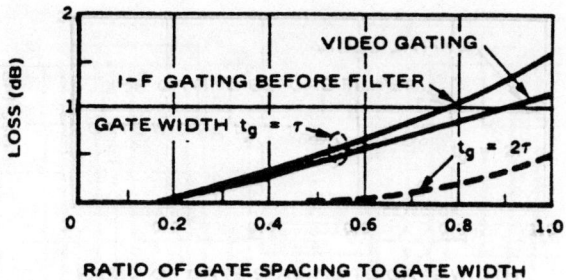

Figure 1-19. Range-Gate Straddling Loss

Range-gate straddling loss is a variable factor, depending on where the signal happens to fall with respect to a range-gate center. As with all variable losses, one should find an average P_D over the range of possible values rather than an average loss. A computerized simulation is the only reasonable way to perform this tedious task. Average loss, shown in figure 1-19, is the best alternative to computer simulation.

Doppler-Filter Straddling Loss

Radars that use comb filters to provide resolution in the frequency domain can suffer a filter straddling loss (L_f) similar to that caused by range-gate straddling. Losses can be reduced at the expense of greater equipment complexity by using more filters to produce overlap in the frequency domain. L_f is a function of filter spacing, filter width, filter shape, and signal-spectrum shape. Figure 1-20 shows typical results for a rectangular filter shape and for a simple single-pole filter. The signal is assumed to have a sin X/X spectrum in both cases. If this loss is computed accurately for a radar, it will include the

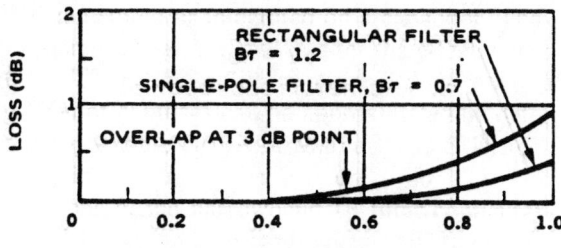

Figure 1-20. Filter-Spacing Loss

Radar Sensor Engineering

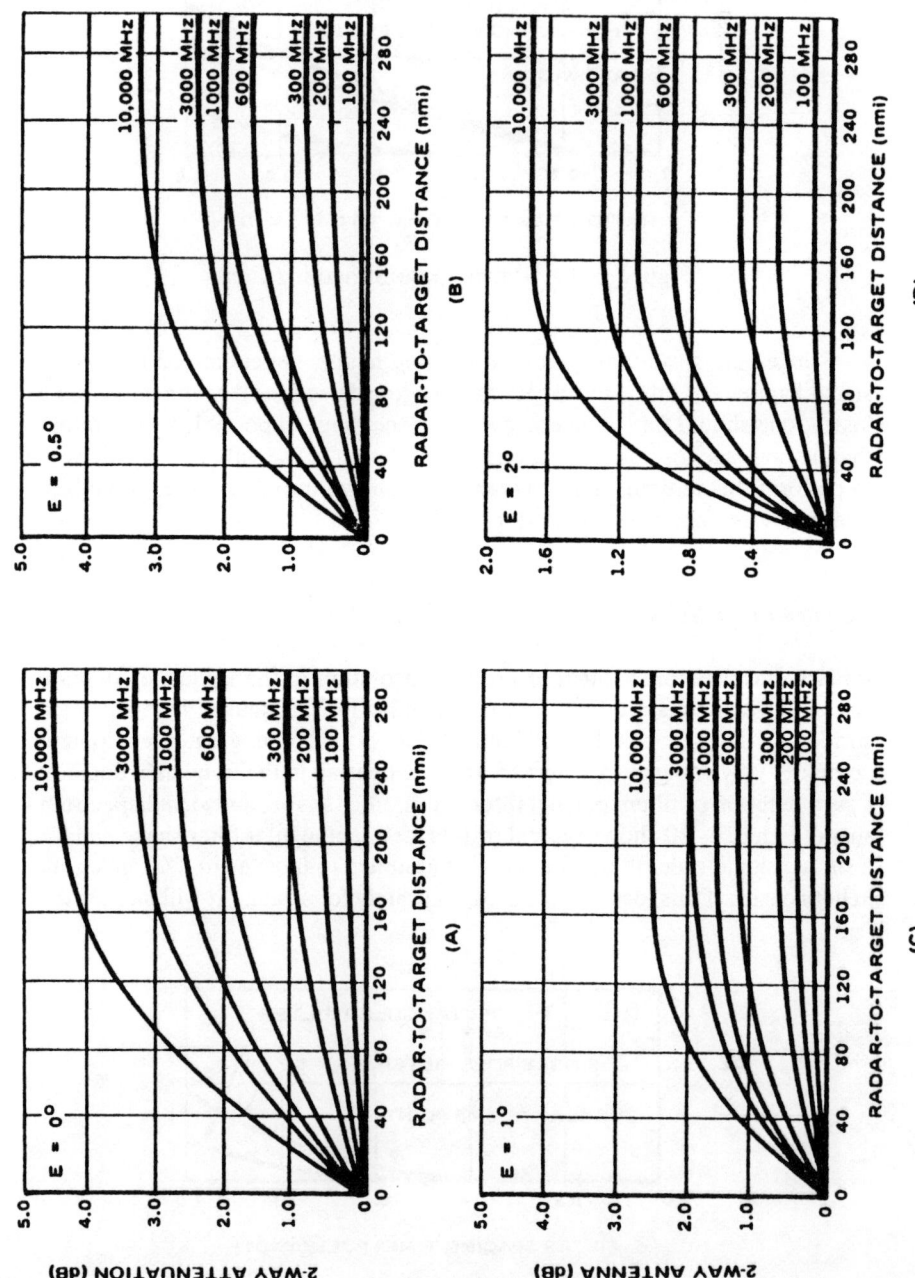

The Radar Range Equation

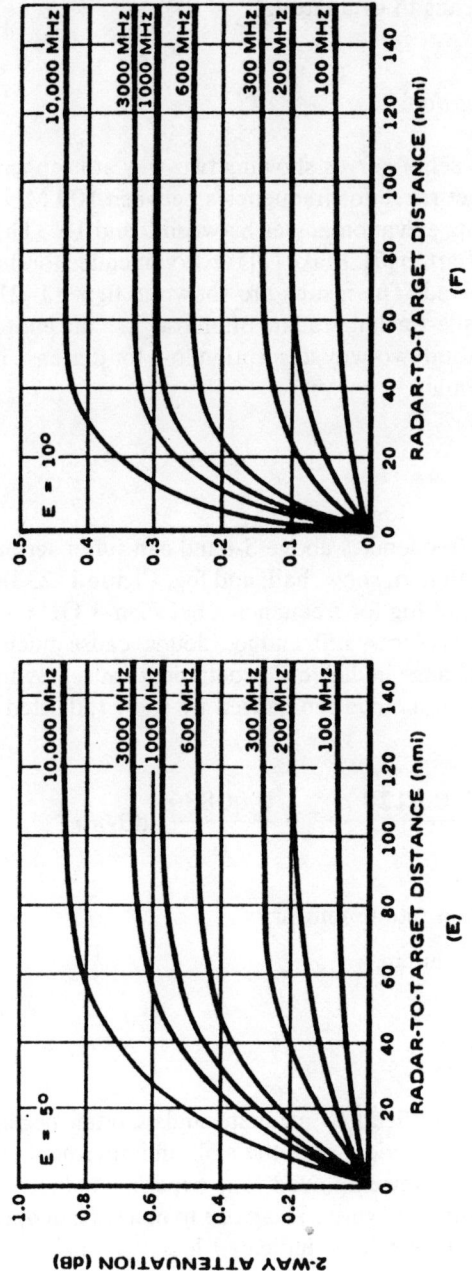

Figure 1–21. Radar Atmospheric Attenuation for Different Elevation Angles

receiver-mismatch loss. As with L_G, a computer simulation to average P_D is the more precise means to evaluate L_f.

Atmospheric-Attenuation Loss

Blake [2] presents a set of curves showing two-way atmospheric attenuation as a function of target range for frequencies between 100 MHz and 30 Ghz and for beam-pointing elevation angles between 0 and 10°. These curves are also reproduced by Barton [7]. Blake [2] is recommended for further detail on the derivations involved. The results are shown in figure 1–21 (a)–(f).

Satellite- or missile-tracking radars often traverse the entire atmosphere. Figure 1–22 shows total two-way absorption loss for this case in a somewhat more convenient format.

Precipitation Loss

Radars operating at frequencies above S-band can suffer serious losses (L_w) in propagating through rain, snow, hail, and fog. Figure 1–23 shows two-way attenuation in rain and fog for frequencies between 3 GHz and 100 GHz. Frozen-water particles (snow, hail, and ice clouds) cause much less attenuation because of the change in dielectric coefficient caused by freezing. Gunn and East [9] show that attenuation caused by snow (adjusted to show two-way loss) is

$$\gamma = \frac{0.0129\, r^{1.6}}{\lambda^2} + \frac{0.0083\, r}{\lambda} \quad \text{dB}/nmi, \qquad (1.18)$$

where $r =$ precipitation rate in mm/hr;

$\lambda =$ wavelength in cm.

Operator Loss

Operator loss (L_o) is difficult to evaluate and is often neglected in radar calcuations. It can vary widely with the skill and altertness of the operator and with the type of display employed. The experimental results reported by Lawson and Uhlenbeck [5; figure 3] lead one to believe that operator loss can approach 0 dB. Other tests have indicated losses as high as 10 dB. These apparent contradictions can perhaps be explained by some remarks made by Skolnick [4, p. 65]. The information bandwidth of a human operator is

The Radar Range Equation

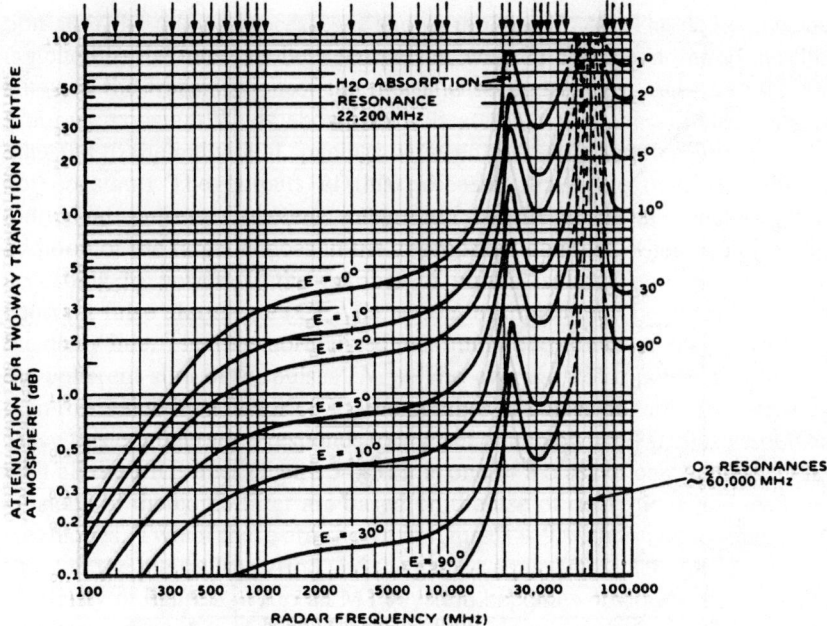

Figure 1-22. Radar Attenuation for Transversal of Entire Troposphere at Various Elevation Angles, Applicable for Targets outside the Troposphere

estimated to be about 20 bits/sec. Most radar displays present data at a much higher rate; thus it is reasonable to assume that a loss must occur. In the tests reported by Lawson and Uhlenbeck, the target was always located at the same azimuth position and could only appear at one of six ranges. Thus the data bandwidth apparently was within the capacity of an alert observer.

Acquisition and tracking radars that receive target designations from other sources should follow the experimental results of Lawson and Uhlenbeck because the operator is alerted and the approximate target location is known. Surveillance radars can approach this level of performance if the threat sector can be confined within narrow limits. If the operator of a surveillance radar has no prior knowledge of target position, it is proper to include a loss, or efficiency factor. Skolnik suggests the following efficiency factor:

$$\rho_o = 0.7 \, (P_D)^2, \tag{1.19}$$

where P_D is the single-scan probability of detection. For $P_D = 0.5$, this relationship produces an operator-loss factor of 7.5 dB.

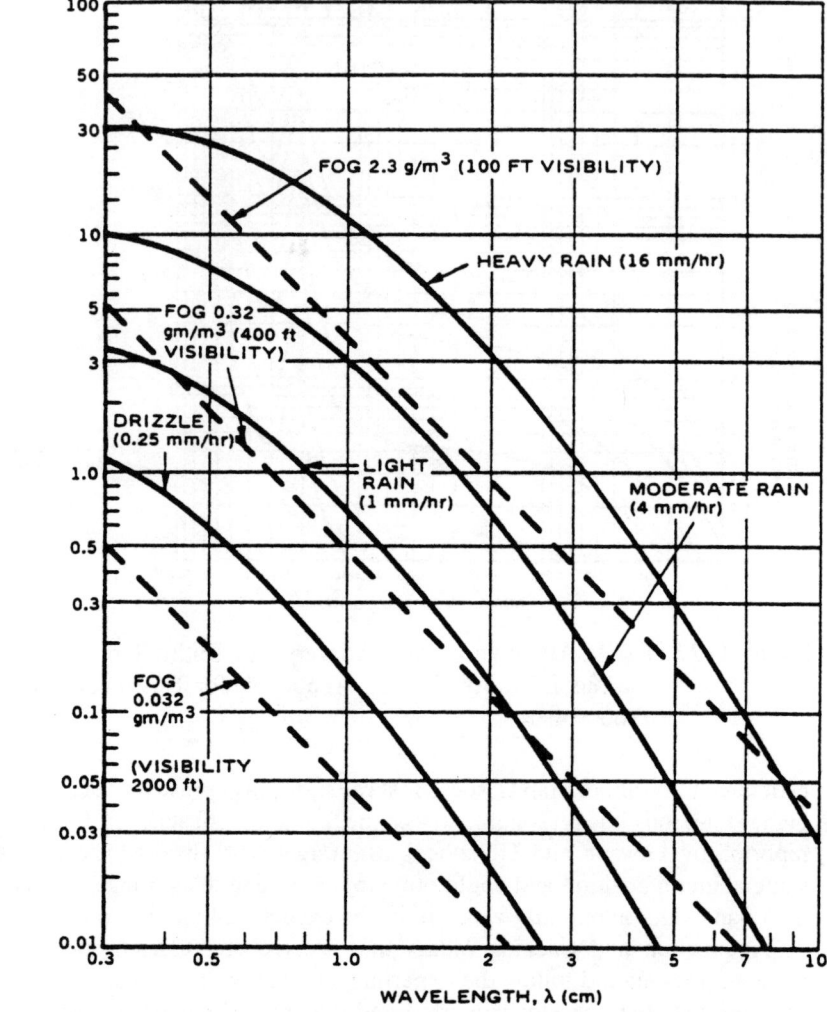

Figure 1-23. Theoretical Values of Attenuation in Rain and Fog

It should be noted that equation 1.19 is applicable to radars that display raw video. Those that utilize threshold circuits prior to displaying the video can reduce considerably the number of noise blips that are presented and thereby reduce the data rate to the operator. Operator loss is reduced in proportion to the threshold value. Typical fan-beam scanning radars that obtain 10–20 hits per scan use thresholds on the order of 0–3 dB and thereby produce only a small reduction in displayed noise. Typical three-dimensional search radars obtain only 1 or 2 hits per scan, with thresholds on the order of

The Radar Range Equation

12–13 dB. Display noise is reduced to the point that an alert operator should suffer no loss in detection probability.

Off-Broadside Loss

Array antennas that perform electronic-beam steering in one or both planes suffer a loss of effective aperture when the beam is scanned away from broadside to the array. The effective aperture is simply the projected aperture along the direction of the beam, varying as the cosine of the off-broadside angle. If the antenna scans in both planes, one merely takes the product of the cosine of the azimuth angle and the cosine of the elevation angle to find the cosine of the absolute angle. Average off-broadside loss is

$$\bar{L}_b = \frac{1}{(\alpha_2 - \alpha_1)(\varepsilon_2 - \varepsilon_1)} \int_{\alpha_1}^{\alpha_2} \int_{\varepsilon_1}^{\varepsilon_2} \cos \alpha \cos \varepsilon \, d\varepsilon \, d\alpha \quad (1.20)$$

where α_1, α_2 = azimuth scan limits;

$\varepsilon_1, \varepsilon_2$ = elevation scan limits.

If the antenna happens to be tilted back from the vertical and it is desired to express beam coordinates in some coordinate system other than that of the antenna (for example, earth coordinates), a different equation must be used. In flat earth coordinates, with the Z axis vertical, the beam position may be expressed as a unit vector (B) as follows:

$$\bar{B} = \bar{X} \sin \alpha \cos \varepsilon + \bar{Y} \cos \alpha \cos \varepsilon + \bar{Z} \sin \varepsilon. \quad (1.21)$$

Antenna broadside, for a tilt-back angle ϕ, is expressed as a unit vector

$$\bar{N} = \bar{Y} \cos \phi + \bar{Z} \sin \phi. \quad (1.22)$$

By taking the dot product of $\bar{B}$ and $\bar{N}$, we obtain the cosine of the absolute off-broadside beam angle:

$$\cos \theta = \bar{B} \cdot \bar{N} = \cos \alpha \cos \varepsilon \cos \phi + \sin \varepsilon \sin \phi. \quad (1.23)$$

Equation 1.23 may be integrated over the scan limits to derive an average value for $\cos \theta$.

Since off-broadside loss is a variable factor, erroneous P_D values will be obtained if $\cos \theta$ is averaged over a large scan volume. A better procedure consists of dividing the scan volume into sectors small enough that $\cos \theta$ is nearly constant over each sector. A different P_D or detection range is then computed for each sector.

2 Radar Resolution and Accuracy

Angular Accuracy

The ultimate angular-measurement accuracy of a radar is a complicated function of its physical and electrical parameters, environmental factors, and the receiver S/N ratio. Each radar is in some way unique, making it impossible to set down a general error-analysis methodology that could apply in all cases. However, it is possible to categorize radars into a few broad classes and to derive limiting accuracy capabilities based on their overall measurement implementations and on received S/N ratio. This section explores these limits for the case of a fan-beam search radar and for two cases of pencil-beam tracking radars.

Fan-Beam Search Radar

Fan-beam search radars normally obtain several returns from a target as the beam scans past it. Displayed on a PPI scope, a point target would appear as a small arc with a width proportional to the antenna beamwidth. If an observer selects the center of this arc as the true target azimuth position, it is possible to obtain angular accuracies that are considerably better than the radar beamwidth. In theory, if the antenna beam shape is accurately known and if nothing acts to alter the amplitude of each signal return, it is possible to determine the target azimuth exactly. In practice, of course, receiver noise and target scintillation will contaminate the signal amplitude and thereby place a limit on the ability of an observer or some electronic beam-splitting device to determine azimuth position exactly.

A paper by Swerling [10] shows that the limiting factors on angular accuracy include scan rate, PRF, beamwidth, beam shape, target-fluctuation characteristics, and S/N ratio. However, for practical beam shapes and for scan rate/PRF combinations that result in at least 10 hits per scan, the errors for various radars tend to converge and become functions of only S/N ratio, the number of hits, and target scintillation. Note that Swerling's case I, III, and V scintillation models do not fluctuate during the observation time of a single scan; measurement accuracy on targets of these types is bounded only by S/N ratio and the number of hits. Figure 2-1 shows the results of Swerling's analysis, normalized with respect to the number of hits (N) and beamwidth (B). Note that β in Swerling's analysis is the angle between beam

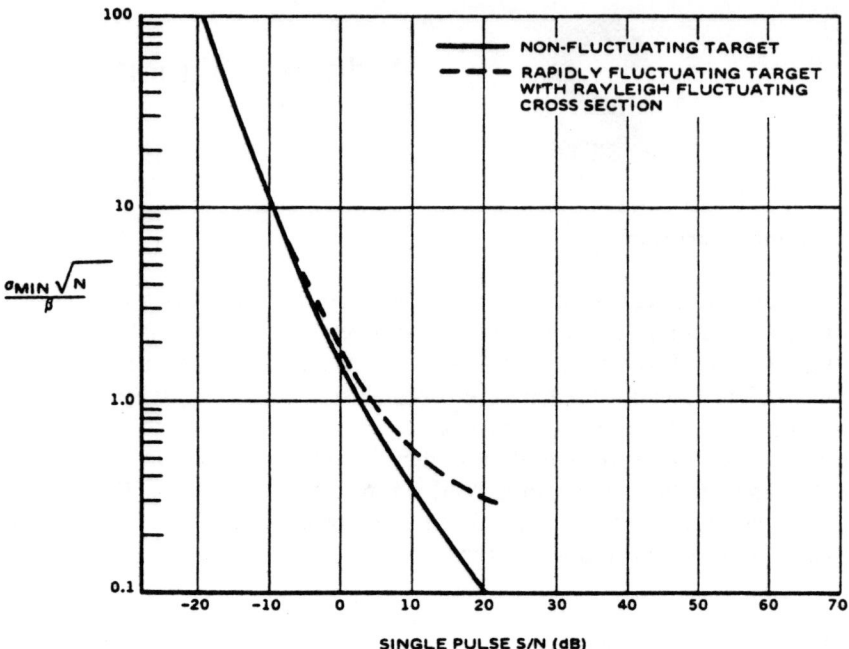

Figure 2-1. σ_{min} as a function of S/N, N, and β for $f(u) = e^{-u}$

center and the $1/e$ point of a two-way Gaussian power pattern. This is related to conventional one-way half-power beamwidth (B) as follows:

$$\beta = 0.425\ B.$$

Also, N is the number of hits integrated as the beam scans through a 2β angle. This is related to the normally quoted N_i as follows:

$$N = 0.85 N_i,$$

where N_i is based on the one-way half-power beamwidth. σ_{min} is the least possible theoretical value for the standard deviation of the error. A Gaussian beam shape is assumed, although the data are applicable with minor errors to other practical beam shapes.

Tracking Radars

Pencil-beam tracking radars are divided into two categories depending on whether lobing is performed on both transmit and receive or only on receive.

Examples of the former are conical-scan and sequential-lobing radars. Monopulse radars fall into the latter category. Some radars have combinations of the two, using monopulse beam forming in one plane and sequential lobing in the other. It is also feasible to perform monopulse beam forming in one plane and to scan in the orthogonal plane. In this case, assuming several hits are obtained as the beam scans past a target, angular accuracy in the scan plane is evaluated as shown in figure 2–1 for a search radar. The factor that differentiates the two classes of tracking radars is two-way beam shape. Monopulse-type radars place the nose of the transmit beam approximately on the target. Conical scanning and sequential lobing radars deliberately squint both the transmit and the receive beams off the target in order to develop an error signal. Considerable loss in S/N can result if large squint angles are employed. In both cases it is possible to develop an error-sensitivity curve that reflects the change in received signal amplitude as the target is moved from the center of the tracking-beam cluster (zero tracking-error position) toward the edge of the cluster. In a noise-free environment the shape of this curve is relatively unimportant. In the presence of receiver noise, however, the slope of this curve near the zero-error position essentially determines measurement error as a function of S/N ratio.

Monopulse. It can be seen from Barton [7, p. 282] that the single-look rms measurement error for monopulse radar is expressed as follows:

$$\sigma_1 = \frac{B}{k_m \sqrt{2\ S/N}}, \qquad (2.1)$$

where B = half-power beamwidth;

k_m = normalized error slope at beam crossover.

Equation 2.1 is based on the previously discussed definition of S/N ratio—that is, the signal energy to noise power density ratio. It does not assume a range gate and filtering arrangement but simply assumes that the peak error signal is measured. If a range gate and filter are employed, the equation should be modified as follows:

$$\sigma_1 = \frac{B}{k_m \sqrt{2\beta\tau(S/N)}}, \qquad (2.2)$$

where β = filter bandwidth;

τ = pulse width.

It is seen that when a matched filter is assumed ($\beta\tau = 1$), the two equations are equivalent.

Most tracking radars employ some form of data smoothing that averages measurement error over a number of signal returns. Equations (2.1) and (2.2) should be modified by multiplying S/N by the number of returns that are averaged.

The value of k_m is determined largely by the squint angle of the monopulse beams. In practical monopulse systems the realizable range of values for k_m is limited by physical factors in the antenna feed and by the need to establish a useful sum beam pattern. Values of approximately 1.2–1.75 are feasible. In Barton [7, p. 273] it is seen that a value of approximately 1.57 is typical.

Conical Scan and Sequantial Lobing. The limits on squint angle that are imposed on a monopulse antenna do not apply to a conical-scan or sequential-lobing antenna. In general, larger squint angles result in steeper error slopes, which reduce system sensitivity to noise and other error sources. However, because the transmit beam is squinted along with the receive beam, a serious loss in S/N ratio occurs for squint angles greater than about one-third beamwidth. Figure 2-2 shows k_s and loss factor (L_k) as a function of squint angle, normalized to unity beamwidth. The error equation for conical scan and sequential lobing [7, p. 278] is:

$$\sigma_t = \frac{1.40}{k_s \sqrt{(S/N)\,(\text{PRF}/\beta_s)}}, \qquad (2.3)$$

where β_s is the tracking servo bandwidth. Note that the factor (PRF/β_s) is equal to the number of pulses smoothed, as for the monopulse case. Because single-look accuracy has no meaning for conical scan or sequential lobing, it is incorporated directly into the error equation.

Angular Resolution

Radar angular resolution is normally quoted as equal to the half-power beamwidth. If two equal-size targets are located at the same range but separated in angle by one beamwidth, most radar observers and automatic acquisition and tracking devices will detect that two targets are present rather than one. However, if one target is considerably larger than the other, it is not generally possible to resolve them at this spacing. Target spacings equal to the null-to-null radar beamwidth (approximately twice the half-power beamwidth) will usually assure resolution in angle. In some extreme cases very large targets may be detected in the pattern sidelobes, but this eventuality does not properly belong under the heading of angular resolution.

Radar Resolution and Accuracy

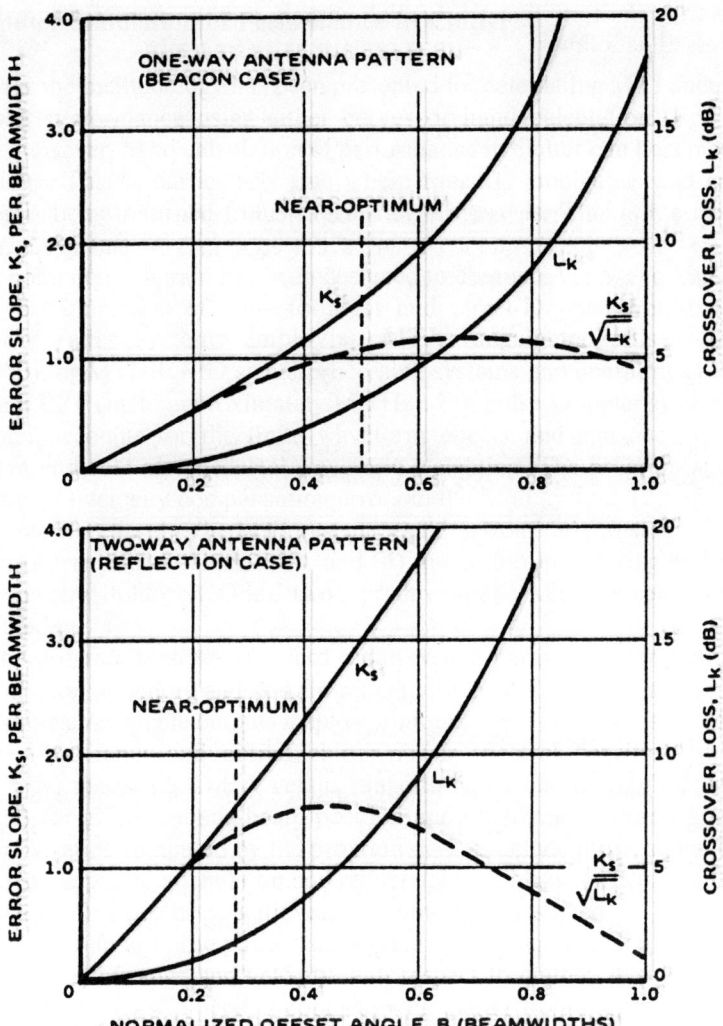

Figure 2–2. Error Slope and Crossover Loss

Range-Measurement Accuracy

The accuracy with which range can be measured by a radar is limited by signal bandwidth and S/N ratio. Many other error sources peculiar to individual radars act to limit measurement accuracy. This section, however, is concerned only with the fundamental error caused by contamination of a signal by thermal noise.

Skolnik [4, p. 464] shows that the limiting rms error associated with measuring the time that the leading edge of a pulse crosses a threshold is expressed as follows:

$$\sigma_r = \frac{t_r}{\sqrt{2S/N}}, \qquad (2.4)$$

where t_r = pulse rise time. If an independent measurement of the pulse trailing edge is made and the two are averaged to find the pulse centroid, the standard deviation is reduced by $\sqrt{2}$:

$$\sigma_r = \frac{t_r}{2\sqrt{S/N}}. \qquad (2.5)$$

Equation 2.5 may be rewritten in terms of bandwidth, pulse width, and signal-to-noise energy ratios by noting the following:

$$t_r = 1/\beta,$$
$$S = E/\tau,$$
$$N = N_0 \beta$$

$$\sigma_r = \left(\frac{\tau}{4\beta E/N_0}\right)^{1/2}. \qquad (2.6)$$

Range-measurement errors can be reduced by smoothing over several successive pulses. In accordance with classical statistics, the error is reduced by $\sqrt{n}$ for n pulses smoothed.

Nonoptimum signal-processing equipment acts to increase the aforementioned rms errors somewhat. For a greater depth of analysis, see Skolnik [4, p. 464] and Barton [7, pp. 351–369].

Range Resolution

The ability to discriminate multiple targets in the range dimension is, for simple pulse radars, an inverse function of signal bandwidth:

$$\Delta R = \frac{C}{2\beta}, \qquad (2.7)$$

where C = speed of propagation in units of R and β. Various pulse-coding schemes and multiple-pulse waveforms must be evaluated by examining their

Radar Resolution and Accuracy

ambiguity functions in detail. Appendix D of Barton [7] shows ambiguity diagrams for several common waveforms.

Doppler-Measurement Accuracy

The equations for Doppler-measurement rms error are similar in form to those for angular and range errors. For a single rectangular pulse, based on a frequency discriminator composed of two adjacent rectangular filters, the rms error is:

$$\sigma_{f_1} = \frac{\beta}{2\sqrt{S/N}}. \qquad (2.8)$$

If n noncoherent measurements are performed (noise and signal not correlated between measurements) the error is reduced to

$$\sigma_f = \frac{\beta}{2\sqrt{n(S/N)}}. \qquad (2.9)$$

A report by Manasse [11] considers velocity-measurement error in greater detail and concludes the following:

$$\sigma_f = \frac{1}{\alpha\sqrt{2\,S/N}}, \qquad (2.10)$$

where α is defined as 2π times the rms division of the signal with respect to its time centroid. For a single rectangular pulse of duration τ,

$$\alpha = \frac{\tau}{\sqrt{12}}(2\pi) = \frac{\pi\tau}{\sqrt{3}},$$

$$\sigma_f = \frac{1}{2.56\,\tau\sqrt{(S/N)n}}. \qquad (2.11)$$

If a coherent pulse train is used, the pulse width (τ) is replaced by the total waveform length (T). The factor n is still included to represent the improvement in S/N ratio that results from integrating n pulses.

$$\sigma_{f_e} = \frac{1}{2.56\,T\sqrt{n(S/N)}}. \qquad (2.12)$$

It should be noted that the improved accuracy of coherent pulse trains is bought at the expense of velocity ambiguities that occur at intervals equal to the waveform PRF. If the PRF is selected high enough, perhaps no real targets will have Doppler frequencies that fall outside the first ambiguous interval; in this case the ambiguities are no problem. If Doppler frequencies greater than the PRF must be considered, however, some means of resolving the resulting ambiguity must be found. The use of two different PRFs on subsequent looks at a target may resolve the problem. This problem is discussed in further detail under the section on pulse-Doppler radar performance.

Doppler Resolution

Two targets may be resolved in the velocity domain if their Doppler frequency shifts differ by more than the signal bandwidth in the case of simple single-pulse waveforms.

$$\Delta V = \frac{\lambda \beta}{2} = \frac{\lambda}{2\tau} . \qquad (2.13)$$

Noncoherent multiple-pulse waveforms improve Doppler measurement accuracy but do not improve resolution.

Coherent pulse trains have velocity resolution proportional to pulse-train length rather than pulse length:

$$\Delta V = \frac{\lambda}{2T} = \frac{\lambda n}{2(\text{PRF})} . \qquad (2.14)$$

3 Radar Performance in Clutter

The purpose of this chapter is to summarize the characteristics of MTI and pulse-Doppler waveforms, and to give the radar engineer the tools for design of the waveforms and calculation of performance, particularly in a clutter environment.

First, the various MTI techniques are discussed along with their limitations in a clutter environment; then the pulse-Doppler waveform will be discussed. The equations for hand calculation of performance are reviewed, as are more accurate methods using computer simulation. This chapter draws mainly on Reilly [12] and Nathanson [27].

MTI Operation and Performance Limits

Clutter is distinguished from receiver noise by its relatively narrow low-frequency spectrum, which implies that its returns are correlated from one pulse to the next. Hence its effects can be reduced by filters that reject energy while passing the Doppler-shifted echoes from targets that have higher velocities than the clutter. The simplest MTI processor, the delay line canceler, subtracts two successive echoes from the same location. The portion of the return from stationary objects will cancel, leaving a residue due to the moving targets.

Various types of MTI processors are possible, distinguished by whether the phase, the amplitude, or both phase and amplitude of the returned signals are processed. The systems that use only phase or amplitude do not match the performance of those that use both. The characteristics of each technique will be discussed briefly.

Phase-Processing MTI System

This system is the so-called *coherent* MTI system since it makes use of the phase information. A block diagram of such a system is given in figure 3–1. The basic phenomenon that is exploited by all MTI systems is illustrated by the vector diagram of figure 3–2. If we observe two successive radar returns from the same region in space, such region containing a fixed clutter target and a moving target within the same radar-resolution cell, the resolutant

43

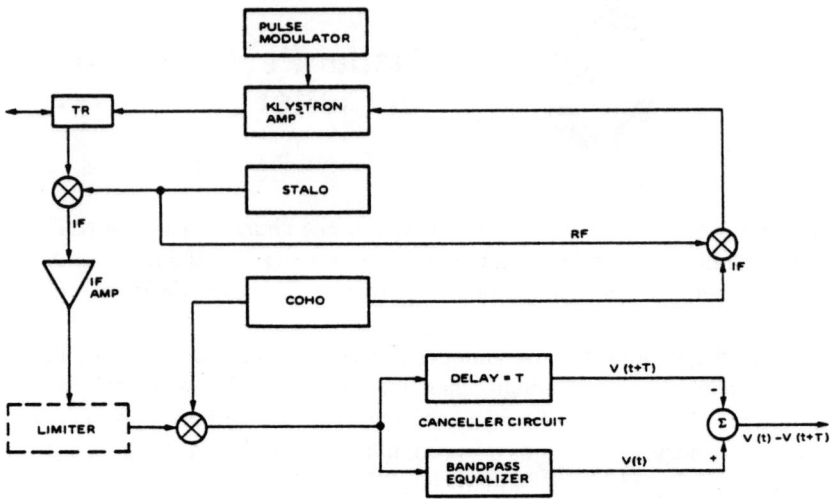

Figure 3-1. Phase-Processing MTI System

signal amplitude and phase will change, as shown by the two sum vectors. The clutter amplitude and phase remain constant from look to look; the target amplitude remains constant, but its phase changes as a result of its motion. Therefore, the sum of the two changes in both amplitude and phase and the difference between the two is finite except for particular relationships between ω_d and PRF. ω_d is the Doppler radian frequency due to target motion, expressed as

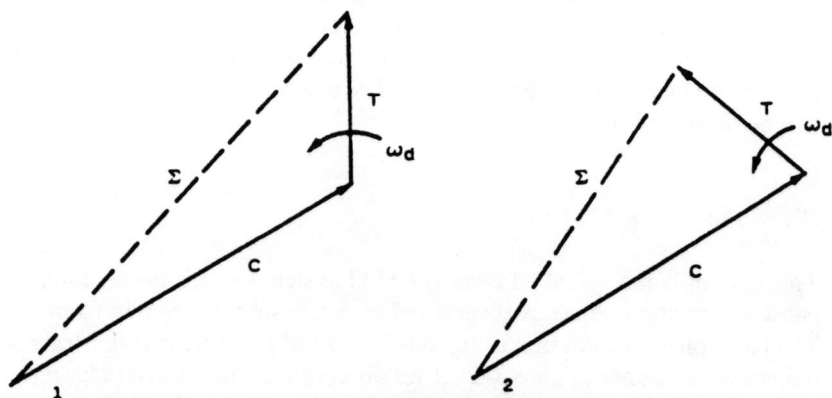

Figure 3-2. Target Plus Clutter-Signal Characteristics

Radar Performance in Clutter

$$\omega_d = 2\pi f_d = 2\pi\left(\frac{2V}{\lambda}\right), \quad (3.1)$$

where target velocity (V) and wavelength (λ) are in compatible units.

In the simplest processor, the phase-detected signal is processed in a delay line canceler that forms the difference between the two signals separated in time by the interpulse period. It is important to note here that a limiting action is performed that keeps the peak amplitude of the two signals the same and makes the output of the canceler proportional only to the phase. The signal output from the subtractor is

$$E_0 = E_1 - E_2 = 2E \sin(\pi f_d/f_r)\cos\left[2\pi f_d\left(t + \frac{1}{2f_r}\right) + \phi_0\right], \quad (3.2)$$

where E_0 = the output of the subtractor:

E_1 = the first signal return;

E_2 = the signal one interpulse period later;

f_r = pulse-repetition frequency;

ϕ_0 = phase shift due to range = $4\pi R_0/\lambda$;

E = amplitude of uncanceled signal;

f_d = Doppler frequency.

As this equation indicates, the difference in voltage is a sine wave at the Doppler frequency whose amplitude depends on the relationship between the Doppler frequency and the pulse-repetition frequency. Figure 3–3 shows the response of the single canceler.
In this figure, S_o and S_i represent the peak output and peak input signal power, respectively. Note that there is a power gain over portions of the PRF interval depending on the phase angle ϕ_0. For optimum ϕ_0 the average gain over the PRF interval is 3 dB. If, in addition to averaging over the Doppler frequency, the average is taken uniformly over all phase angles, the average signal power gain is 1 rather than 2; that is, there is an average of 3-dB signal loss for a phase-processing MTI system. A disadvantage of the phase-processing system is that a high degree of phase stability is required. Also, blind speeds exist at multiples of the PRF, or

$$f_d \text{ blind} = nf_r.$$

The corresponding blind speeds are

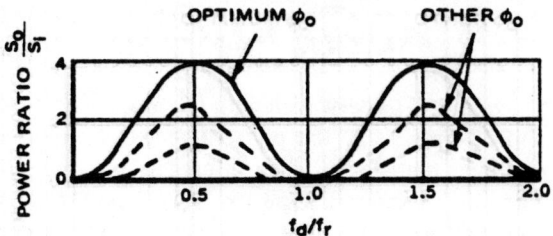

Figure 3–3. Power Response of Single-Delay-Line Phase Processor

$$V_d \text{ blind} = \frac{\lambda}{2} nf_r.$$

Amplitude-Processing MTI

The amplitude-processing, or envelope-processing MTI, a noncoherent system, is diagramed in figure 3–4. The main advantage of this system is that the local oscillator stability is not as crucial as in other systems. A disadvantage is that clutter must be present in relatively large amounts to detect moving targets.

The returns from each pulse are:

$$E_1^2 = E_c^2 + E_s^2 - 2 E_c E_s \cos(\phi_0 - \Delta\phi/2),$$

$$E_2^2 = E_c^2 + E_s^2 - 2 E_c E_s \cos(\phi_0 - \Delta\phi/2), \qquad (3.3)$$

where E_c, E_s = clutter and signal voltage, respectively;

E_1, E_2 = magnitude of signal plus clutter voltage for the two pulses;

ϕ_0 = average phase of the signal relative to the clutter (range dependent);

$\Delta\phi$ = relative phase change of signal during the interpulse period.

Assuming square-law detection, the output of the canceler is:

$$E_o^2 = E_1^2 - E_2^2 = 16 E_c^2 E_s^2 (\sin^2 \phi_0)(\sin^2 \Delta\phi/2), \qquad (3.4)$$

but

$$\Delta\phi = 2\pi f_d/f_r,$$

Radar Performance in Clutter

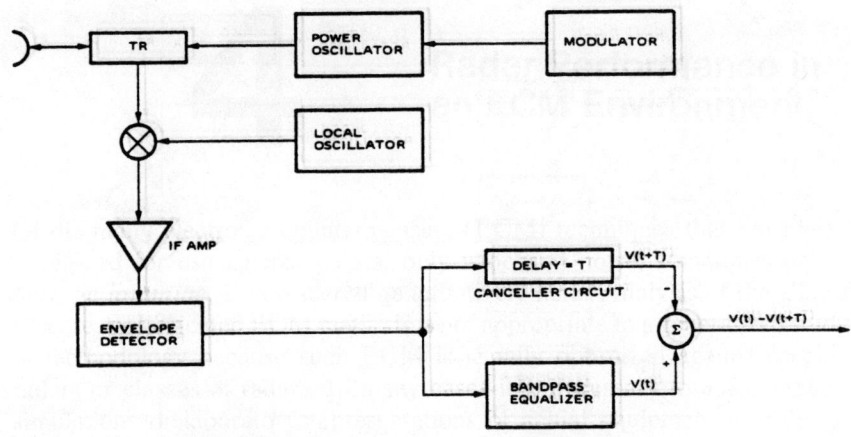

Figure 3–4. Envelope-Processing MTI System

then

$$E_o^2 = 16E_c^2 E_s^2 \ (\sin^2 \tau f_d/f_r) \ (\sin^2 \phi_0).$$

Hence, as in the phase processing MTI system, both blind speeds and blind phases exist. As shown here, there must be a clutter voltage present in order to obtain an output.

Vector-Processing (or IF) MTI

This version of MTI makes use of both the amplitude and phase information of the echo signals. This technique can be implemented at either *IF* or video. The *IF* technique is block diagramed in figure 3–5. The returns from each pulse are:

$$E_1 = E \sin [2\pi(f_{IF} + f_d) t + \phi_0],$$

$$E_2 = E \sin [2\pi(f_{IF} + f_d) (t + T) + \phi_0],$$

where T = interpulse period;

E = amplitude of the *IF* signal;

f_{IF} = intermediate frequency

f_d = Doppler frequency

ϕ_0 = initial phase of echo.

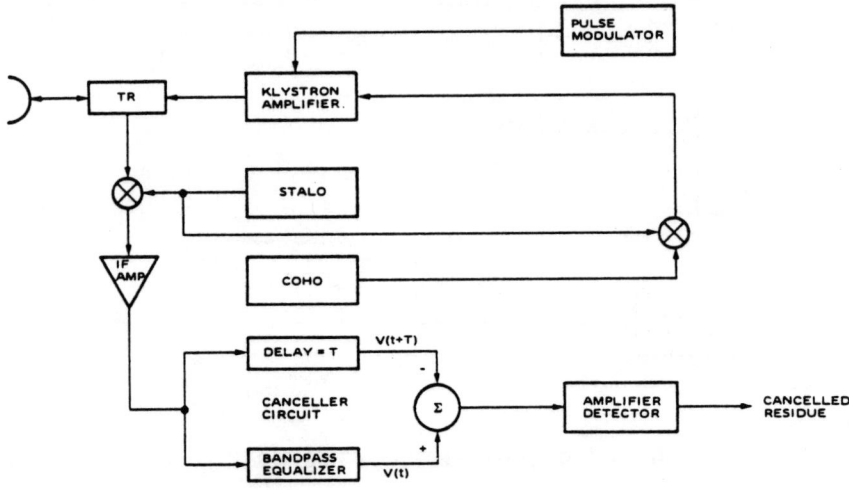

Figure 3–5. Vector-Processing MTI

The output of the canceler is

$$E_o = 2E \sin[\pi(f_{if} + f_d)] \cos\left[2\pi(f_{IF} + f_d)\left(t + \frac{T}{2}\right) + \phi_0\right] \quad (3.5)$$

To ensure that the difference signal is zero when the Doppler frequency is zero, it is required that $f_{IF} = nf_r$ where f_r is the repetition frequency. If this is the case, the response follows the curves of figure 3–3. An improved version of this system uses two channels in quadrature phase following the IF amplifier. The squares of the outputs of the two canceler circuits are then added, and the response of the system is always that of the solid curve of figure 3–3. Wainstein and Zubakov [21, p. 213] have shown that this quadrature processing optimizes the signal to clutter ratio. The transfer function is:

$$E_o = 2E \sin[\pi(f_{IF} + f_d)T].$$

Clutter-Locked MTI System

The presence of a mean-clutter Doppler component other than zero can drastically degrade MTI performance. The mean component can originate either from the average motion of the clutter itself or from the motion of the radar platform (as in a moving ship or aircraft). Therefore, it is desirable to remove the average clutter frequency from the radar signal. One method, known as the *clutter-locking technique,* is illustrated by the block diagram of

Radar Performance in Clutter

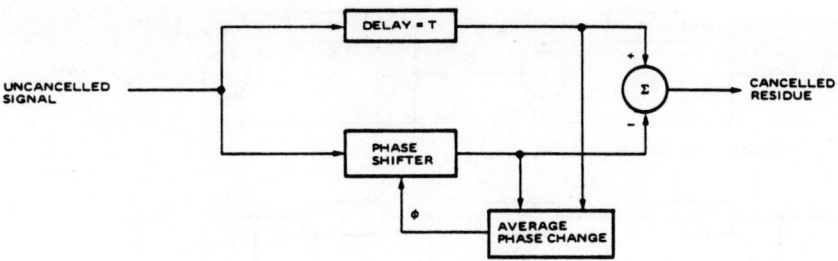

Figure 3-6. An Example of a Clutter-Locking Circuit

figure 3-6. With this technique the average phase change in each interpulse interval is determined and compensated for by the phase shifter.

The manner in which MTI performance degrades as a result of inexact determination of the mean Doppler is discussed in a later section. Since the phase-averaging process involves a finite number of samples, the mean velocity at the range of interest can only be approximately estimated. Further, if more than one type of clutter is present, this clutter-locking technique would not be effective if there were a wide disparity in their mean velocities. Another disadvantage is that it would reject strong target signals in the absence of clutter since the phase compensation would be made for the target alone. Therefore, in such a system it is desirable to have an adaptive technique whereby the clutter-locking circuits can be bypassed when the clutter power is small.

Multiple-Canceler Systems

The multiple cancelers provide greater clutter rejection than the single-canceler circuits described previously. One form of multiple canceler is composed of cascade sections of single-canceler circuits, as illustrated in figure 3-7(a). The configuration using n cascaded sections is equivalent to the weighted sum of $(n + 1)$ pulses, as shown in figure 3-7(b), where the weighting follows the binomial coefficients of $(1 - X)^n$.

Peak power output for an n-stage canceler is:

$$(S_o/S_i)n = s^{2n}\sin^{2n}(\pi f_d/f_r). \qquad (3.6)$$

This equation illustrates that the blind speeds are independent of the number of canceler stages.

Signal power, when averaged over all possible Doppler frequencies for an n-stage canceler, is:

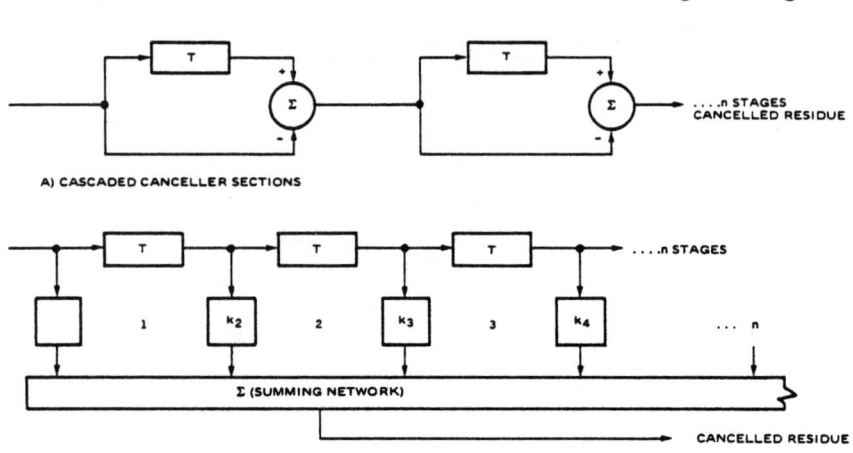

Figure 3–7. Multiple-Delay-Line Cancelers

$$(S_o/S_i)_1 = 2 = 3 \text{ dB},$$
$$(S_o/S_i)_2 = 6 = 7.8 \text{ db},$$
$$(S_o/S_i)_3 = 20 = 13 \text{ dB}.$$

If combined phase and amplitude processing (vector or *IF* processing) were not used, the multiple canceler would experience a 3-dB loss in performance because of the "blind phases," as explained for the single canceler. One disadvantage of the cascaded *n*-section canceler is that because of the $\sin^n$ responses, the rejection of target signals near the blind frequencies becomes more severe as *n* is increased. It is possible to obtain other MTI response curves with multiple-canceler systems that use various feedback and feed-forward circuits. A summary of the literature on the topic is given by Skolnik [4, pp. 132–135].

Staggered-PRF Systems

The blind speeds inherent in the previously described MTI processors can pose serious limitations on target detection. The use of a varied pulse-repetition frequency provides a technique for extending the first blind speed to any desired velocity. For a system that has two repetition intervals available, T_1 and T_2, the canceler may form the successive differences:

$$E_o = [E(t) - E(t + T)] - [E(t + T) - E(t + 2T + \Delta T)],$$

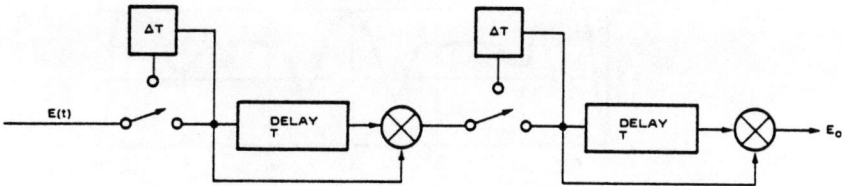

Figure 3–8. Canceler Configuration Useful in Staggered-PRF Systems

which is equivalent to the weighted sum

$$E_o = E(t) - 2E(t + T) + E(t + 2T + \Delta T). \qquad (3.7)$$

This is the output of the canceler shown in figure 3–8.

The response characteristics of the three-pulse canceler for different values of T_2/T_1 are shown in figure 3–9.

Digital MTI

Recent advances in digital circuitry make it practical to implement the video storage and cancellation digitally. Radars using digital MTI have been built and demonstrated. Figure 3–10 shows the basic digital canceler for two-pulse cancellation. In this example the radar receiver's phase-detected output is sampled at discrete range intervals and converted to a binary number. The analog/digital (A/D) output for each range interval is sent to a digital store, and after one interpulse period the stored digital words are read out sequentially and digitally subtracted from the current A/D-converter output for each range interval. Higher-order cancelers can be implemented by simply increasing the digital storage and performing the necessary computations. The loss due to processing only one coherent video channel is equal in both analog and ditigal systems. A/D converters have been built with 10-MHz bandwidth. The theoretical cancellation capability depends on the resolution of the A/D converter and is equal to about 6 dB per bit or, for a nine-bit converter, about 54 dB.

Limitations on MTI Performance

Having discussed the various techniques for attenuating clutter magnitude, we will review the limitations in performance. The *clutter attenuation* (*CA*) is defined as the ratio of input-clutter power, Ci, to output-clutter power, Co:

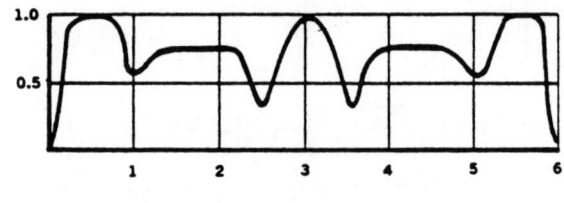

(A) $T_2/T_1 = 5/7$

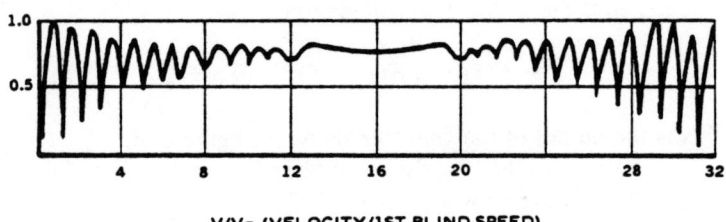

(B) $T_2/T_1 = 63/65$

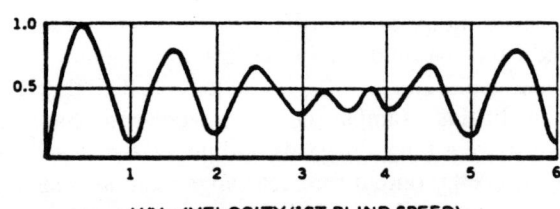

(C) $T_2/T_1 = 7/8$

Figure 3-9. Relative-Power Response Curves for Staggered-PRF System

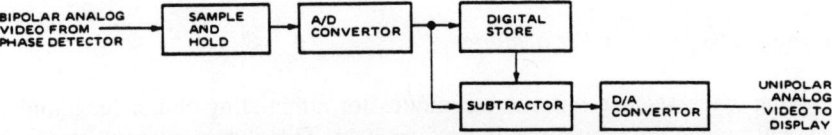

Figure 3-10. Basic Digital MTI Canceler

Radar Performance in Clutter

$$CA = \frac{Ci}{Co}.$$

The *clutter-improvement factor* can be defined as the signal-to-clutter ratio at the output of the MTI system compared with that at the input, where the signal is understood as that averaged uniformly over all radial velocities; that is:

$$I = \frac{\bar{S}o/Co}{Si/Ci},$$

which can be expressed as:

$$I = \frac{So}{Si} CA. \qquad (3.8)$$

This improvement factor is limited mainly by system instabilities, antenna motion, and clutter fluctuations.

Limitation from Clutter Statistics

The improvement factor in this case is dependent on the correlation function of the clutter signal evaluated at a single point—the interpulse time.

Approximate expressions for $I_1 > 10$ dB are as follows. For the single canceler,

$$I_1 = \frac{\lambda^2 f_r^2}{8\pi^2(\sigma_v^2 + V_0^2)}, \qquad (3.9)$$

where

σ_v = velocity spread of clutter, in meters per second.

For the double canceler,

$$I_2 = \frac{1}{2} I_1^2 = \frac{\lambda^4 f_r^4}{128\pi^4 \sigma_v^4} \text{ (for } V_0 = 0\text{)}. \qquad (3.10)$$

Curves for I_1 and I_2 are shown in figures 3–11 and 3–12 as a function of Doppler spread when $V_0 = 0$. These curves apply to either clutter-locked systems or ones that receive clutter of zero mean Doppler. The effects of a

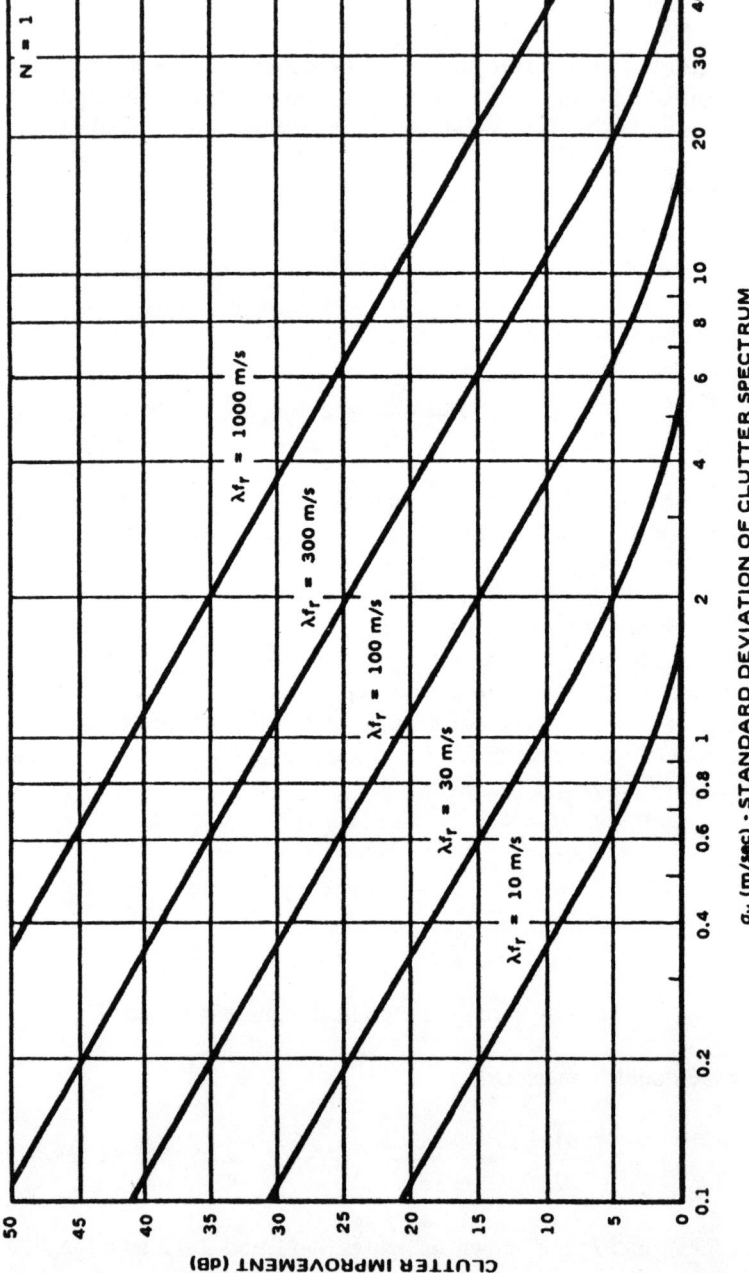

Figure 3-11. Clutter Improvement for a Single-Canceler System as Limited by the Spectrum Width of Clutter

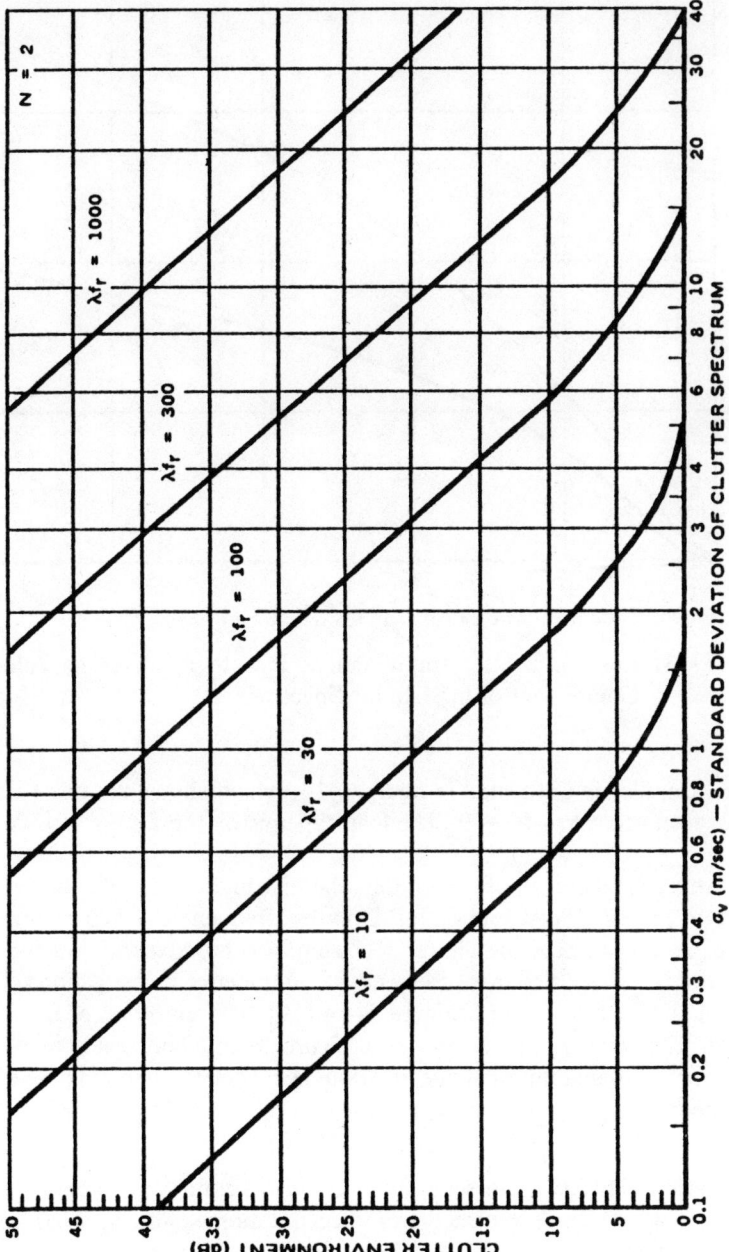

Figure 3–12. Clutter Improvement for a Double-Canceler System as Limited by the Spectrum Width of Clutter

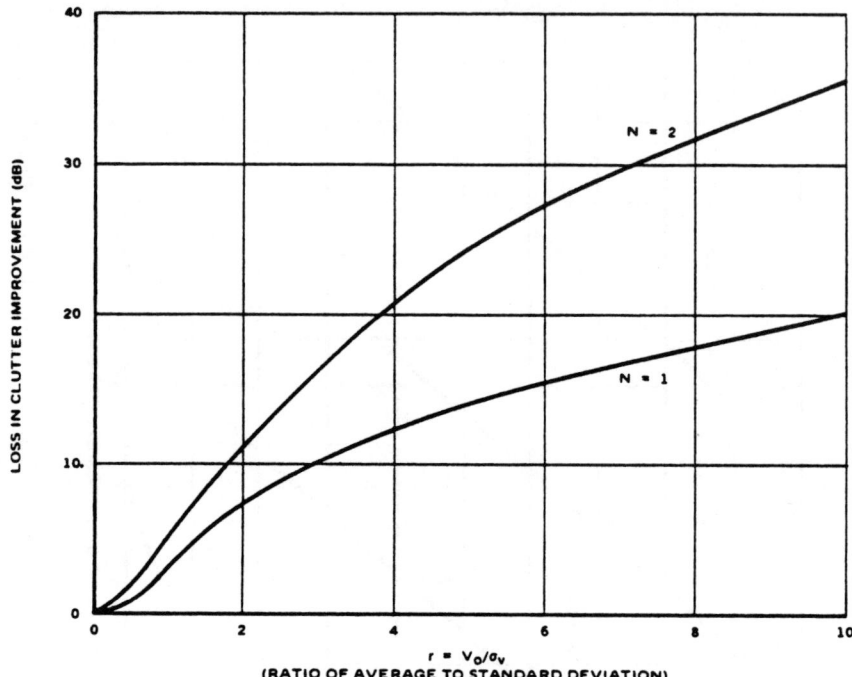

Figure 3–13. Loss in Clutter Improvement Due to an Average-Velocity Component of the Clutter Spectrum

nonzero Doppler is illustrated in figure 3–13, which shows the loss (in dB) from a system that has $V_0 = 0$. The loss depends on the ratio $r = V_0/\sigma_r$.

Sea Clutter. In the case of sea clutter, the distribution of velocities of the scatterers causes a distribution of Doppler frequencies. When the sea becomes more agitated, the clutter spectrum also broadens and hence is a function of sea state. The measurements of a number of investigations of the spectrum spread are shown in figure 3–14 [19]. The variation of the mean Doppler with wind speed is shown in figure 3–15. The resultant clutter improvement obtainable for single and double cancelers, including the effect of the mean Doppler component, is shown in figures 3–16 and 3–17.

Precipitation Clutter. Four separate mechanisms contribute to the Doppler spectra of radar signals returned from precipitation—namely, wind shear, beam broadening, turbulence, and fall-velocity distribution. If we assume that these effects are independent, then the variance of the Doppler velocity

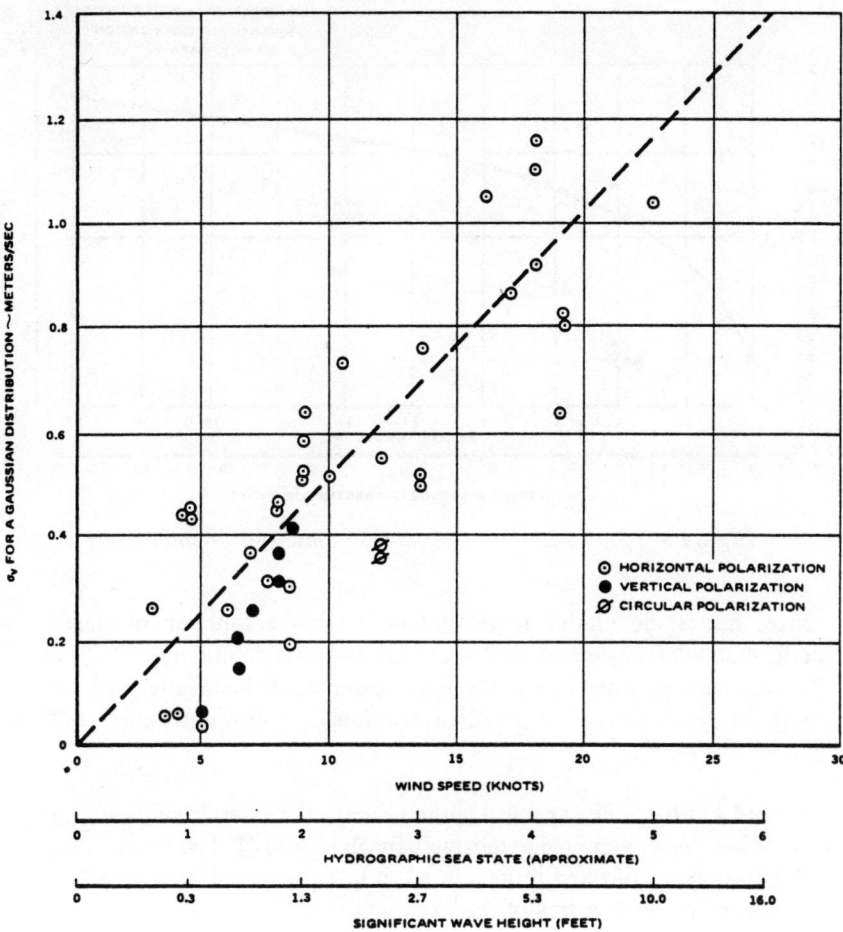

Figure 3-14. Velocity Spread for Coherently Detected Sea-Clutter Signals

spectrum, σ_v^2, can be represented as the sum of the variance of each contributing factor:

$$\sigma_v^2 = \sigma^2_{\text{shear}} + \sigma^2_{\text{beam}} + \sigma^2_{\text{turb}} + \sigma^2_{\text{fall}}.$$

Of these, the spread due to shear and turbulence will have the dominant effect.

The combined effect of these factors is given in figure 3-18 for the wind-speed profile shown in figure 3-19. The spectrum width of precipitation clutter is a function of range because the wind-shear effect is a function of

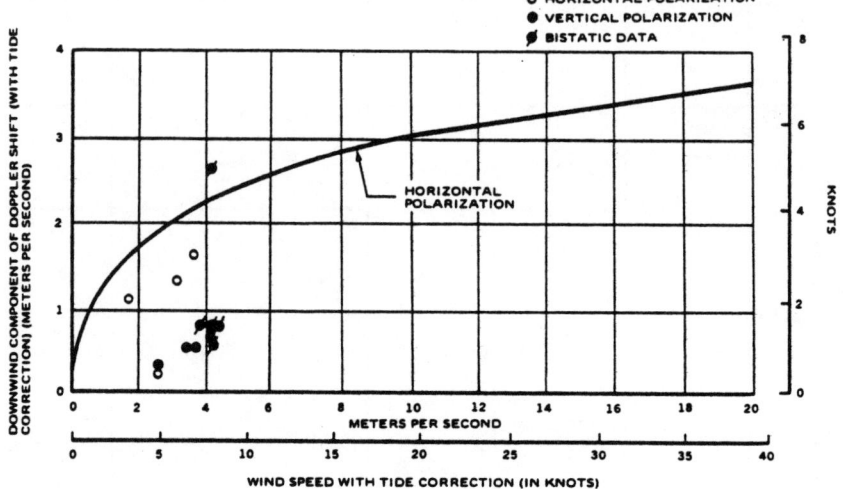

Figure 3-15. Variation of Mean Doppler with Wind Speed

range; hence the clutter improvement is also a function of range. The performance of single and double cancelers is shown in figures 3-20 and 3-21, respectively, when clutter locking is assumed. When clutter locking is not used, the performance is degraded accordingly, as shown in figures 3-22 and 3-23.

Ground Clutter. The spectral spread due to the internal motion of ground clutter has been examined extensively by Barlow [14]. The frequency spread of various types of fixed targets is given in figure 3-24 and table 3-1. The frequency spread in terms of the exponent a is

$$\sigma_c = \sqrt{8a \ \sigma_v} \qquad (3.11)$$

The effect on clutter attenuation for a single canceler and fixed targets is given as a function of the clutter exponent a by Grisetti, Santa, and Kirkpatrick [18] in figure 3-25. For a double canceler that can be determined from the relation $I_2 = 1/2I_1^2$.

Chaff. The theory pertaining to chaff clutter is almost exactly the same as that for weather clutter. The total cross-section of chaff weighing W pounds has been shown to be

$$\sigma \simeq 3{,}000 \frac{W}{f_0} \ \text{m}^2, \qquad (3.12)$$

Radar Performance in Clutter

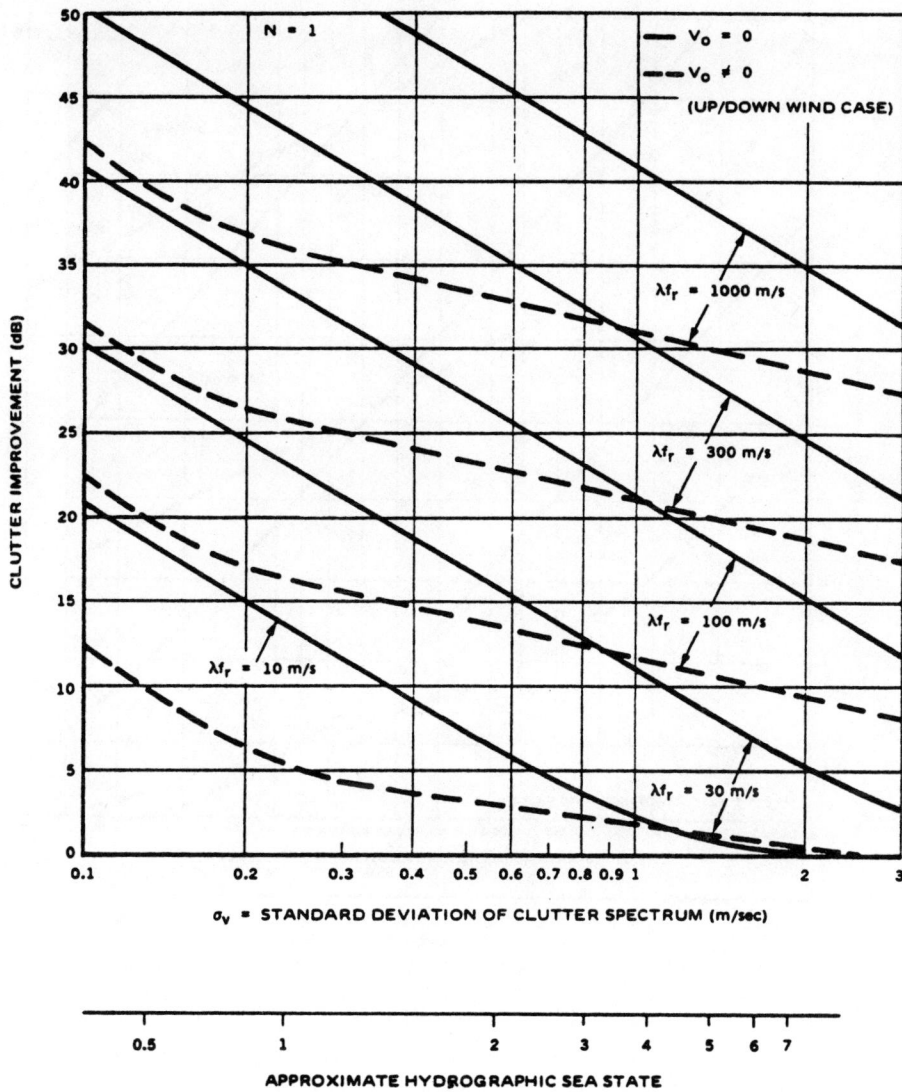

Figure 3-16. Single-Canceler Clutter Improvement as Limited by Sea Clutter

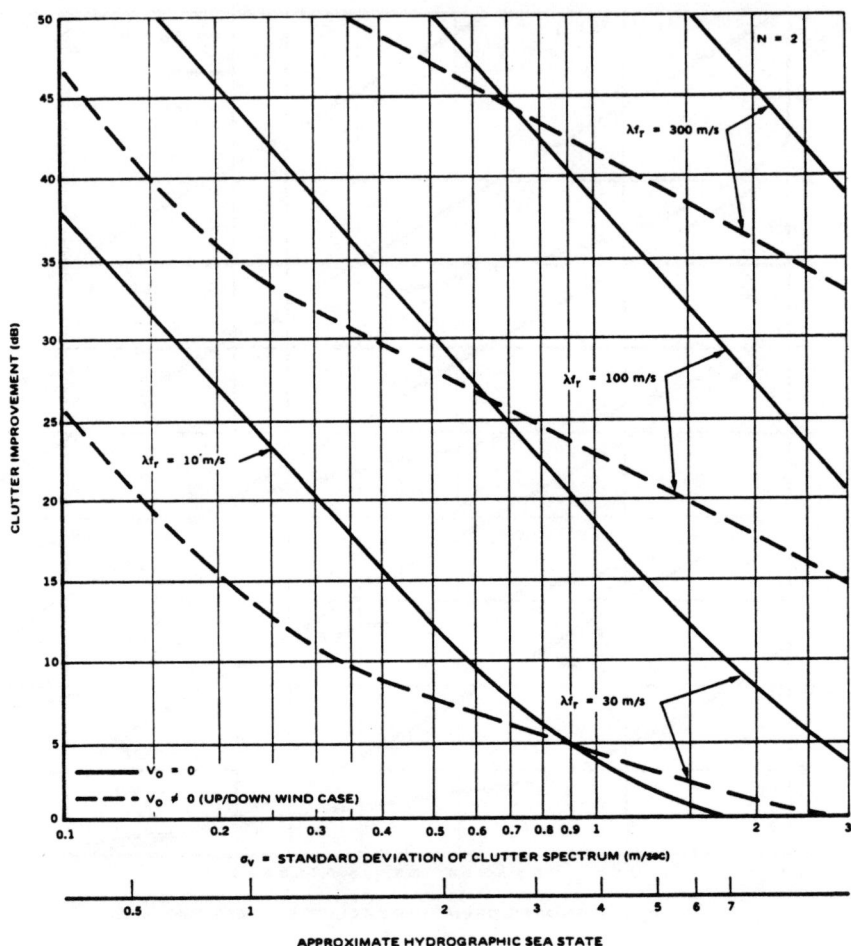

Figure 3-17. Double-Canceler Clutter Improvement as Limited by Sea Clutter

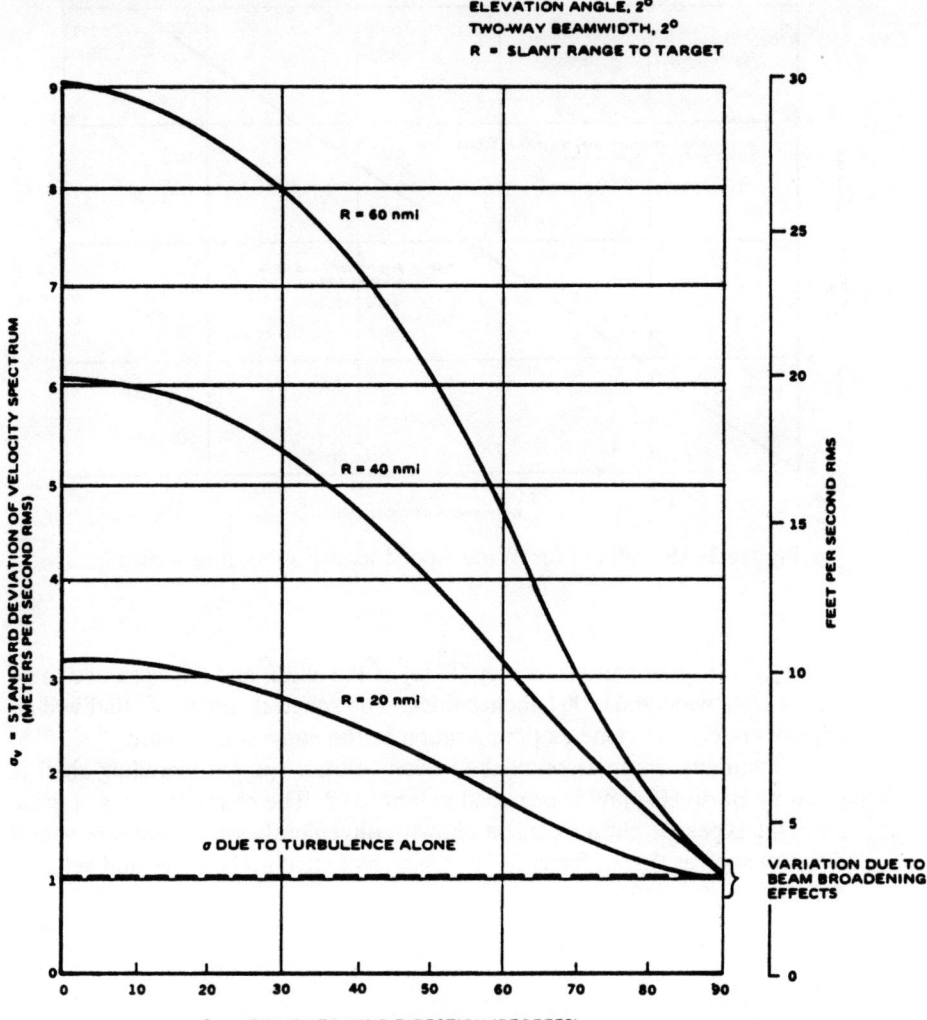

Figure 3–18. Standard Deviation of Doppler Spectrum for Precipitation as a Function of Azimuth Angle

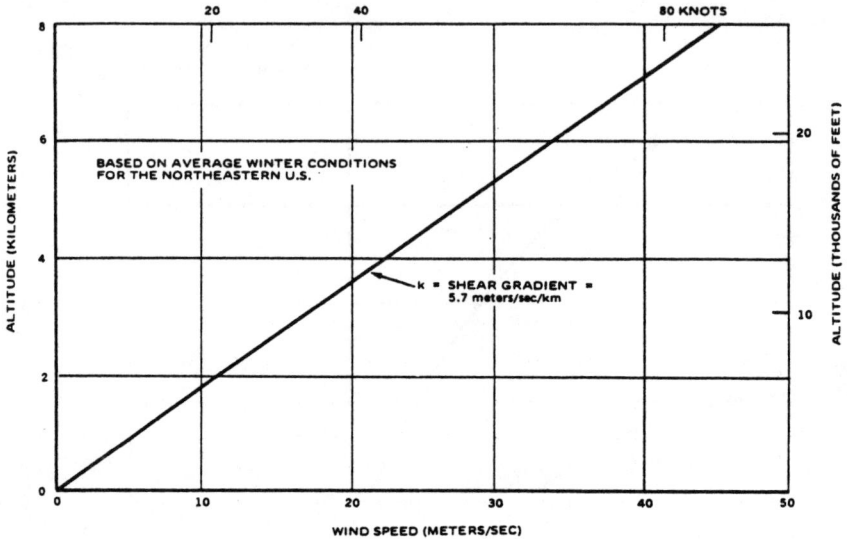

Figure 3-19. Model for Wind Speed as a Function of Altitude

where f_0 is the resonant frequency (kHz) of the chaff, and it is assumed to be 0.01 inches wide and 0.001 inches thick. The spectral spread of chaff will be approximately the same as precipitation in the same atmosphere.

A summary comparison of the various clutter spectra including chaff is given by Barton [7] and is provided in table 3-2. The characteristics of these different types of clutter can be shown conveniently on an environmental diagram such as that of figure 3-26, giving its extent in the range and velocity domain.

Table 3-1
Target Spectra

Target	a	Curve in Figure 3-24	σ S-Band (Hz)	σ C-Band (Hz)
Heavily wooded hills, 20-mph wind blowing	2.3×10^{17}	1	4.4	8.8
Sparsely wooded hills, calm day	3.9×10^{19}	2	0.34	0.68
Sea echo, windy day	1.41×10^{16}	3	17.9	35.8
Rain clouds	2.8×10^{15}	4	40.0	80.0
Window jamming (chaff)	1×10^{16}	5	21.2	42.4

Radar Performance in Clutter

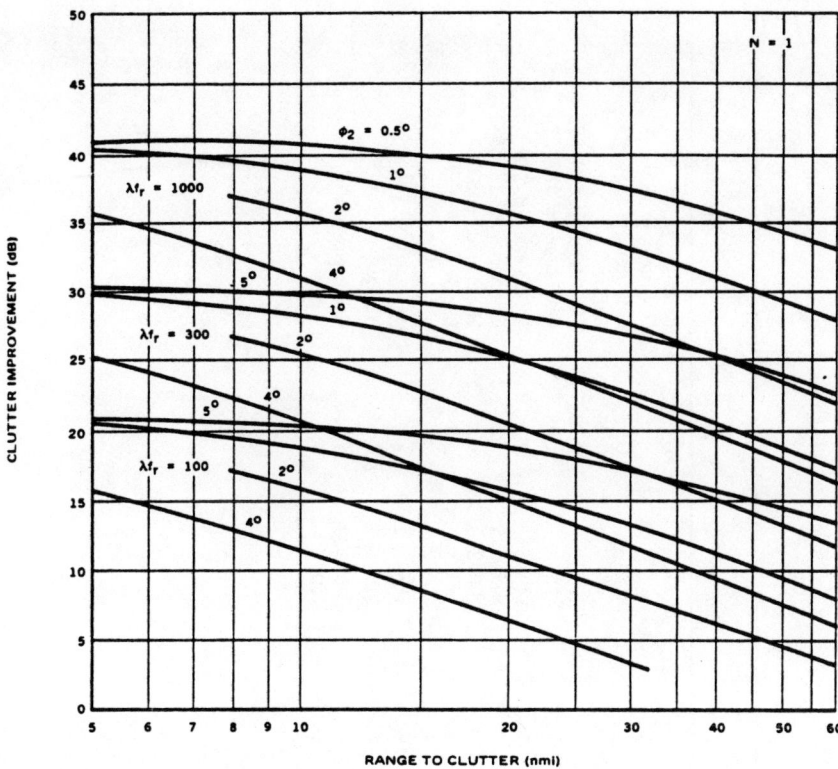

Note: Radar looking up/down wind.

Figure 3–20. Single-Canceler Clutter Improvement as Limited by Precipitation Clutter—Clutter Locked

Limitation from Errors in Estimating Mean Clutter Velocity

As shown previously, the mean-velocity component of clutter can seriously degrade MTI performance. The system that compensates for the mean-velocity component in the clutter-locked system was described earlier. However, since the estimate of the mean clutter frequencies involves a finite number of samples, it will have a statistical fluctuation. Since the loss in clutter improvement depends on the ratio $r = V_0/\sigma_v$, the rms ratio for the clutter-locked system can be shown to be $\bar{v} = 1/\sqrt{n}$, where n is the number of samples. For instance, for a clutter-locked system that uses two samples, from figure 3–13 the loss for a single canceler and a double canceler would be 1.5 and 3.0 dB, respectively.

Table 3-2
Characteristics of Clutter Spectra

Source of Clutter	Wind Speed (Knots)	Ratio m^2	Barlow's a	$\sigma_c \lambda$ (cm/sec)	σ_v (ft/sec)	Reference
Sparse woods	(calm)	30	3.9×10^{19}	3.5	0.057	Barlow
Rocky terrain	10					Goldstein, p. 583
Wooded hills	10	5.2	7.2×10^{18}	8	0.13	Goldstein, p. 583–585
Wooded hills	20		2.3×10^{17}	45	0.74	Barlow
Wooded hills	25	0.8	9×10^{17}	23	0.38	Goldstein, pp. 583–585
Wooded hills	40	0	1.1×10^{17}	65	1.06	Goldstein, pp. 583–585
Sea echo			2.4×10^{16}	140	2.3	Wiltse et al., p. 226
Sea echo			$(1-2) \times 10^{16}$	165–205	2.5–3.3	Goldstein, pp. 580–581
Sea echo	8–20	0	$(0.6–2.6) \times 10^{16}$	100–220	1.5–3.5	Hicks et al., p. 831
Sea echo	(windy)		1.4×10^{16}	183	3.0	Barlow
Chaff	0–10	0	$(1.4–8) \times 10^{16}$	75–180	1.2–3.0	Goldstein, p. 472
Chaff	25	0	7×10^{15}	250	4.1	Goldstein, p. 472
Chaff			10^{16}	215	3.5	Barlow
Rain clouds		0	$(0.7–3) \times 10^{15}$	370–800	6–13	Goldstein, p. 576
Rain clouds			2.8×10^{15}	410	6.7	Barlow

Basic relationships:

$$\sigma_v = \frac{\sigma_c \lambda}{2} \text{ (cm/sec)} = \frac{\sigma_c \lambda}{61} \text{ (ft/sec)};$$

$$\sigma_v = \frac{\sigma_c}{\sqrt{8a}} = \frac{10^9}{\sqrt{8a}} \text{ (ft/sec)} \qquad a = \frac{\sigma_c^2}{8\sigma_v^2};$$

$$\sigma_c = (0.85) \text{ (half-power spectrum width)}.$$

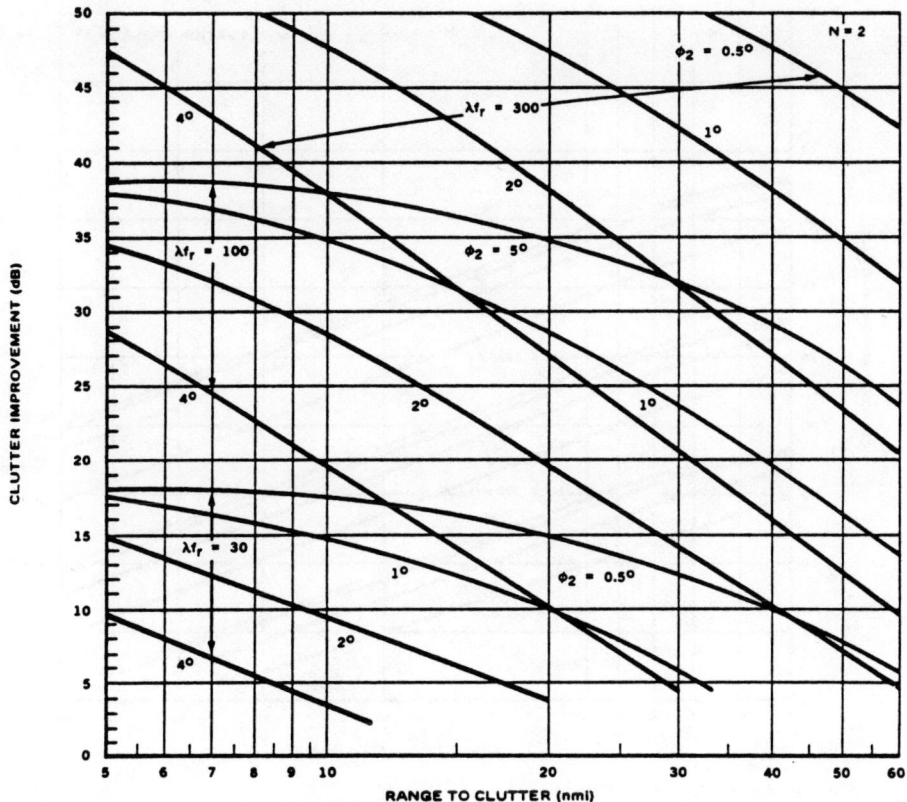

Note: Radar looking up/down wind. Mean velocity assumed to be zero.

Figure 3–21. Double-Canceler Clutter Improvement as Limited by Precipitation Clutter—Clutter Locked

Limitations from System Instabilities

Phase Instabilities. Unfortunately, an MTI system is unable to distinguish phase changes caused by system instabilities from those caused by the motion of targets. These phase instabilities may cause the canceler circuit to produce a residue for fixed-target echos.

For small values of phase error, $\sin \Delta\phi = \Delta\phi$ ($\Delta\phi$ is the phase error over the interpulse period), and the improvement limitation is given by the following equations:

single canceler:

$$I_1 = \frac{2}{\Delta\phi^2} \ . \tag{3.31}$$

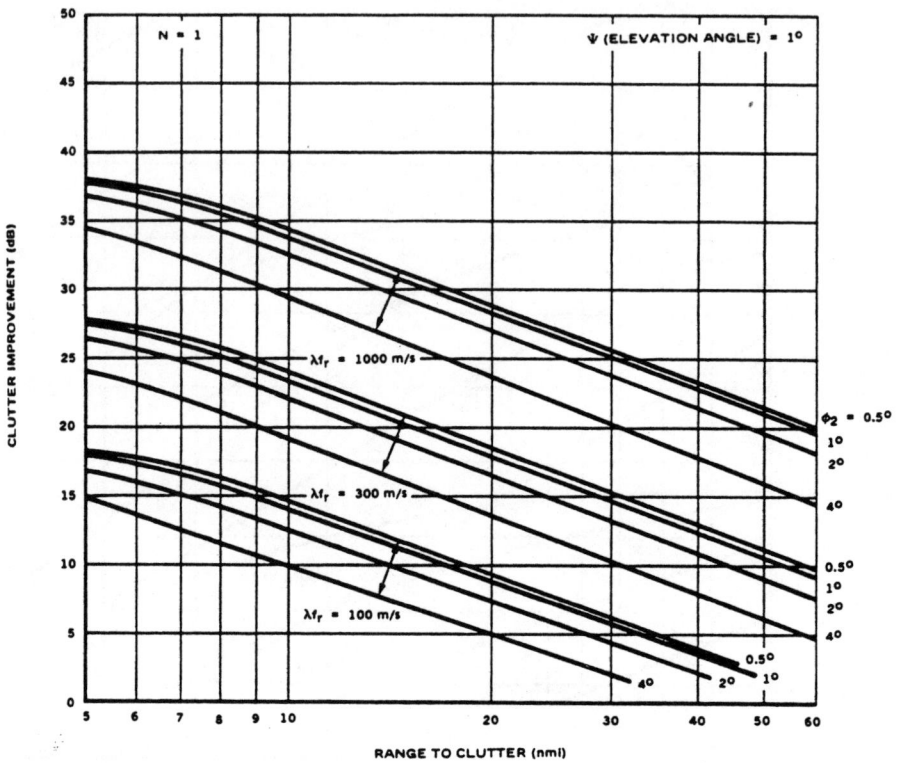

Note: Radar looking up/down wind.

Figure 3–22. Single-Canceler Clutter Improvement as Limited by Precipitattion—No Clutter Locking

Double canceler:

$$I_2 = \frac{6}{\Delta\phi^4} \quad \text{(for } \Delta\phi_1 = \Delta\phi_2\text{)}, \tag{3.14}$$

$$I_2 = \frac{3}{\Delta\phi^2} \quad \text{(For } \Delta\phi_1 \text{ independent of } \Delta\phi_2\text{)}, \tag{3.15}$$

where $\Delta\phi_1$ and $\Delta\phi_2$ are the phase errors during the first and second interpulse periods, respectively.

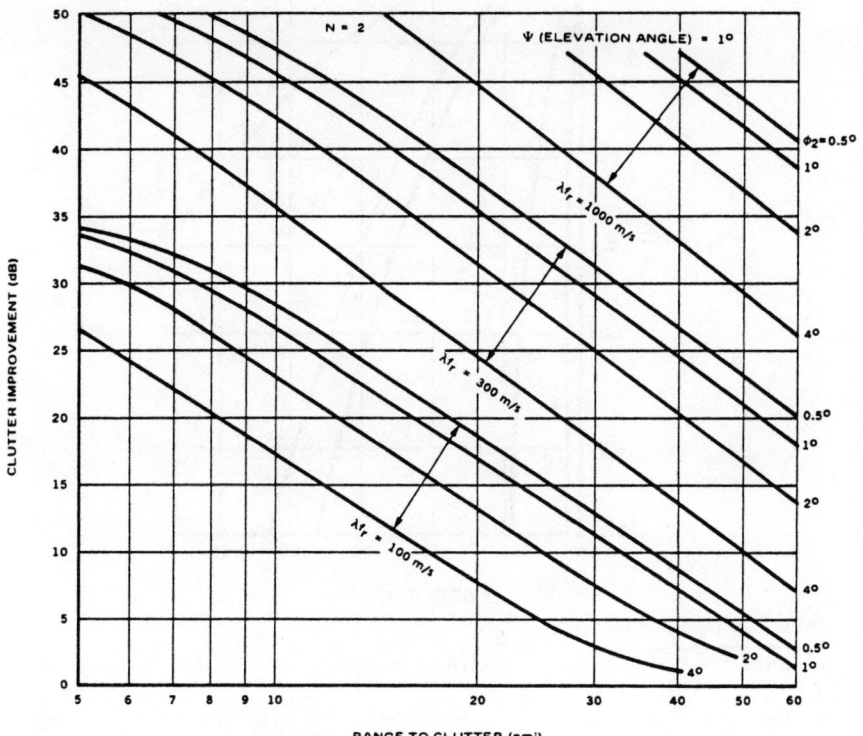

Note: Radar looking up/down wind.

Figure 3-23. Double-Canceler Clutter Improvement as Limited by Precipitation—No Clutter Locking

Limitations from Pulse Jitter. If the echoes from a target have time jitter, the output of the canceler will consist of two spikes of width equal to the difference between the timing errors for the two pulses. The improvement ratio for a single or double canceler is:

$$I_1 = I_2 = \frac{\tau^2}{2\sigma_\varepsilon^2}, \qquad (3.16)$$

where τ is the pulse length and σ_ε is the variance of the leading time of the pulse. Also, for pulse-width variations:

$$I_1 = I_2 = \frac{\tau^2}{\sigma_\delta^2}, \qquad (3.17)$$

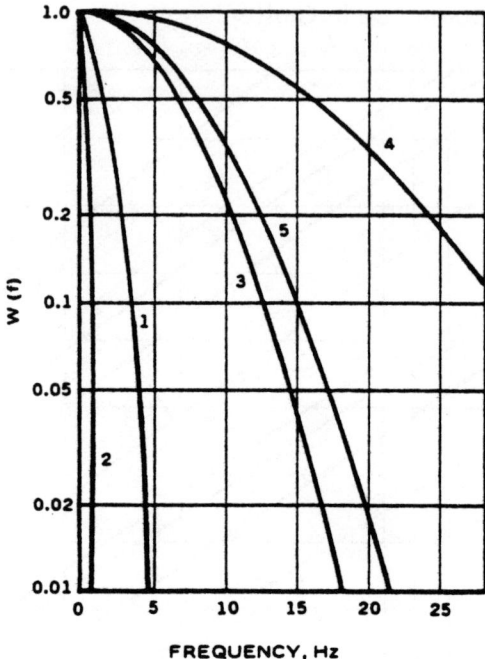

Note: See table 3–1 to identify curves.

Figure 3–24. Frequency Spectra of Various Types of Fixed Targets

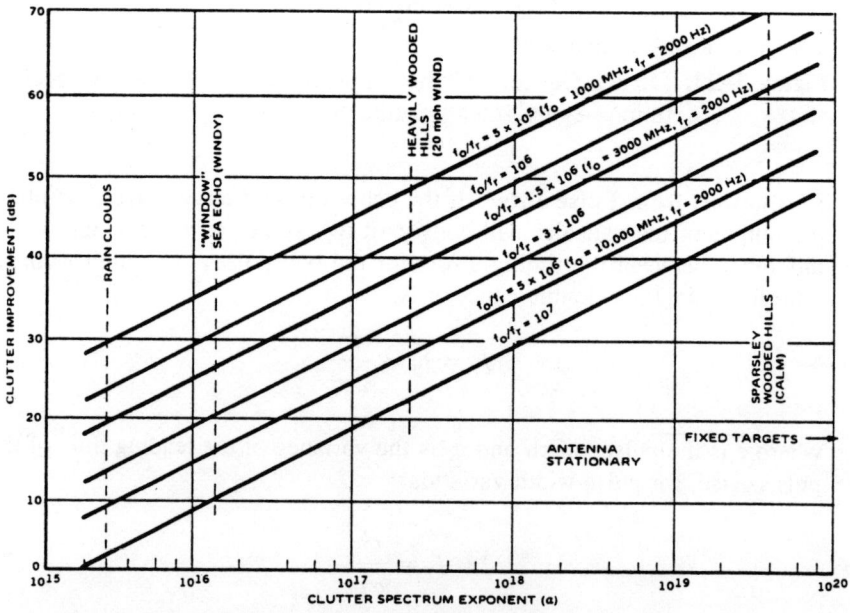

Figure 3–25. Effect of Internal Fluctuations on Clutter Attenuation

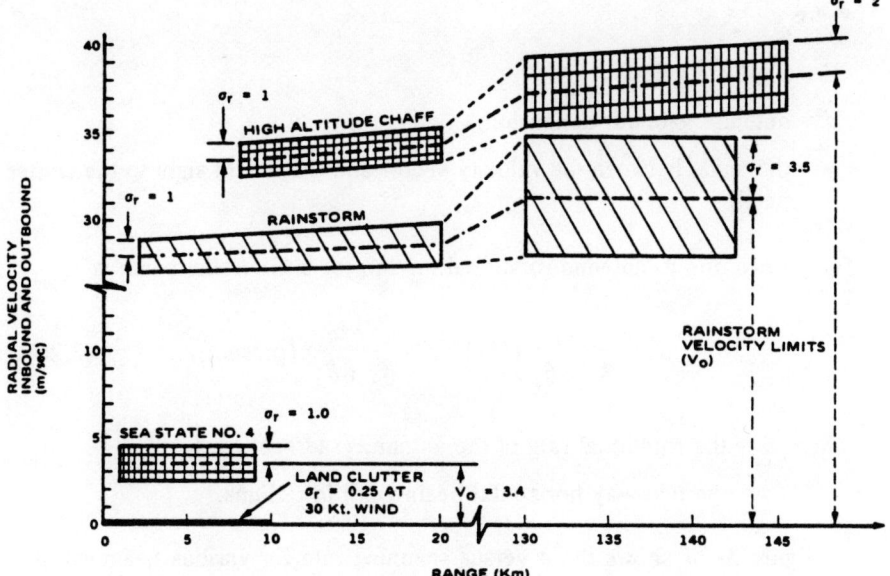

Figure 3–26. Environmental Diagram for Air-Defense Radar

where σ_δ^2 is the variance of the variation in pulse width.

Limitations from Amplitude Instabilities. A cancelor residue can result from variations in signal amplitude due to pulse-to-pulse fluctuations in transmitted power or signal gain. This ratio is:

$$I_1 = I_2 = \frac{E^2}{\sigma_E^2}, \qquad (3.18)$$

where E is the signal voltage and σ_E is the variance. This limitation does not apply to the phase-processing system, since it is insensitive to amplitude fluctuations.

Limitations from Antenna Motion. The motion of the radar antenna can degrade MTI performance because of the spectrum broadening of the clutter echoes. This results from translational motion of the antenna as well as rotational motion. The translational motion is important for a rapidly moving object such as an aircraft. The spectral spread due to platform motion is

$$\sigma = \frac{0.625}{d} V \sin \gamma \text{ (Hz)}, \text{ or, } \sigma = \frac{0.312 \lambda V \sin \gamma}{d} \text{ (m/sec)}, \qquad (3.19)$$

where

V = platform velocity;

d = antenna azimuth dimension;

γ = the angle between the velocity vector and the line of sight to the clutter patch.

The spread due to antenna rotational motion for a Gaussian beam is

$$\sigma = \frac{\alpha}{3.78\theta_2} \text{ (Hz) or} = \frac{\lambda\alpha}{7.6\theta_2} \text{ (m/sec)}, \qquad (3.20)$$

where α = the rotational rate of the antenna (rad/sec);

θ_2 = the two-way horizontal beamwidth in radians.

Figure 3–27 shows the σ versus scanning rate for various beamwidths. This value as well as the foregoing value for platform motion can be used with figures 3–11 and 3–12 to determine the limitation on clutter improvement.

Limitations in Staggered-PRF Systems. Blind speeds inherent in MTI operation pose a serious limitation to target detection. Figure 3–28 shows the loss due to a staggered PRF for various stagger ratios.

Limitations from Second-Time-Around Clutter. Although the second-time-around signal is often smaller than the first (depending on main-beam pointing angle), it is completely uncorrelated with the first signal and receives no attenuation. Hence it places a limitation on the maximum achievable clutter attenuation. For this case,

$$I_1 = \frac{2(R_1 + cT/2)^2}{R_1^2 \, \Delta E_2}, \qquad (3.21)$$

where R_1 = the range from which the first pulse was received;

T = the interpulse interval;

ΔE_2 = the percentage of the beam volume that is filled by clutter at a range of $R_1 + cT/2$.

Radar Performance in Clutter

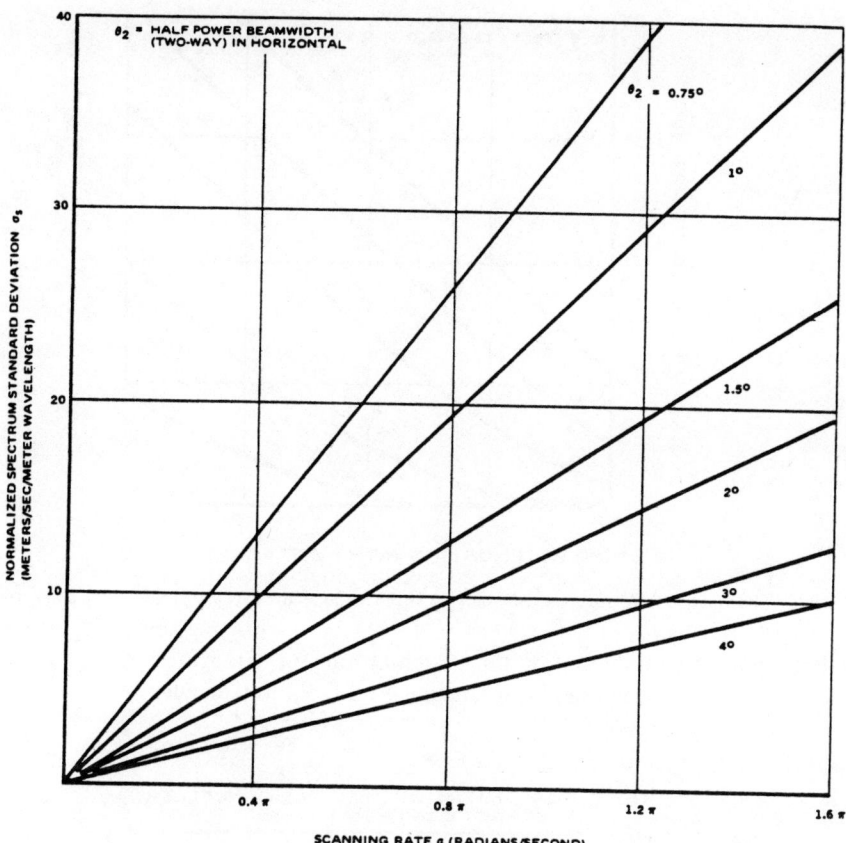

Figure 3-27. Normalized Standard Deviation of Clutter Spectrum Due to Antenna Scanning

For the double canceler:

$$I_2 = \frac{6(R_1 + cT/2)^2}{R_1^2 \, \Delta E_2}. \tag{3.22}$$

The improvement factor for a double canceler is shown in figure 3-29.

Note that the second-time or nth-time-around clutter can be canceled if sufficient pulses are transmitted before processing begins to establish steady-

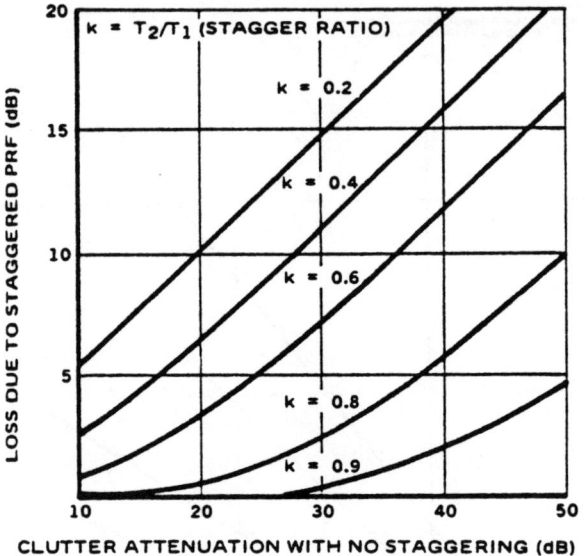

Note: The period of the unstaggered canceller is the same as the average period in the staggered canceler.

Figure 3–28. Loss in Clutter Attenuation for Staggered-PRF System Compared with Unstaggered Double Canceler

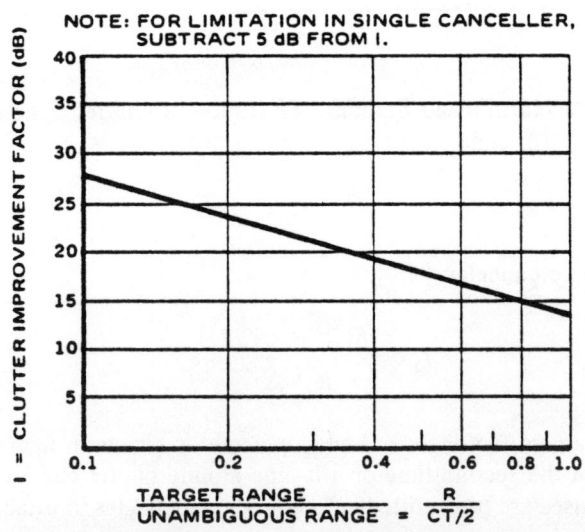

Note: Clutter is assumed to fill the beam completely at the ambiguous range.

Figure 3–29. Limitation on Clutter Improvement Due to Second-Time-Around Clutter for Double Canceler

Radar Performance in Clutter

state signal returns over all n ambiguous range intervals. To some extent this occurs automatically with scanning search and track radar that transmit many pulses while the antenna moves through an angle equal to the beamwidth. Step-scanning radars, such as fixed arrays with electronic-beam steering, must transmit additional pulses at each beam position before processing begins in order to establish steady-state signal conditions. This constitutes a direct loss of radar power and time that cannot be avoided in nth-time-around echoes are to be canceled.

MTI-System Improvement Factor

The previous section discussed the factors that limit the clutter-improvement ability of MTI radars. These individual limitations may be combined to give the system clutter improvement. The system improvement ratio can be determined from the individual ratios by:

$$\frac{1}{I_T} = \frac{1}{I_a} + \frac{1}{I_b} + \frac{1}{I_c}, \qquad (3.23)$$

where I_T is the total-system improvement factor, and I_a, I_b, I_c, and so forth are the individual improvements.

To illustrate the relative importance of the various limitations, consider a hypothetical radar with the following parameters:

- $\lambda = 50$ cm (transmitted wavelength);
- $f_r = 0.6$ kHz (pulse-repetition frequency);
- $\tau = 3$ μs (pulse width);
- $\phi_2 = \theta_2 = 1.5°$ (two-way vertical and horizontal beamwidths, respectively);
- $\alpha = 0.4\pi$ radians/sec (antenna scanning rate);
- $n = 2$ (number of cascaded cancelers) (type of cancellation scheme: IF canceler)

Assume the system instabilities are as follows:

- $\sigma_\varepsilon = 20$ nsec (timing jitter);
- $\sigma_\delta = 15$ nsec (pulse width variation);
- $\Delta\phi = 10$ milliradians, pulse-to-pulse independence (phase instability at the canceler frequency);
- $\sigma_E = 0.005$ (amplitude instability pulse-to-pulse variation).

Table 3–3
System Limitations for Example MTI

Source of Limitation	I_2 (dB)
A. Clutter motion	
1. Spectrum width $\sigma_v = (\sigma^2_{shear} + \sigma^2_{turb})^{1/2} = 4.1$ m/s	34
2. Antenna scanning ($\sigma_s = 3.25$ m/s)	38
3. Total σ_v (root sum square of 1 and 2) = 5.2 m/s	30
Additional loss due to mean-velocity component	
a. No clutter locking ($V_0/\sigma_v = 2.9$)	
Loss = 16 dB	
b. With clutter locking ($V_0/\sigma_v = 0.707$)	
Loss = 3 db	
4. Total limitation due to motion	
a. No clutter locking	14
b. With clutter locking	27
B. Timing jitter	40
C. Pulse-width variation	46
D. Phase instability	45
E. Amplitude instability	46
F. Second-time-around clutter	—[a]
G. Total-system-improvement factor	
1. No clutter locking	13.8
2. With clutter locking	24.6

[a] In this example the radar beam is completely out of the clutter at the first ambiguous range. Therefore, second-time-around clutter causes no limitation. (Altitude is determined with the aid of a "4/3 earth" chart.)

Let the source of clutter be precipitation at 50 nmi, where $\beta = 45°$ (azimuth with respect to wind direction) and $\Psi = 2°$ (elevation angle). The precipitation is assumed to exist up to 15,000 ft altitude. Table 3–3 lists the clutter-improvement limitation due to each individual cause. The additional loss due to the mean-velocity component is given when no clutter locking is used or, alternatively, when clutter locking is provided on the basis of two samples. Table 3–3 shows that the overall-system improvement is 13.8 dB when clutter locking is not used and 24.6 dB when it is used. In this example the primary limitation is the motion of the clutter.

Calculation of Clutter Magnitude

To calculate the clutter magnitude, first determine the area or volume of the clutter cell and then multiply by the backscatter coefficient σ_0, which is the radar cross-section per unit area. The geometry of the clutter cell for area and volume clutter is shown in figures 3–30 and 3–31. As noted here for area illumination, there is a transition between the beamwidth-limited case and the pulse-length-limited case depending on the angle of incidence. Representative values of σ_0 for sea echo illustrating the variation with grazing angle

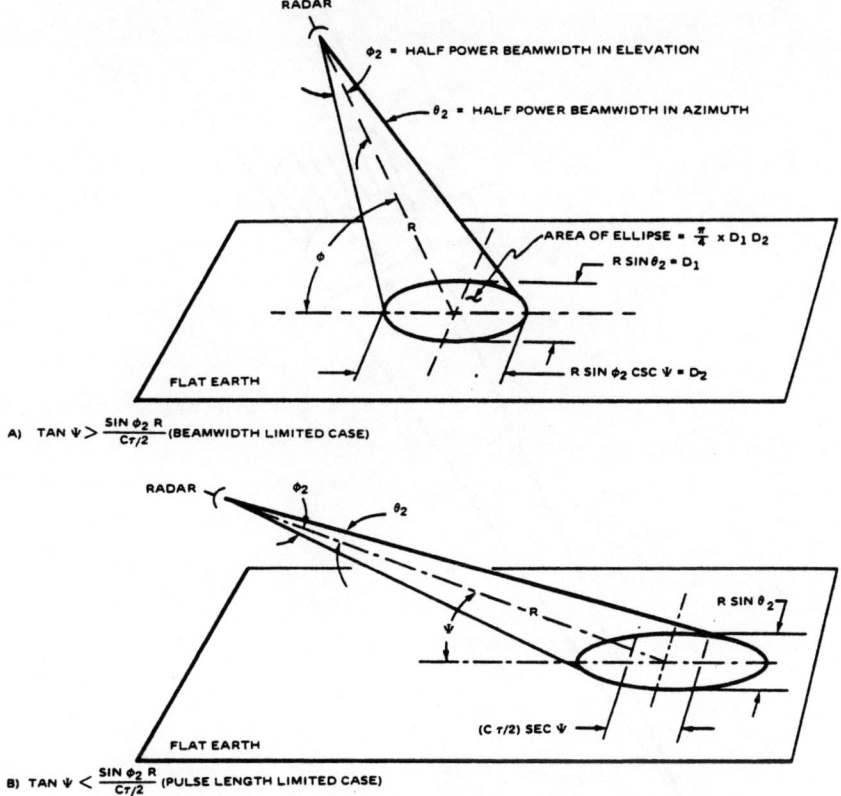

Figure 3–30. Area Illumination for Pulse Radar

and frequency are shown here from Skolnik [4]. Additional data that are probably more accurate at low grazing angles are shown in figure 3–33. The data of figures 3–32 and 3–34 illustrate a frequency dependence of between $f^{1.4}$ and $f^{2.1}$. There is also a polarization dependence that depends on the radar frequency and the roughness of the sea. At S-band the echo signal for smooth sea and vertical polarization can be 20–30-dB greater than with horizontal polarization. For wave heights of 3–5 ft, this polarization dependence disappears.

For precipitation, which is volume clutter, the radar reflectivity can be used. A plot of η versus precipitation rate is shown in figure 3–35, from Barton [7]. As indicated previously, chaff can be considered as similar to rain, and its cross-section has already been given.

For ground clutter the data of Krason and Randig [17] are most representative and are shown in figure 3–36. In this case γ is plotted, which is

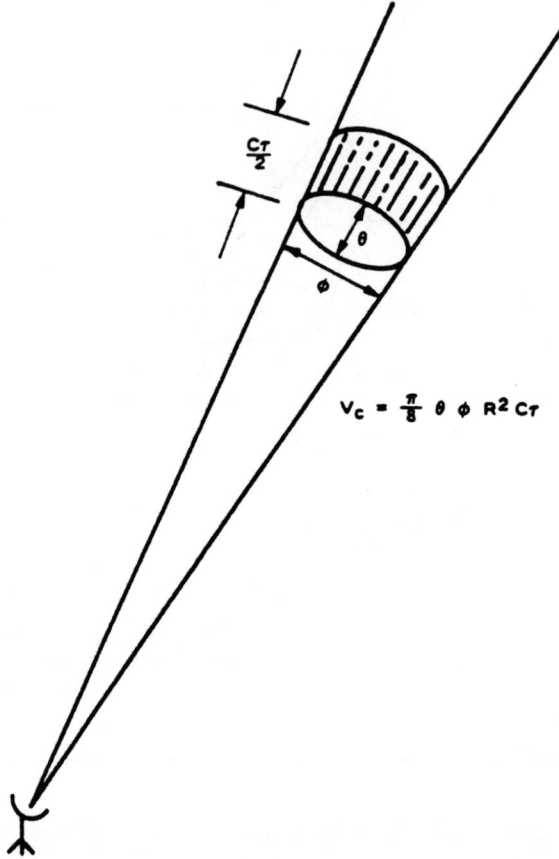

Figure 3-31. Volume Illumination for Pulse Radar

equal to $\sigma_0/\sin \Psi$, where Ψ is the depression angle. The value of γ at or near grazing tends to be independent of frequency, whereas at large angles there is an f-to-f^2 relationship. Clutter also exhibits a statistical fluctuation, as illusrated in figure 3-37 in data taken at X-band (from Pidgeon [16]).

To obtain the clutter cross-section at a particular range, σ_0 (η in the case of precipitation) is multiplied by the area or volume, respectively, of the clutter cell.

Referring back to figures 3-30 and 3-31, for area clutter the cross-section in the pulse-length-limited case is ($\sigma_0 \equiv \sigma^0$)

$$\sigma_c = \sigma_0 \ R = \frac{c\tau}{2} \sin \theta_2 \sec \Psi \qquad (3.24)$$

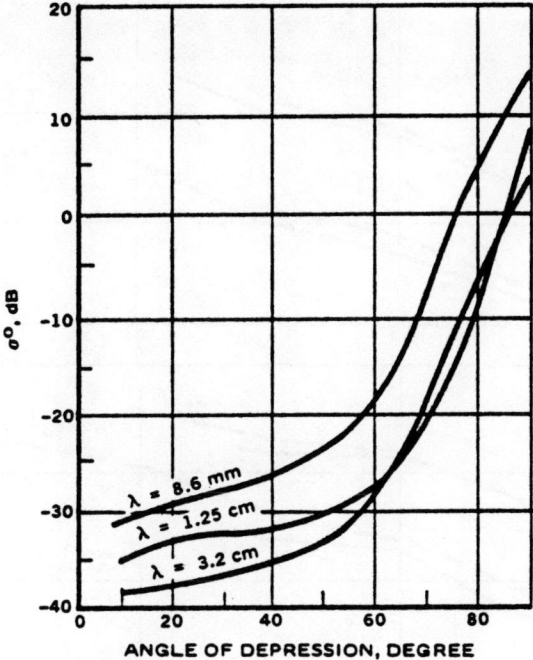

Figure 3–32. Variation of σ^0 for Sea Echo as a Function of Grazing Angle and Frequency

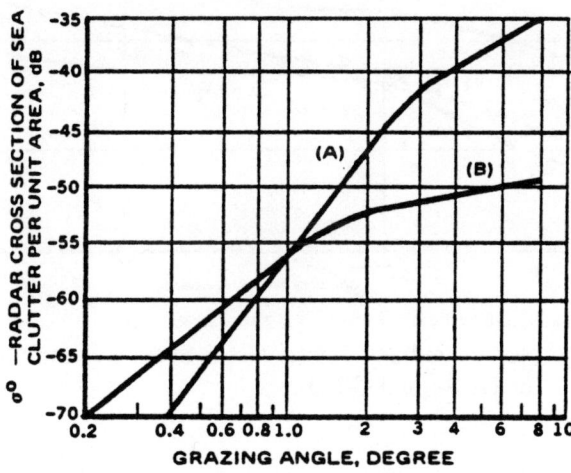

Note: (a) Slight sea; (b) moderate sea.

Figure 3–33. Variation of σ^0 for Sea Echo as a Function of Grazing Angle at Low (Small Angles)

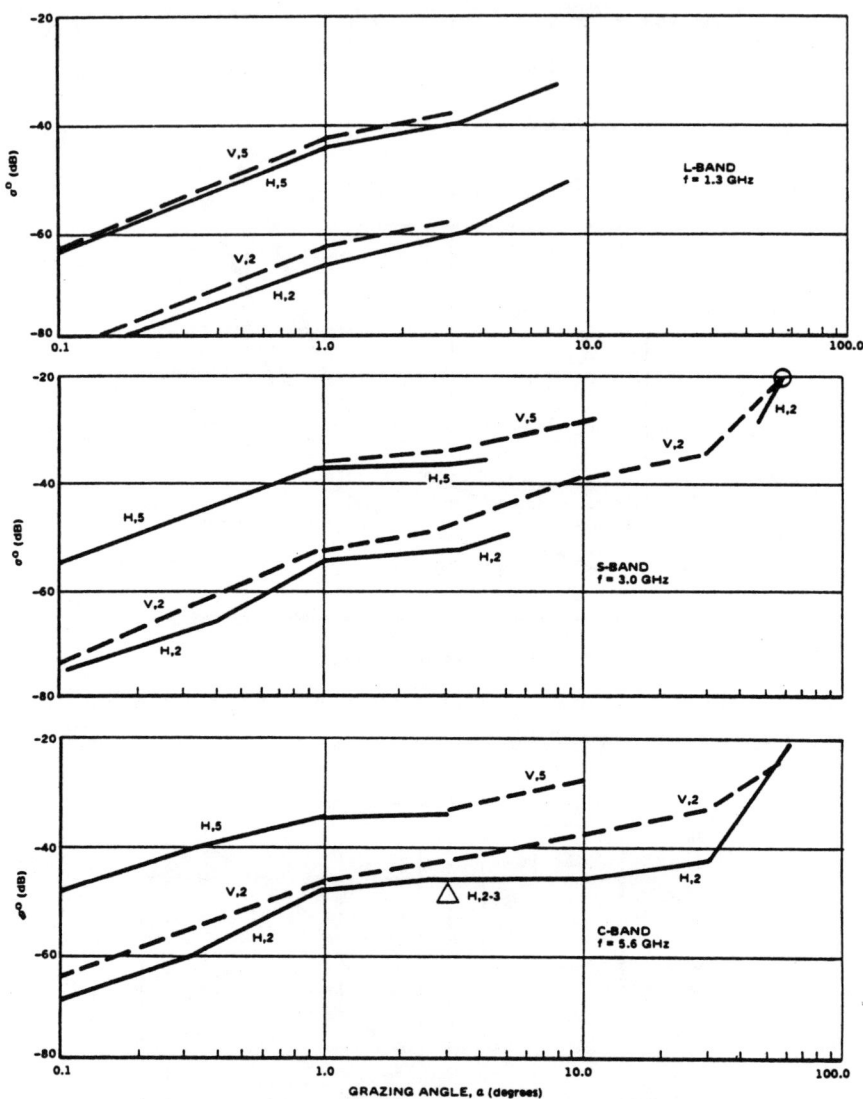

Note: for horizontal and vertical polarization at sea states 2 and 5.

Figure 3–34. Normalized Sea-Clutter Cross-Section *versus* Grazing Angle for L-, S-, and C-Bands

Radar Performance in Clutter

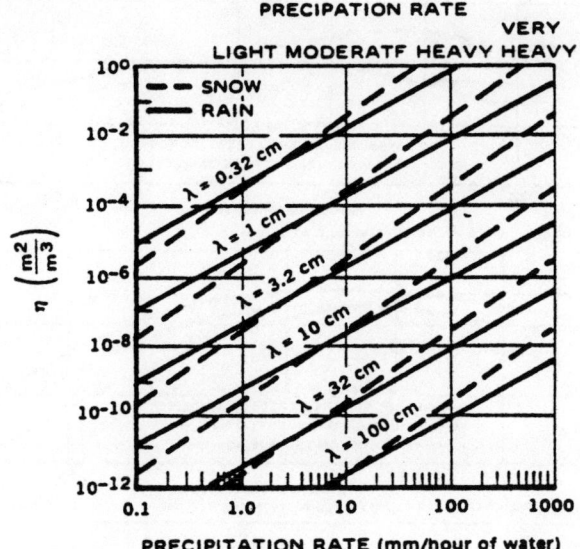

Figure 3–35. Radar Reflectivity *versus* Precipitation Rate

where σ_0 = the backscatter coefficient (m²/m²);

R = the range to the interfering clutter;

τ = the pulse length;

θ_2 = the half-power beamwidth in azimuth;

Ψ = the angle of incidence;

η = the reflectivity coefficient;

c = the velocity of light.

For the beamwidth-limited case

$$\sigma_c = \frac{\tau}{4}\, \sigma_0\, R^2 \sin\theta_2 \sin\phi_2 \csc\Psi, \qquad (3.25)$$

where ϕ is the half-power elevation beamwidth.

For volume clutter,

$$\sigma_c = \eta\, \frac{\pi}{8}\, \theta_2\, \phi_2 R^2\, \tau c, \qquad (3.26)$$

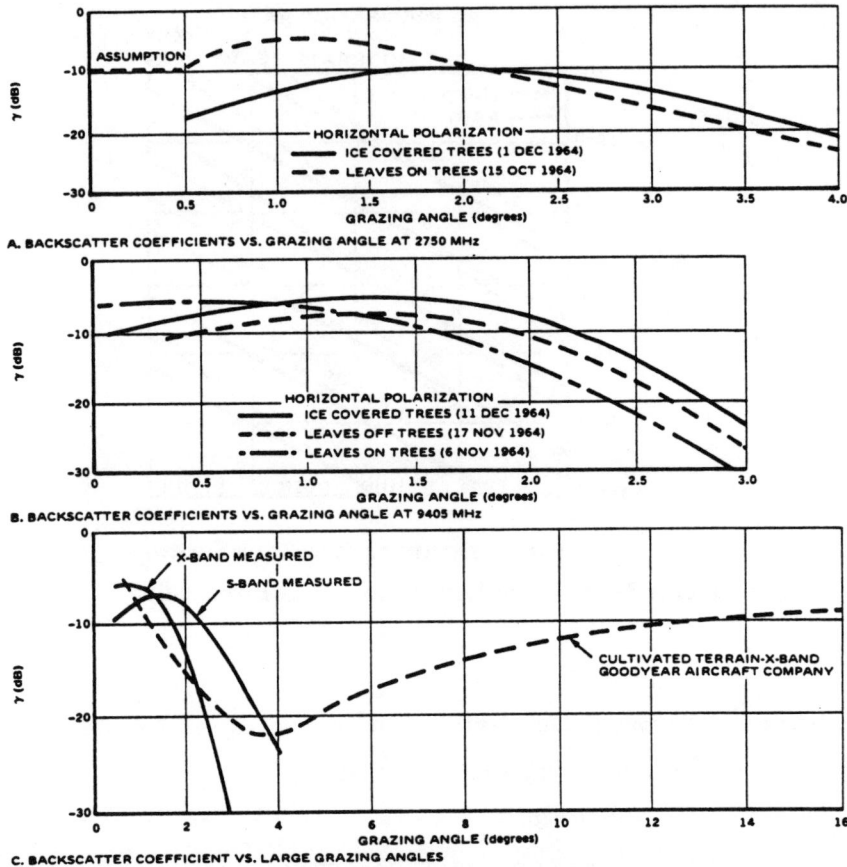

Figure 3-36. Backscatter Coefficients as a Function of Grazing Angle

where η = the reflectivity coefficient (m^2/m^3).

Clutter is also received in the sidelobes; and although the relative antenna gain is reduced at this point, the illuminated area is larger since clutter is being received from all angles. The expression for sidelobe-clutter cross-section is (assuming area clutter):

$$\sigma_{SL} = \pi R \ c\tau \ \sec \Psi \ \bar{G}_{SL} \ \sigma_o \qquad (3.27)$$

where σ_{SL} = sidelobe clutter magnitude relative to a target in the mainbeam;

$\bar{G}_{SL}$ = the average two-way antenna sidelobe gain relative to two-way mainbeam gain.

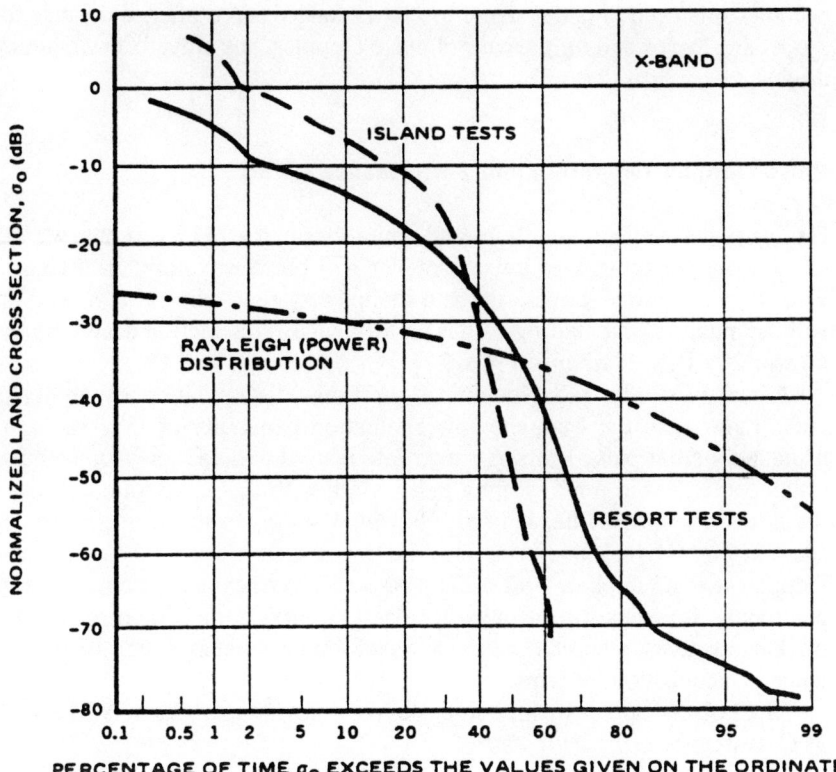

Figure 3-37. Probability Distribution of Land Cross-Section per Unit Area, σ_0

If range is ambiguous, clutter reflected from the shorter ranges competes with target energy at longer ranges, and the expression is modified to:

$$\sigma_{sc} = \pi R_c c\tau \sec \Psi \, \bar{G}_{SL} \left(\frac{R_t}{R_c}\right)^4 \sigma_0 \qquad (3.28)$$

where σ_{sc} = sidelobe-clutter magnitude relative to a target in the main beam at range R_c interfering with a target at range R_t. This expression is approximate since clutter is returned from a series of clutter rings at all ambiguous range intervals. More accurate calculations of this clutter can be obtained in a computer simulation, which is discussed later.

The clutter-to-target ratio is obtained by dividing the clutter cross-section thus obtained by the target cross-section. The improvement factor required becomes this ratio plus the signal-to-clutter ratio desired after allowance for processing losses and, in the case of pulse Doppler, any waveform weighting used.

Pulse-Doppler Operation and Performance Limits

Having reviewed the capabilities and limitations of the MTI systems, we will discuss the advantages of pulse Doppler and burst waveform in the same vein. Representative waveforms and Doppler systems will be reviewed, as will the means of calculating clutter performance. This section draws mainly on work by F.E. Nathanson [13, 27].

A pulse-Doppler radar combines the range-discrimination capability of pulse radar with the frequency-discrimination capability of CW radar by using a coherent pulse train. A range gate is used to select only those pulse trains coincident in time within a pulse width with the pulse-train echoes of the target. A narrow filter is used following the range gate to select a single spectral line. Burst waveforms may be considered a special case of pulse Doppler with an implied limit of 50–100 pulses per beam position. The truly continuous transmission of pulses (pulse Doppler) is more appropriate for tracking systems, whereas the burst waveforms are better suited to surveillance or acquisition systems.

The pulse-Doppler technique has several advantages over CW-pulse or MTI systems:

1. the ability to measure range and velocity unambiguously over a large region of the ambiguity plane;
2. the ability to reject unwanted echoes in the Doppler and range domains;
3. coherent rather than noncoherent integration of the returns from many pulses, with the attendent reduction in the required (S/N) or (S/C) per pulse;
4. less sensitivity than MTI systems to the mean velocity and spectral width of the unwanted clutter if the target velocity is separated from the center of the clutter spectrum by more than $1/T_d$, where T_d is the coherent integration time;
5. much greater rejection of signal leakage directly from transmitter to receiver, and often greater close-in clutter rejection than with CW systems.

The only truly general limitation is that pulse-Doppler receiver circuitry is usually more complex than that found in pulse, CW, or simple MTI receiver systems.

The Ambiguity Diagram

Since the pulse-Doppler technique implies many pulses per beamwidth, ambiguity is difficult to avoid. This is particularly true with the advent of large-phased arrays with long detection ranges and narrow antenna beams for which the time per beam position is considerably reduced. The ambiguity problem is most acute in search systems for air defense, where the PRF and pulse length must be chosen to accommodate a broad span of target ranges and radial velocities. The carrier frequency and PRF determine the Doppler ambiguities (blind speeds) from the usual Doppler equation:

$$f_d = \frac{2V}{\lambda} = \frac{2Vf_0}{C}, \qquad (3.29)$$

where f_d = Doppler frequency shift;

λ = the transmit wavelength = C/f_0;

V = the echo radial velocity.

For an unambiguous velocity response, it is required that $f_d <$ PRF or $1/T$, where T is the interpulse period, and that $R_{max} \leq C/2\,(T - \tau)$, where τ is the pulse duration. The solution of these inequalities is shown in figure 3–38. The dotted line shows that for ±300-knot velocity coverage at S-band (3,000 MHz), the unambiguous range is less than 27 nmi. It becomes obvious that for many short-wavelength applications, unambiguous operation is impractical and parameters should be chosen by careful study of the ambiguity diagram. The ambiguity function has proved useful in interpreting the performance resulting from the transmission of finite pulse trains in either multiple-target or clutter environments. The ambiguity diagram will now be examined for transmitted trains of rectangular pulses that are samples of a constant-frequency carrier.

Constant Interpulse Period

The ability of a particular waveform and processor to separate a moving target from stationary clutter is shown by its ambiguity function. Figure 3–39 shows an example of the ambiguity function associated with a stationary target. It has a maximum amplitude response at the point labeled $\Psi(0,0)$. If a moving target exists in the same radar-resolution cell, it will have an identical ambiguity function that is shifted along the Doppler velocity axis. Thus the

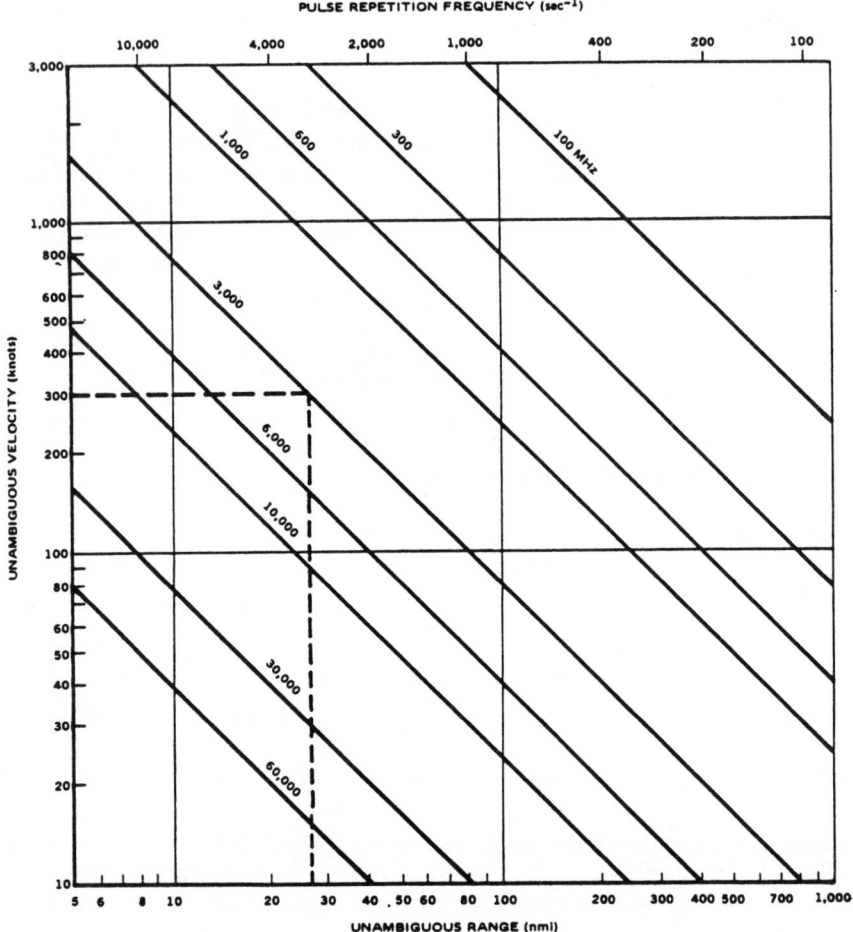

Figure 3–38. Unambiguous Velocity *versus* Unambiguous Range

maximum response of the target will coincide in time and frequency with some smaller response of the stationary target. By range gating in the time domain and filtering in the frequency domain, one may obtain an output signal that contains the peak signal from the moving target summed with a lesser response from stationary targets. Effectively, target-signal-to-clutter-signal ratio has been improved.

The uniform pulse train consisting of N rectangular pulses of length τ and interpulse period T is the simplest and most widely used pulse-Doppler waveform. Its utility lies in the complete absence of time ambiguities in the region between the transmit pulses. This waveform has been examined

Radar Performance in Clutter

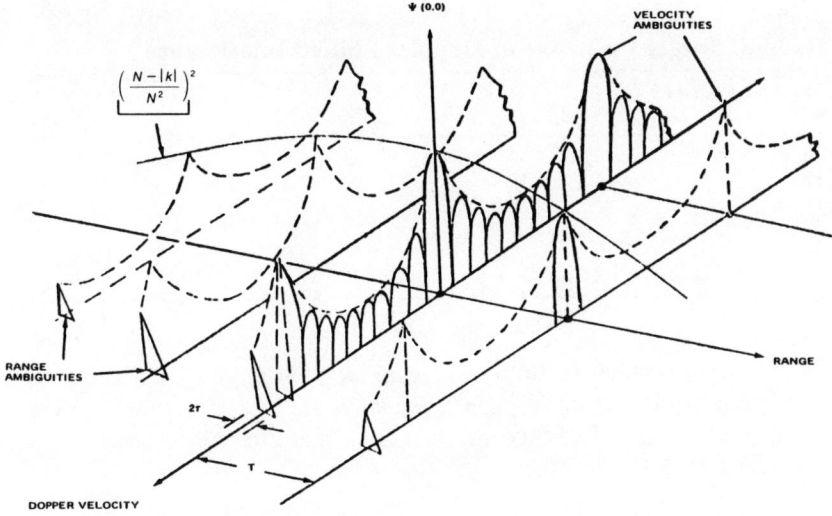

Figure 3-39. Ambiguity Diagram for Uniform Pulse Train

extensively by Resnick [22]. Figure 3-39 gives the ambiguity diagram of this type of waveform for $N = 9$. The central response is labeled $\Psi(0,0)$, and the velocity response near the range axis where $\tau = 0$ is of the form:

$$\Psi(o,\omega) = \frac{\sin^2 (1/2\ N\ \omega T)}{N^2 \sin^2 (1/2\ \omega T)} . \qquad (3.30)$$

The range response on the Doppler axis is the square of the autocorrelation function of the individual rectangular pulses repeated at intervals of T. For a finite pulse train the envelope of the range ambiguity peaks of period T decrease as

$$\frac{(N - |k|)^2}{N^2}, \qquad (3.31)$$

with k representing integer values of the interpulse period over $-(N-1) \leq k \leq (N-1)$. If the number of pulses is large, the sinusoidal sidelobes rapidly decrease from the values on the range axis at multiples of $1/T$. The Doppler response envelope is minimum at $1/2T$, where the peak sidelobe is roughly equal to $1/N$ times the peak amplitude ($1/N^2$ in power). Table 3-4 gives the average values of the sidelobe power response at other regions along the Doppler axis for various values of N and at various percentages of the unambiguous Doppler frequency. These values can be interpreted as the

Table 3-4
Ratio of Target Response to Doppler-Shifted Interference

Number of Pulses (N)	0.1/T, 0.9T (dB)	0.3/T, 0.7/T (dB)	0.5/T (dB)
8	—	19.1	21.0
12	15.4	22.5	24.6
16	16.5	24.7	27.0
20	18.1	27.3	29.0
24	20.2	29.0	30.6
32	23.0	31.5	33.1

average improvement in clutter-to-signal ratio for point clutter at the target range, but separated in Doppler frequency, and illustrate the interference rejection of the matched filter receiver for a uniform pulse train.

Staggered Pulse Train

Since in the time-frequency plane of the ambiguity diagram the area between range ambiguities has identically zero sidelobes, the uniformly spaced pulse train provides maximum freedom from extended clutter echoes. When the target is relatively small and the clutter is widely distributed or the interfering targets are large, this would be the best burst waveform to use. With the uniform pulse train, amplitude and pulse taper can be used, and each individual pulse can be modulated to increase range resolution.

The main disadvantages of the uniform pulse train are that range ambiguities will be close in unless the PRF is low, whereas Doppler ambiguities will be close in unless the PRF is high. The staggered pulse train, on the other hand, provides maximum freedom from significant ambiguities.

To construct a suitable waveform, the interpulse period should be varied so that a maximum number of time-sidelobe locations are generated. Resnick [22] has shown that

$$\frac{\Psi}{N^2}(0,0) \geq \Psi(\tau,0) \text{ for } \tau > \tau'. \qquad (3.31)$$

That is, the central spike of the ambiguity function is N^2 greater than any zero-Doppler time sidelobe in the region outside twice the pulse width. The ambiguity function for this type waveform is shown in figure 3-40. The waveform merits use where major range ambiguities cannot be tolerated and the maximum expected target-Doppler velocities are within the first few ambiguous regions, as illustrated in figure 3-41. For further details on this type of waveform, see Resnick [22].

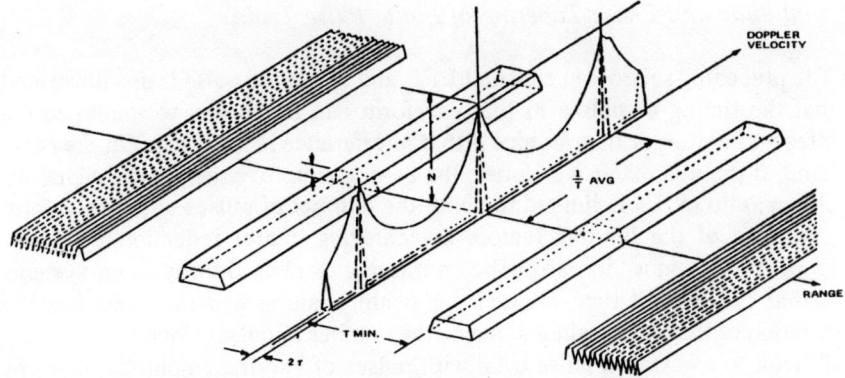

Figure 3–40. Envelope of the Correlation Function for a Nonuniformly Spaced Pulse Train of N Pulses

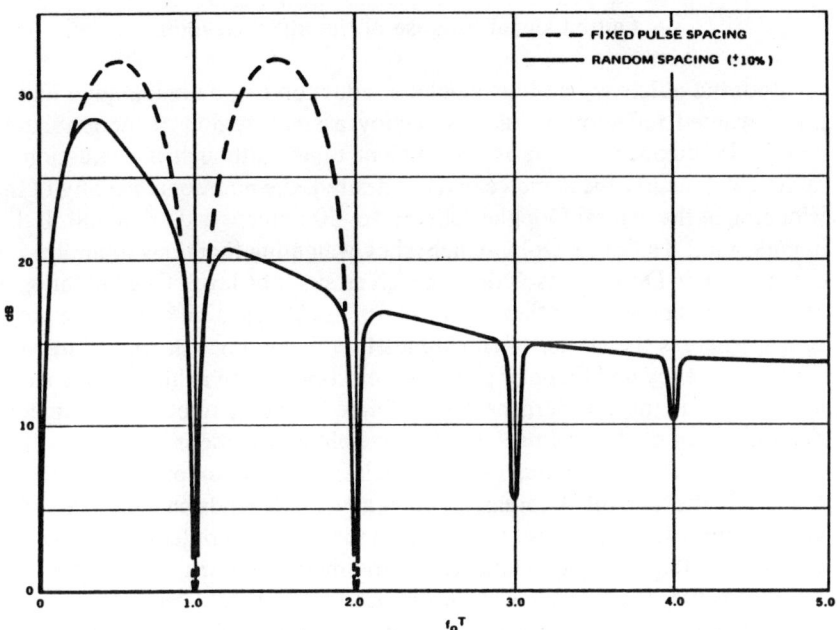

Figure 3–41. Pulse-Doppler Point-Clutter Improvement Factor

Amplitude and Phase Tapering of Finite Pulse Trains

The preceding section on constant PRF and staggered pulse trains illustrated that the timing of pulses in the waveform can be chosen to minimize the effects of clutter or interference if that interference is localized. On the other hand, if there is extensive clutter, the average improvement in the signal-to-clutter ratio (S/C) is limited to about the number of pulses in the waveform (N). One of the limiting factors in achieving clutter reduction by pulse-Doppler techniques in narrow-beam rotating or phased-array radar systems is that the limited time allotted per beam position generally restricts the duration of the pulse train and hence the number of pulses. Since it is usually difficult to transmit a pulse train with pulses of varying amplitude, it seems best to consider first the optimiation of the receiver for the constant-amplitude, uniformly spaced pulse train of finite extent. Taylor, Cosine, and Hamming amplitude weighting of the received pulse train can improve the signal-to-clutter ratio in many situations. A figure of merit for these receivers would be efficiency (E) for a given Doppler sidelobe reduction, defined as:

$$E = \frac{\text{Output signal-to-noise of weighted processor}}{\text{Output signal-to-noise of the matched filter}}.$$

Both the efficiency and the sidelobe reduction can be made quite high if the unwanted reflectors in targets occupy a small region of the ambiguity plane. Hamming weighting is an efficient taper, although if evaluation of targets with nearly the same velocity is desired, the additional penalty of the widening of the central Doppler lobe by 40–50 percent is detrimental. If this is important, Taylor or Dolph-Chebyshev weighting functions minimize the degradation in Doppler resolution for a given sidelobe level. One limitation of the antenna tapers is that they optimize for clutter rejection only at the target range, whereas the clutter often extends into the ambiguous range intervals. Better efficiency and Doppler sideband rejection can be achieved for specific separations if the interference has a limited range extent. Unfortunately, determination of the optimum receiver weighting functions usually involves variational calculus techniques that require a digital computer for solution. Generally the amount of optimization obtainable depends on the range extent of the clutter. If the extent of the clutter in range and Doppler is small, and its location in the ambigity plane is approximately known, substantial S/C improvements can be achieved on the use of a matched filter. If the range extent of the clutter increases to one-half the length of the total waveform, and the Doppler extent exceeds one-third the unambiguous velocity, then the pulse trains of 20–40 pulses. If the range extent of the clutter is comparable to the length of the waveform, the optimization yields only 2.6–3.0-dB improvement.

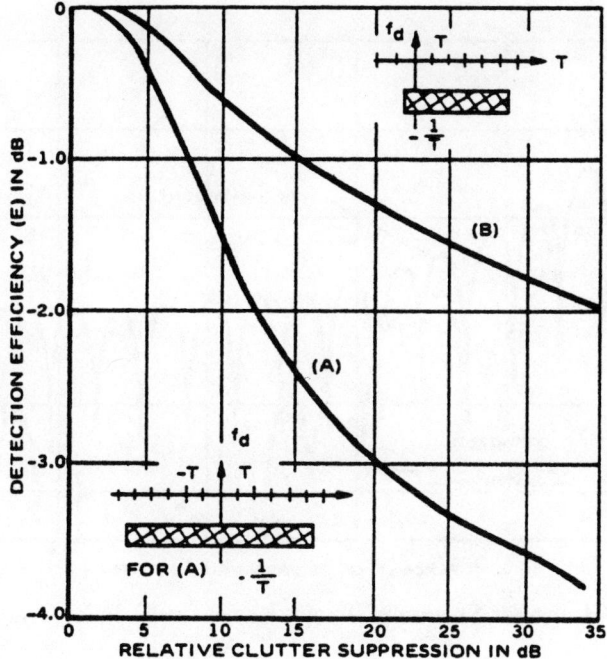

Note: Clutter cloud distributed from $f_d = 5/6T$ to $f_d = 1/2T$ in range: (a) from $n = 15$ to $n = 5$; (b) from $n = 0$ to $N = 5$.

Figure 3–42. Detection Efficiency *versus* Relative Clutter Suppression with a 21-Pulse Train

Figure 3–42 is an example of the improvement in clutter reduction as a result of optimum waveforms for moderate clutter extents. For curve *a* the range extent is 10 interpulse periods, and for curve *b* it is 5 interpulse periods. out of a 21-pulse train. It can be seen that even for moderately extensive clutter, significant relative improvement can be obtained with only 1–2-dB loss in efficiency. A similar optimization is sketched in figure 3–43. The zero cut of the ambiguity diagram is shown for the unweighted and optimum pulse trains [$\Psi(0,\omega)$].

In situations where it is necessary to use a burst of only a few pulses, the signal-to-clutter improvement is inadequate for the environment. If the transmitter train is sufficiently linear to weight the amplitude of the pulses in the transmission as a function of time, an additional improvement may be obtained. Ares [23] has shown that if the clutter extends throughout the two $N - 1$ range ambiguities, a filter matched to the tapered transmission yields a signal-to-clutter improvement that is within 3 dB of the optimum mismatched receiver.

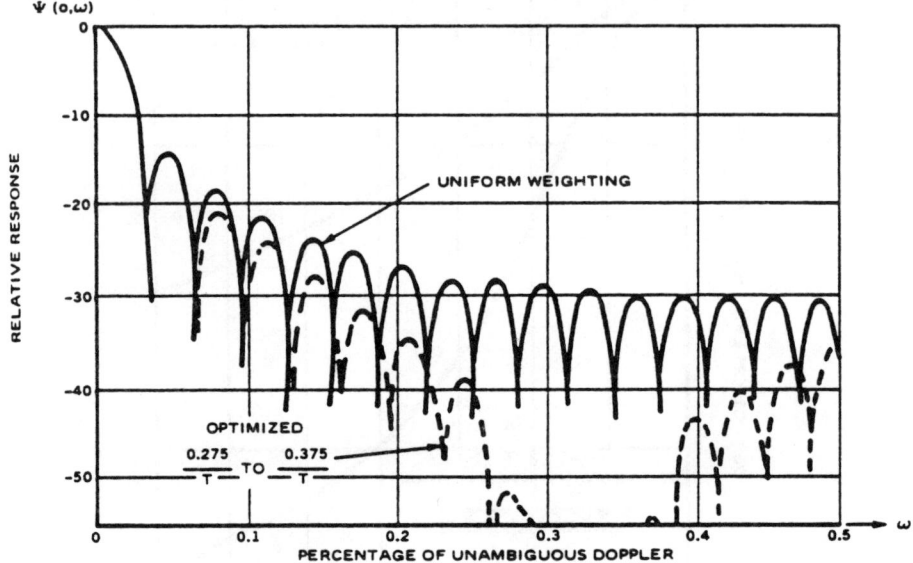

Figure 3-43. Power Spectra at Target Range for 17 Pulse Transmissions

Pulse-Doppler Receivers

There are numerous configurations for pulse-Doppler receivers for both search and tracking systems. The choice of configuration is based on the function of the radar. The pulse-Doppler search radar usually requires more sophisticated circuitry than the tracking radar since more range gates and Doppler filters are needed to "find" the target.

Emphasis will be on coherent integration in the range, Doppler, and angle circuits. The range and angle circuits may involve coherent or incoherent processing depending on the clutter environment, or the range and Doppler processing may be coherent and the angle processing incoherent to reduce the total amount of circuitry. A discussion of these hybrid coherent pulse trackers is given by Barton [7, pp. 385-389].

A block diagram for a pulse-Doppler receiver based on narrowband IF filters is given in figure 3-44. The parameters of the waveform are shown at the top of the figure, with the square pulses repeating samples of the coherent constant-frequency carrier. The spectrum of a continuous pulse train with interpulse period T and N approaching infinity is shown as the series of impulses centered about the carrier frequency f_0 in figure 3-45. If the receiver is to integrate many of the target echo pulses coherently, the second or Doppler bandpass filter must have a width $b \ll 1/T$. In the example of a 5-kHz PRF, the bandwidth would be about 100 Hz for 50-pulse coherent integration. If the filters are not perfectly rectangular, there is a loss of

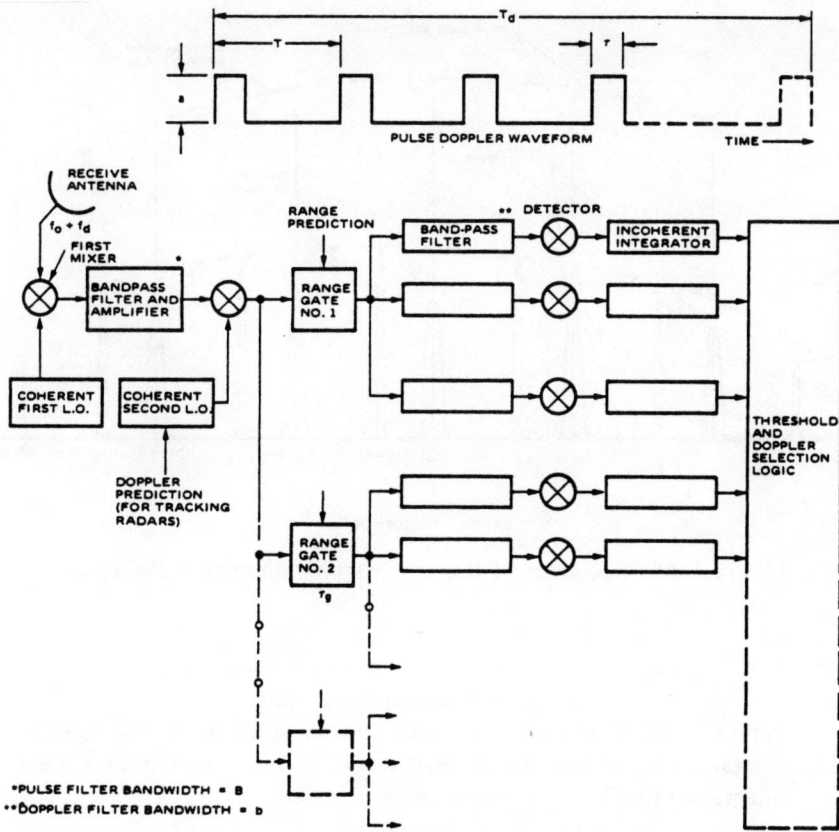

Figure 3–44. Pulse-Doppler Waveform and Basic Receiver Block Diagram

response in the crossover region between the filters. On the other hand, a rectangular filter is not a matched filter for a uniform pulse train. After the Doppler filtering, the sinusoidal output of the filters is envelope or square-law detected and is often stored or integrated on a capacitor. The integration at this point is incoherent for all the pulses after the effective time constant of the Doppler filter. The reduction in per-pulse S/N below that required for coherent integration of N pulses can be approximated by determining N/N_c and entering the appropriate curves of Marcum and Swerling. Curves from Marcum and Swerling [3] are shown in figure 3–46.

Several limitations of the narrowband IF filter configuration for pulse-Doppler receivers have led to the zero-frequency IF (homodyne) configuration. These limitations include:

1. The coherent integration time is fixed at the values of the original design and cannot be adapted to the spectrum of different types of targets or better knowledge of the target velocity.

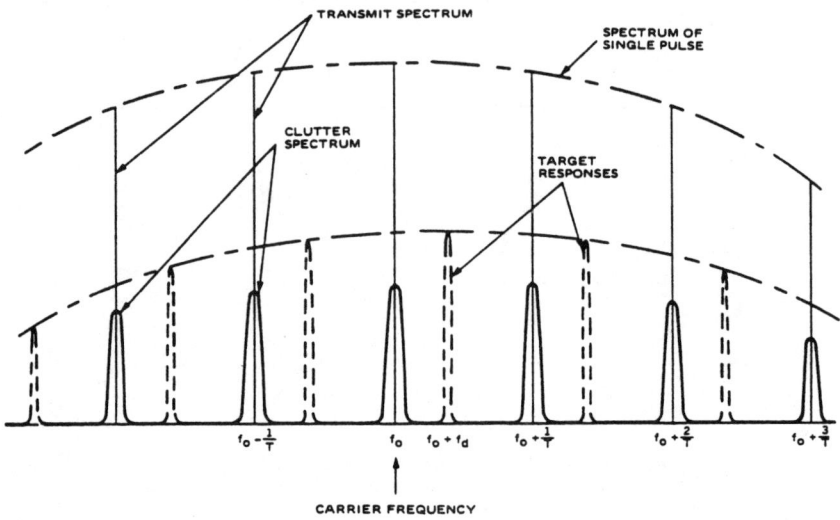

Figure 3-45. Spectrum of Returns from a Coherent Pulse Train

2. The range-gate switch is not easily designed when there is only about one cycle of the second IF during the gate time.
3. With phased arrays or other rapidly scanning antennas, the signals in crystal filters cannot be dumped instantaneously to permit coherent integration to start at a new location in space.

A basic block diagram of the homodyne pulse-Doppler receiver is shown in figure 3–47. Direct conversion from RF to zero frequency is illustrated here, although it is not necessary. The mixers are of the single-sideband type to attenuate the Doppler images and to provide the in-phase (I) and quadrature (Q) components of the signal. These bipolar pulses then allow reconstruction of the Doppler signal.

Using this technique, the integrator time constant can be much greater than the minimum desired integration time without loss of detectability or signal-to-noise ratio and can be varied over a wide range. Also in this system, the range gating can be performed at an IF where there are many carrier cycles per range gate. As the required number of Doppler filters becomes large, increasing the number of Doppler predictions, mixers, integrators, and so forth, it may be useful to derive the Dopplers in a digital computer. This implementation is more amenable to pulse doppler systems with lower PRFs.

If the number of pulses to be coherently integrated is relatively small, the tapped delay-line implementation can be used. The portion of this configura-

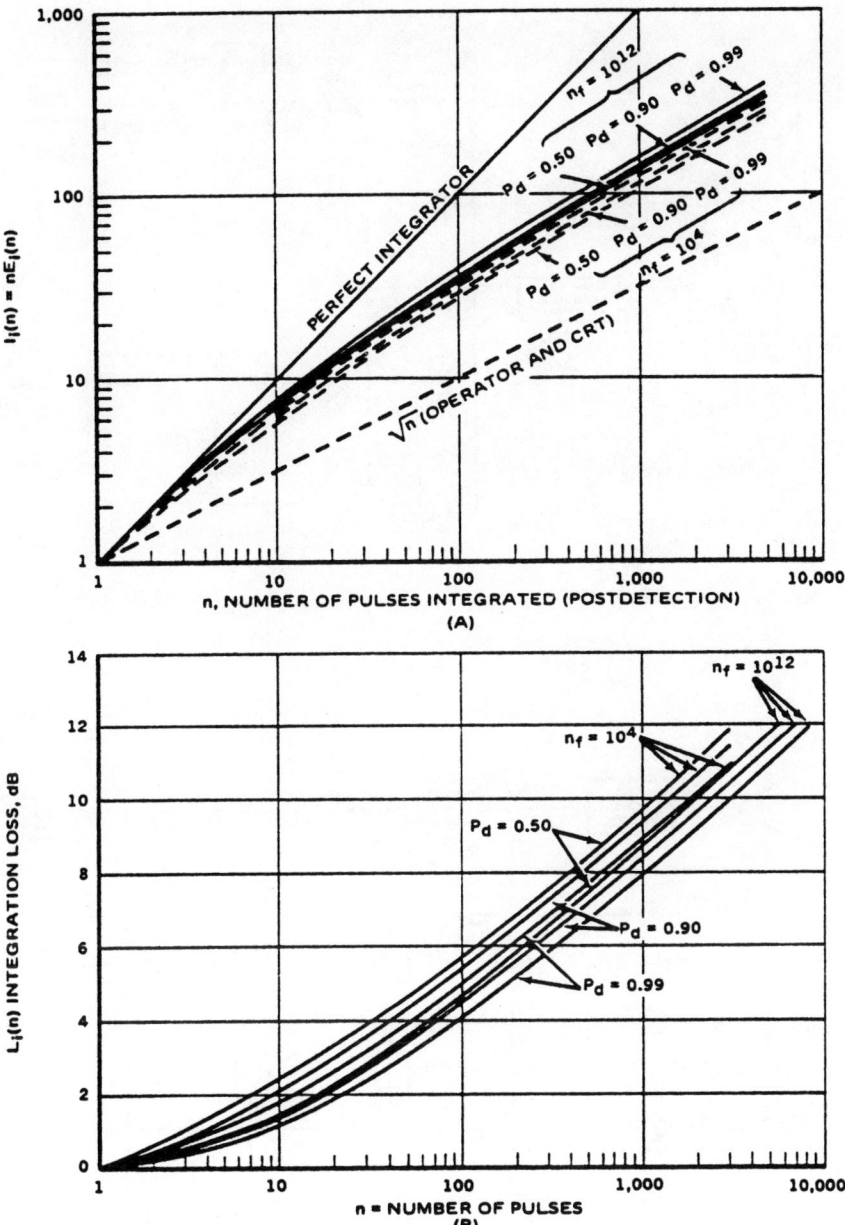

Figure 3–46. (a) Integration-Improvement Factor, Square-Law Detector, P_d = Probability of Detection, n_f = False-Alarm Number; (b) Integration Loss as a Function of n, the Number of Pulses Integrated, P_d, and n_f

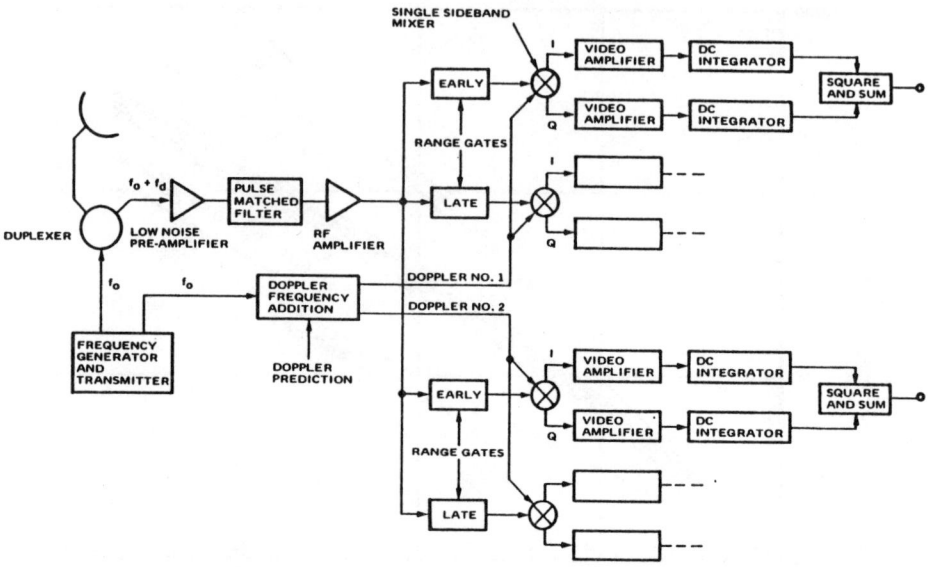

Figure 3–47. Zero *IF* (Homodyne) Pulse-Doppler Receiver

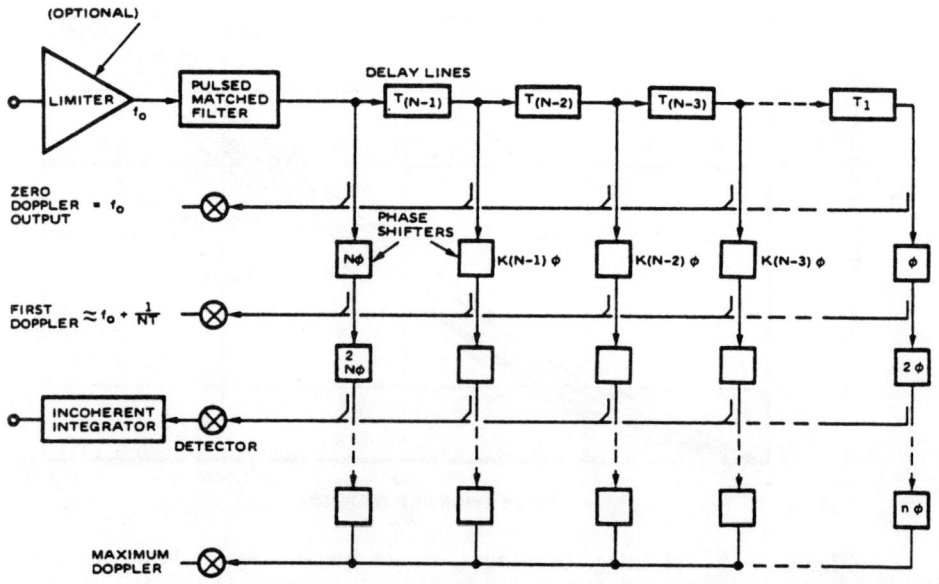

Figure 3–48. Tapped Delay-Line Receiver for Coherent Pulse Train

tion after the mixing processes is shown in figure 3–48. The delay lines are matched to the intervals between pulses, with Ti corresponding to the time between the transmission of the first and second pulses, and so forth. The arbitrary constant K_n is shown to indicate that the phase shifts must be adjusted to the individual spacings between pulses if the interpulse period is not constant. The outputs of this processor have the bandwidth of the individual pulses. Coherent integration results from the vector-voltage addition of the target echoes in the appropriate channel, whereas noise adds according to power. If the number of pulses is large enough to achieve constant false alarm rate (CFAR) action, a limiter may be placed ahead of the delay line. Then the normalized maximum output power is obtained from the coherent sum of N_c pulses (N_c^2). The average noise power level is less than this value by a factor of N_c, the number of pulses coherently integrated. Since this configuration responds to any set of narrowband signals associated with all the PRF lines (figure 3–45) it is one of a class of comb filters. Other versions of the comb filter are based on the use of delay lines and feedback techniques. Two of these are shown in figure 3–49 with the delay-line length equal to the interpulse period and the gain A equal to the delay-line loss. The similarity of figure 3–47 to an MTI system becomes obvious if the amplifier (A) has a gain of -1. A null in the response occurs at zero Doppler and multiples of the PRF. Coherent integration occurs only at odd multiples of one-half of the PRF. Figure 3–49(b) is more appropriate to a pulse-Doppler system if the loop gain is near unity. The response characteristics of feedback filters, including higher-order feedback network, are given by Flesher and Cohn [24] and by Galejs [25]. In general, the feedback filter may be at either IF or video. The circuit used in figure 3–49(c) has been widely used to extract Doppler information and is commonly known as the Coherent Memory Filter (Registered trademark of Federal Scientific Corporation, New York).

Energy Considerations

The basic purpose of the pulse-Doppler system is to integrate coherently many (N_c) pulses of length τ, and it is almost sufficient to assume for range computations that the energy is enhanced by the factor N_c if the transfer characteristic of the narrowband filters following the range gate optimally integrates N_c pulses. In the simplest processor, such as the tapped delay of figure 3–49, N_c pulses are added voltage-wise (N_c^2 in power), and N_c noise samples are added with random phase (N_c in power). In effect, the S/N required per pulse for a given probability of detection and false alarm is reduced by N_c after allowance is made for the apropriate number of false-alarm opportunities.

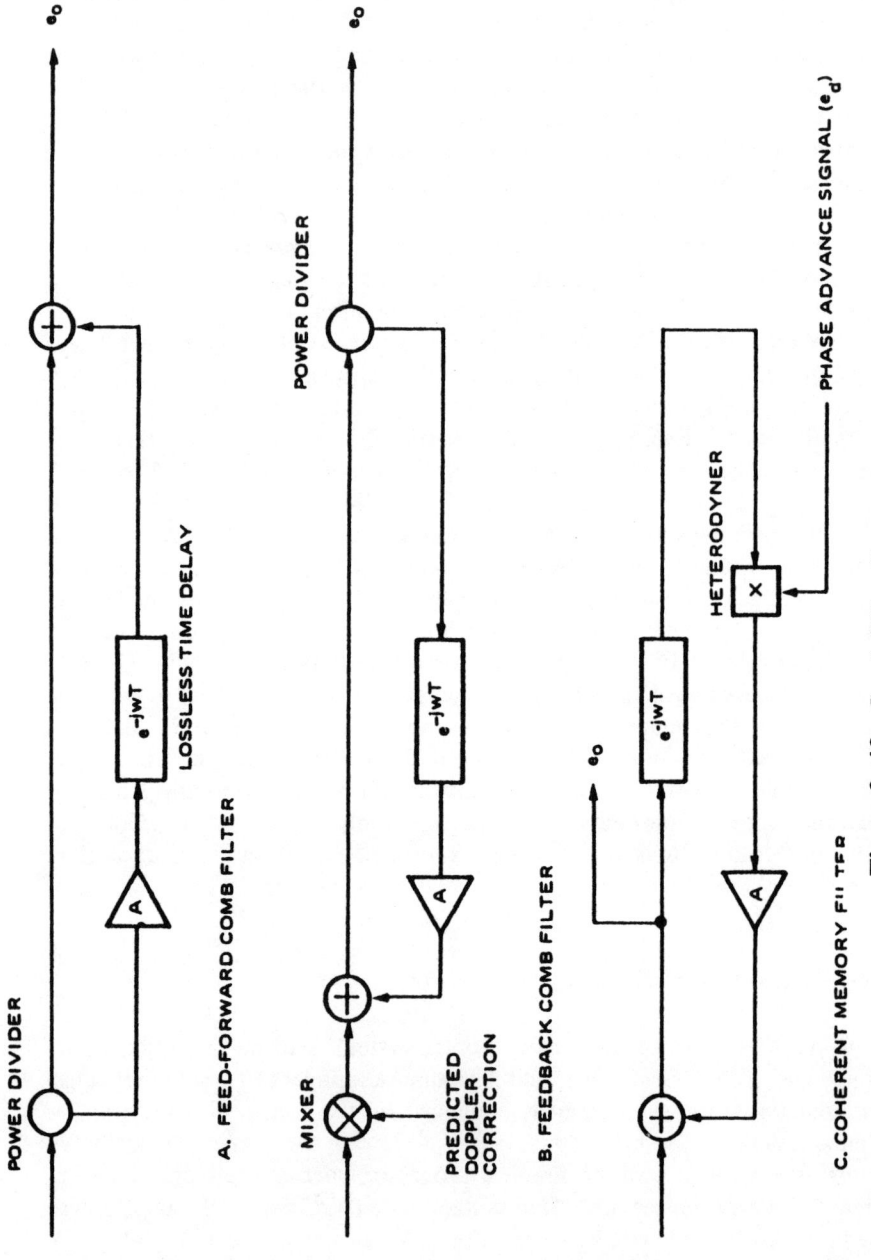

Figure 3-49. Comb-Filter Processors

In many practical cases the Doppler filter bandwidth is deliberately made greater than the matched bandwidth for the number of available pulses in the transmission. This occurs where: (1) the target spectrum is broader than $1/NT$, (2) the target is accelerating or decelerating in the time NT sufficiently to broaden the spectrum beyond $1/NT$, or (3) the construction of the optimum number of Doppler filters is too costly. In these cases coherent integration exists for N_c pulses; and if the detected output of the doppler filters is linearly integrated, there is incoherent integration of N/N_c "bursts."

The calculation of the efficiency factor (L_s) for pulse Doppler must account for some potential added losses:

1. The pulse-matched filter may deviate from the optimum value.
2. The range gate may be too narrow, too wide, or not centered on the target return.
3. The Doppler-matched filter shape (or bandwidth) may deviate from the optimum.
4. The target response may straddle more than one Doppler filter or fall in the "notches" between them.
5. There may be a significant Doppler dispersion loss if

$$\frac{2T_d\ V/C}{\tau} \geq 1.$$

6. Doppler image noise may be "folded" onto the desired doppler spectrum.
7. Phase- and time-delay errors reduce the sensitivity, as shown in the section on MTI.
8. FM and AM transmitter noise imposed on nearby clutter may exceed receiver noise.

Analysis of the improvement obtained from coherent integration for various filter shapes has been reported by North [26], Galejs [25], and others. For the case of a uniform pulse trian of rectangular pulses, and a uniform, video-comb filter with an idealized square passband, it has been shown that

$$\text{Improvement} = N_c\ L_s = 10 \log (0.45\ N_c) \text{ if } \tau < T/4.$$

By altering the gain and response of the "teeth" of the comb filter, additional improvement can be obtained on the order of 2–4 dB.

Galejs [25] has extended this analysis to physically realizable comb filters and to the case where the pulse-train envelope is a $\sin x/x$, which would result from an antenna beam scanning past a target.

He obtained the following S/N improvements, above $10 \log 0.45 \, N_c$, for practical filters and rectangular pulse trains:

Pulse Shape	Optimum Filter	Cascaded Delay-Line Filter	Feedback Filter
Rectangular	3.8 dB	1.7 dB	1.84 dB
Sin x/x	1.7	1.7	1.84

The exact values quoted here are for special cases. They should not be used without study of the detailed assumptions in the references, including the discussions on when the output is sampled. In practice, the avarage loss in the range gates and Doppler filters is 1–2 dB.

Clutter Computations

The signal-to-clutter ratio can be written, after simplification, as

$$\frac{E_s}{E_c} = \left[\frac{\sigma_t}{(\Sigma\sigma) \, c\tau/2 \, \theta\phi} \right] \left[\frac{1}{R_t^2 \Sigma_k \, (R_t/R_k)^2} \right] \left[\frac{N}{K} \right] \qquad (3.32)$$

where Σ_σ = clutter-scattering cross-section density (usually expressed in square meters per cubic meter for volume clutter);

σ_t = the target cross-section;

τ = the pulse length;

θ and ϕ = the one-way half-power beamwidths;

R_k = range to the kth ambiguous range;

R_t = range to the target;

$K = \frac{1}{\pi} [(\sin \pi \, bT)/(\cos \pi \, bT - \cos 2\pi f_c \, T)]$;

b = the filter bandwidth;

T = the interpulse spacing;

f_c = the average clutter-Doppler shift;

N = the number of pulses in the train.

Radar Performance in Clutter

If the extent of the clutter is small compared with the interpulse period (less than $C\,T/2$) and only located at the target, then

$$\frac{E_s}{E_c} = \left[\frac{\sigma_t}{(\Sigma\sigma)\,(C\tau/2)\,\theta\phi}\right]\left[\frac{1}{R_t^2}\right]$$

$$\left[\frac{N/K}{\dfrac{1}{\pi}\left(\dfrac{\sin \pi bT}{\cos \pi bT - \cos 2\pi f_c T}\right)}\right] \quad (3.33)$$

For clutter that extends quite close to the radar, the first ambiguous range dominates, and the ratio becomes:

$$\frac{E_s}{E_c} = \left[\frac{\sigma_t}{(\Sigma\sigma)\,(C\tau/2)\,\theta\phi}\right]\left[\frac{R_1^2}{R_t^4}\right]\left[L_a(R_t - R_1)\right]\frac{N}{K}, \quad (3.34)$$

where R_1 is the nearest ambiguous range.

For $bT = 1/N$ and N large, the maximum clutter attenuation for a Doppler separation of $f_c\,T = 1/2$ can be approximated by:

$$CA_{max} = 2N^2, \quad (3.35)$$

near the optimum Doppler separation. Alternatively, for an average overall Doppler velocity (except for the ambiguous regions), the improvement factor is $I = N^2$. As indicated by the foregoing equations, the summation of the clutter echoes from many ranges provides a limit to the signal-to-clutter ratio improvement in partially extended clutter. Ares [23] has shown that for a given target-to-clutter velocity ratio and interpulse period T there exists for each target range (R_t) a value of N that maximizes the signal-to-clutter ratio. The optimum number of pulses found by Ares is

$$N = \frac{1}{\sqrt{3}}\frac{R_t}{(CT/2)}, \quad (3.36)$$

and the optimum clutter attenuation is

$$CA_{opt} = \frac{3}{2N}, \quad (3.37)$$

or the improvement is

$$I_{\text{opt}} = \frac{2}{3}N = \frac{2}{3\sqrt{3}} \frac{T_t}{(CT/2)}, \qquad (3.38)$$

if the interpulse period is not fixed by other considerations, or double optimization involving the period and the number of pulses can be performed. An example is given in table 3–5. For a given range R_t (in μsec) and target Doppler frequency f_d, the optimum number of pulses is shown; also shown in the table is the length of the burst, $T_d \approx 2R/c \sqrt{3}$ and the Doppler resolution b in kHz. The improvement factor for these cases is shown in table 3–6. Ares also considered receiver-amplitude weighting of the truncated pulse train and showed that there was little additional improvement.

Limitations Due to System Instabilities

Various system instabilities impose limitations on the pulse-Doppler radar as they do on MTI performance. These take the form of phase, frequency, time, and amplitude instabilities. They produce unwanted energy in the region between spectral lines. These can be enumerated as follows:

Periodic Interpulse Phase Modulation. This modulation gives rise to sidebands at $f_c + f_m$ and $f_c - f_m$ with magnitudes of $\theta/2$. Each sideband energy relative to that at the carrier is $\theta^2/4$, where θ is the phase modulation in radians.

Periodic Interpulse Amplitude Modulation. This also gives sidebands with a relative sideband energy of $m^2/4$.

Random Interpulse Phase Modulation. When considering a pulsed waveform of N pulses with coherent integration, the coherent energy will add as N^2, and the random variation will add as N. Then, if σ is the variance from pulse to pulse, the relative power level of the random phase-modulation noise is σ^2/N.

Random Interpulse Amplitude Modulation. Following the preceding argument, the relative power level of random amplitude modulation noise is σ^2/N, where N is the number of pulses in the waveform.

Time Jitter. Lack of periodicity in a pulsed waveform will result in noise-frequency sidebands due to the time jitter. This will be $2\sigma_{\Delta\tau}^2/(N^2)$ where $\sigma_{\Delta\tau}$ is the variance of the time jitter.

Intrapulse Noise. This noise is reduced in coherent systems by the number of pulses in the waveform.

Table 3-5
Optimum Number of Uniformly Spaced Pulses for Extensive Clutter

Target Range R_t (μsec)	Dwell Time T_d (μsec)	Doppler Resolution b(kHz)	N_{opt} for Doppler Frequency Offset (kHz) of:									
			4	5	8	10	12	15	20	25	30	35	40
400	231	4.33	2	2	4	5	6	7	9	12	14	16	18
450	260	3.85	2	3	4	5	6	8	10	13	16	18	21
500	289	3.46	2	3	5	6	7	9	12	14	17	20	23
550	318	3.15	3	3	5	6	8	10	13	16	19	22	25
600	346	2.89	3	4	6	7	8	10	14	17	21	24	28
650	375	2.67	3	4	6	8	9	11	15	19	23	26	30
700	404	2.48	3	4	7	8	10	12	16	20	24	28	32
750	433	2.31	3	4	7	9	10	13	17	22	26	30	35
800	462	2.17	4	5	7	9	11	14	18	23	28	32	37
900	517	1.94	4	5	8	10	12	16	21	26	31	36	42
1,000	577	1.73	5	6	9	12	14	17	23	29	35	40	46
1,100	635	1.58	5	6	10	13	15	19	25	32	38	44	51
1,200	693	1.44	6	7	11	14	17	21	28	35	42	49	55
1,300	751	1.33	6	8	12	15	18	23	30	38	45	53	60
1,400	808	1.24	6	8	13	16	19	24	32	40	49	57	65
1,500	866	1.16	7	9	14	17	21	26	35	43	52	61	69
1,600	923	1.08	7	9	15	18	22	28	37	46	55	65	74

Table 3–6
Optimum Improvement Factor I (in dB) for Uniformly Spaced Pulses and Extensive Clutter

Target Range R_t (μs)	Doppler-Frequency Offset (kHz)									
	4	5	8	10	12	15	20	25	30	35	40
400	1.3	1.3	4.3	5.2	6.0	6.7	7.8	9.0	9.7	10.3	10.8
450	1.3	3.0	4.3	5.2	6.0	7.3	8.2	9.4	10.3	10.8	11.5
500	1.3	3.0	5.2	6.0	6.7	7.8	9.0	9.7	10.5	11.3	11.9
550	3.0	3.0	5.2	6.0	7.3	8.2	9.4	10.3	11.0	11.7	12.2
600	3.0	3.0	6.0	6.7	7.3	8.2	9.7	10.5	11.5	12.0	12.7
650	3.0	4.3	6.0	7.3	7.8	8.7	10.0	11.0	11.9	12.4	13.0
700	3.0	4.3	6.7	7.3	8.2	9.0	10.3	11.3	12.0	12.7	13.3
750	3.0	4.3	6.7	7.8	8.2	9.4	10.5	11.7	12.4	13.0	13.7
800	4.3	5.2	7.3	7.8	8.7	9.7	10.8	11.9	12.7	13.3	13.7
900	4.3	5.2	7.8	8.2	9.0	10.0	11.5	12.4	13.2	13.8	13.9
1,000	5.2	6.0	8.2	9.0	9.7	10.5	11.9	12.9	13.7	14.3	14.4
1,100	5.2	6.0	8.7	9.4	10.0	11.0	12.2	13.3	14.0	14.7	14.9
1,200	6.0	6.7	9.0	9.7	10.5	11.5	12.7	13.7	14.5	15.1	15.3
1,300	6.0	7.3	9.0	10.0	10.8	11.9	13.0	14.0	14.8	15.5	15.6
1,400	6.0	7.3	9.4	10.3	11.0	12.0	13.3	14.2	15.1	15.8	16.0
1,500	6.7	7.8	9.7	10.5	11.5	12.4	13.7	14.6	15.4	16.1	16.4
1,600	6.7	7.8	10.0	10.8	11.7	12.7	13.9	14.9	15.6	16.4	16.9

Computer Simulation of Performance in Clutter

Thus far the methodology has been presented for calculating MTI and pulse-Doppler performance using either curves or hand calculation. For some cases, however, hand calculation of performance in clutter can become quite laborious. This is particularly true where range is ambiguous and ground clutter is being returned from several range intervals. This is the case for airborne systems.

A computer program has been designed to compute the S/C or $S/C + N$ ratios as a function of range-to-target and target-Doppler frequency for airborne radar systems employing MTI, range-gated pulse Doppler (RGPD), or MTI/RGPD signal-processing systems and operating at low or medium PRF (up to approximately 15 kHz). For pulse-Doppler systems, the program accounts for the finite waveform effects and can accommodate uniform and $\cos^2$ plus pedestal-waveform weighting. The program can also simulate (1) velocity compensation errors in the MTI system, (2) maximum realizable rejection by the MTI system resulting from radar instabilities, (3) effects of transmitter instabilities, and (4) quadrature processing that eliminates spectral foldover. One-, two-, or three-stage MTI cancelers can be simulated.

The program requires standard radar-system parameters for inputs; the antenna-gain function and terrain-reflectivity models are in the form of subroutines. Available outputs include printouts of the characteristics of each clutter ring, input and output clutter-power spectral densities, and the $S/(C + N)$ matrix ($S/C + N$ versus range-to-target and target-Doppler frequency) for given antenna-beam position and other radar parameters. The $S/C + N$ ratios computed by the program do not include target signal losses due to atmospheric attenuation and reduced antenna gain (when target is off antenna axis), since no specific target geometry is assumed.

Consider an airborne pulse-radar system operating at a PRF such that it is ambiguous in range and velocity with respect to clutter. A signal from a target at range R_T will need to compete with clutter from ranges $R_T + nR_{un}$, where n is an integer and R_{un} is the unambiguous range ($c/2$ PRF) as illustrated in figure 3–50. A single line of the transmitted spectrum will result in received clutter spectral densities from the clutter rings at ranges $R_T = R_{un}$ of the type illustrated in figure 3–51. The relative spectral extent of the clutter rings will depend on the depression angle. The relative magnitudes will depend on the range, reflectivity, atmospheric attenuation, and antenna gain along the clutter rings.

For an infinite waveform consisting of line spectra, adjacent frequency sidebands of the transmitted spectrum will yield similar clutter spectral densities displaced by $\pm n$ PRF. The total spectral density over a PRF interval will be the sum of the contributions, in that interval, of adjacent

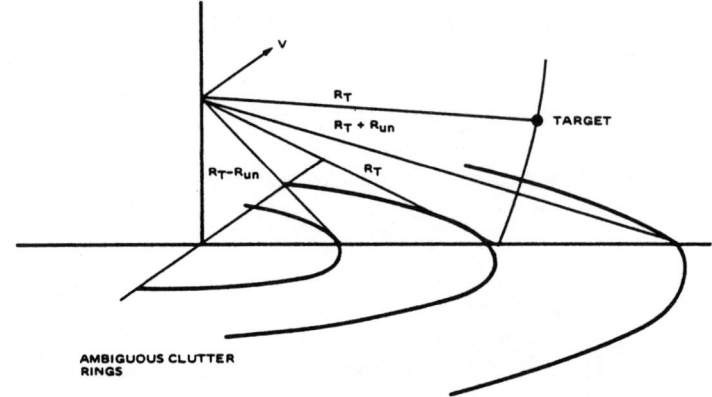

Figure 3–50. Clutter Rings

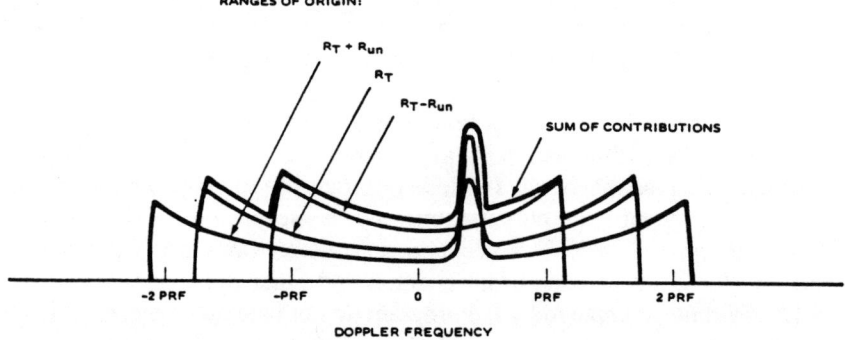

Figure 3–51. Effective Clutter Spectral Densities Due to One Sideband

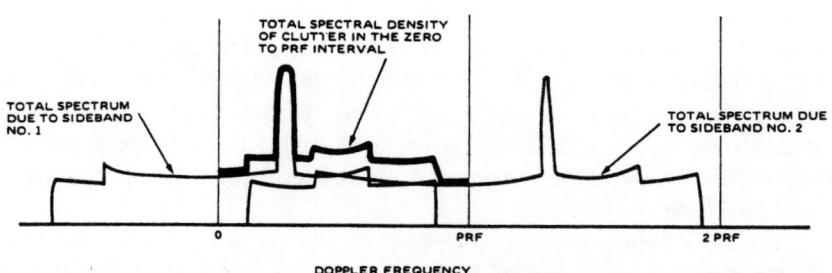

Figure 3–52. Total Spectral Density of Clutter

Radar Performance in Clutter

transmitted sidebands, as illustrated in figure 3–52. Since the adjacent sidebands will be of nearly equal magnitudes, the clutter spectral density will be periodic, with the period equal to PRF. The composite spectrum indicated in figure 3–52 represents the total input-clutter spectral density of an infinite waveform.

The output-clutter spectral density from an MTI canceler can be determined by applying the MTI gain function to the input-clutter spectral density. For an MTI canceler with n stages, the gain function is given by

$$G(f) = \left[2 \sin\left(\frac{\pi}{\text{PRF}} (f - f_s) \right) \right]^{2n}, \qquad (3.39)$$

where f is the doppler frequency and f_s is the compensation frequency that locates the notch of the canceler at the main lobe spectrum (f_s is necessary to realize maximum rejection with MTI). Conversely, the input-clutter spectral density could be shifted by f_s. The residue at the output of the MTI canceler is illustrated in figure 3–53. The MTI-only systems S/C is determined by the ratio of the magnitude of the signal (including MTI gain) and the integral of the clutter residue.

For MTI range-gated pulse-Doppler systems (MTI/RGPD) the corresponding residue of a finite waveform must be found. The procedure for this is to convolve the infinite waveform residue with the Fourier transform of the finite-waveform weighting function; for a rectangular weighting function the Fourier transform is of the form $(\sin x)/x$. The effect of this process will be to smear the clutter spectral density (residue) over the PRF interval. When the resulting spectral density is relatively uniform, contiguous rectangular filters

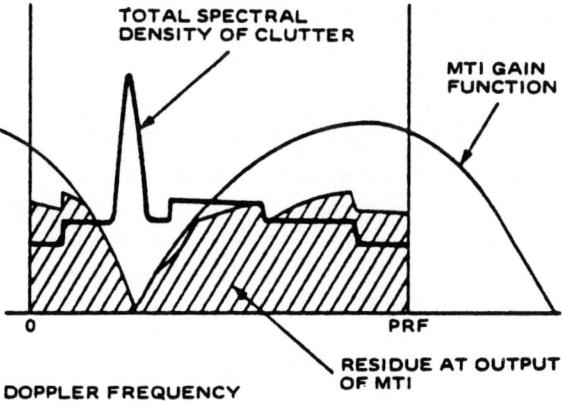

Figure 3–53. Clutter Rejection by MTI

can be assumed in the model of the filtering operation. The spectral smearing or spectral spreading effect of the finite waveform will apply to the target signal as well as clutter. Thus, in general, the target signal may be split among adjacent filters, thereby complicating the evaluation of the S/C and P_d. A similar straddling problem will also occur when range gating. The most tractable way to deal with straddling losses is to compute the S/C ratios assuming no straddling losses and afterward to modify the S/C values by an appropriate signal-loss factor. The idealized S/C ratios can be determined by computing the signal magnitude (including MTI gain) and integrating the finite waveform residue of the MTI canceler over the bandwidth of the filter containing the target signal. When these calculations are carried out over a PRF interval with the target signal at the centers of successive filters, S/C versus Doppler frequency is obtained.

Iteration of the foregoing calculations at other ranges will yield S/C versus Doppler frequency and range to target. Because of the periodicity of the S/C function, complete calculations for a given set of radar parameters are required only over one unambiguous range interval and one PRF interval in Doppler frequency. The S/C function will be identical at other *PRF* intervals; in range, the S/C at other unambiguous range intervals will vary as R^4 and can be readily extrapolated. Since the antenna-gain loss and atmospheric attenuation of the target signal are functions of the target geometry, these losses are not included in the computed S/C and must be added subsequently.

Although the foregoing discussion pertains strictly to the evaluation of S/C ratios, receiver noise and effects of transmitter instabilities can be introduced, thus yielding $S/C + N$ ratios. Both are modeled as contributing uniform spectral densities and are added to the input-clutter density. The transmitter instabilities are assumed to be spurious transmitter sidebands occurring at the center of each filter (worst case) and to produce noise proportional to the input clutter. Since by far the greatest input-clutter contribution is due to the antenna mainlobe, the transmitter instability model essentially places an averaged mainlobe spectral density proportional to the magnitude of the spurious sideband at the center of each filter.

The most direct method of including receiver noise is by means of the S/N ratio. This approach requires that losses peculiar to the target signal but not to clutter be extracted from the S/N ratio. Losses of this type include range-gate and filter straddling losses, atmospheric attenuation (usually differs from the attenuation suffered by the clutter signal), and antenna-gain loss when the target is off axis. The resulting S/N ratio defined for a target of known cross-section at a specific range can be used to determine the equivalent clutter corresponding to the receiver noise. With the receiver noise in terms of equivalent clutter magnitude, $C + N$ and $S/C + N$ can be

Radar Performance in Clutter

determined; the actual $S/C + N$ ratio will be obtained by applying the losses, previously extracted from S/N, to $S/C + N$.

Note that the calculations within the program are in terms of effective-clutter radar cross-sections and target radar cross-sections (m^2, m^2/Hz, dB above 1 m^2, and so forth). That is, the computations are carried in the units of radar cross-section magnitude and density referenced at the range of the target. The computations could just as well have been carried out in terms of watts of clutter power and signal power in the receiver. The relationship between these quantities is a constant and is given by the radar range equation. Since the primary output is a ratio $(S/C + N)$, both approaches yield the same result. The approach used here has the advantage that the significance of the clutter or clutter plus noise is appreciated more readily when expressed in terms of m^2 at the range of the target, since the target radar cross-section is a familiar parameter; on the other hand, conversion of receiver noise to equivalent clutter magnitudes seems inappropriate, even when the mechanics are valid.

Typical outputs for an MTI/RGPD system are given later on. System parameters are listed in table 3–7. For each target range at which detailed

Table 3–7
System Parameters

Program Parameters	
System Parameters	
Platform altitude	35,000.0 ft
Platform velocity	165.00 m/sec
Clutter extent	229.87 nmi
Radar parameters	
Radar frequency	3,550.30 MHz
Signal bandwidth	2.00 MHz
PRF	2,000.00 PPS
Beam elevation	−1.00°
Beam azimuth	52.80°
El beamwidth	4.95°
Az beamwidth	0.90°
Transmitter instabilities	−60.00 dB
Receiver noise, S/N	100.00 dB at 230.00 nmi
Signal loss	0.0 dB
Radar cross-section of target	5.00 m^2
Processor Parameters	
MTI/RGPD—critical channels	
MTI stages	2
Max MTI rejection	0.000100
Filters	8
Waveform weighting—none	

Table 3–8
Clutter Geometry

Target Range = 10.82 nmi
Clutter Rings

Range (nmi)	Depression Angle(°)	Grazing Angle(°)	Reflectivity	Effective Clutter (m^2)	Input/Output Clutter	
10.82	32.22	32.10	0.03321	0.00		
51.29	6.76	6.13	0.00978	0.01	Clutter	= 0.195E 00
91.76	4.17	3.03	0.00872	0.13	TR noise	= 0.156E-05
132.23	3.32	1.67	0.00754	0.05	Receiver noise	= 0.201E-13
172.70	2.99	0.83	0.00358	0.01	output $(C+N)$	= 0.386E-03
213.17	2.88	0.22	0.00053	0.00	MTI rejection	= 27.04 dB

calculations are made, the computer program prints out three blocks of data. The first block (table 3–8) describes the input-clutter characteristics: slant range and depression angle to the clutter ring, grazing angle and reflectivity (σ_0) at the clutter ring, and effective magnitude of the clutter at the range of the target. The second block of output data (table 3–8) is denoted by the heading *"Input/Output Clutter."* The first value gives the total effective clutter magnitude contributed by the clutter rings (sum of last column of first block of data). "TR Noise" denotes the magnitude of the input clutter due to transmitter instabilities. Note that in modeling transmitter instabilities, sidebands due to instabilities are assumed at each filter to yield worst-case $S/C + N$ data; for this case the clutter due to transmitter instabilities is $0.195 \times 8 \times 10^{-6} m^2$ or $1.56 \times 10^{-6} m^2$. "Receiver Noise" denotes the magnitude of the equivalent input clutter due to receiver noise. "Output $(C + N)$" is the magnitude of the residue at the output of the MTI unit. "MTI Rejection" is the rejection provided by the MTI—that is, the ratio of the sum of inputs (the first three quantities) to the residue, expressed in dB.

The third block of output data (table 3–9) describes the Doppler-processor outputs. *"Frequency Shift"* denotes the frequency by which the clutter spectrum was shifted. The first two columns number the filters and give the Doppler frequencies at the centers of the filters; the bank of filters extends from zero to PRF (0–2,000 Hz for these particular data). The next two columns give the MTI gain at the centers of the filters and the target signal that is the product of MTI gain and the radar cross-section of the target. The last two columns give the magnitude of the clutter in each filter and the $S/C + N$ ratio for a target signal at the center of that filter ($S/C + N$ is just the ratio of the signal and $C + N$, columns 4 and 5). The $S/C + N$ values given in this block must be corrected by signal losses. Part of these losses, including filter and range-gate straddling loss (0 dB for this particular

Table 3-9
Doppler-Processor Outputs

Range to target = 10.82 nmi
Frequency shift = 2,368.20 Hz

Filter	Center Frequency (Hz)	Gain for signal (Numeric)	Target Signal (mi^2)	Total $C + N$ (mi^2)	$S/C + N$ (dB)
1	125.00	0.233E-01	0.166E 00	0.657E-04	32.48
2	375.00	0.152E 01	0.762E 01	0.144E-03	47.23
3	625.00	0.765E 01	0.382E 02	0.244E-04	61.96
4	875.00	0.148E 02	0.740E 02	0.798E-04	59.67
5	1,125.00	0.148E 02	0.740E 02	0.196E-04	65.77
6	1,375.00	0.765E 01	0.382E 02	0.143E-04	64.29
7	1,625.00	0.152E 01	0.762E 01	0.702E-05	60.36
8	1,875.00	0.233E-01	0.166E 00	0.257E-04	36.56

case) are subtracted from the $S/C + N$ values when printing out the $S/C + N$ summary matrix at the end of the cmputations. Thus the $S/C + N$ values in the summary matrix need to be corrected only by the atmospheric attenuation and antenna gain loss off beam center.

The summary matrix for this sample is given in table 3-10. Detailed calculations were carried out only for the first eight ranges in increments of

Table 3-10
Doppler-Processor Output Summary

Range (nmi)	$S/C + N$ (dB) Filter Center Frequency (Hz)				(Frequency Shift = -2368.20 Hz)			
	125.00	375.00	625.00	875.00	1,125.00	1,375.00	1,625.00	1,875.00
10.82	32.48	47.23	61.96	59.67	65.77	64.29	60.36	36.56
15.87	25.42	40.24	55.08	57.01	56.65	57.42	50.60	28.16
20.93	20.36	35.59	52.17	54.37	55.70	53.13	47.41	22.04
25.99	16.74	32.18	48.91	51.92	53.73	51.98	45.81	21.02
31.05	14.51	29.84	47.65	50.83	51.57	50.29	43.58	18.27
36.11	11.93	27.09	45.10	48.75	49.56	48.51	41.35	15.79
41.17	9.52	24.28	42.44	46.58	47.44	46.33	39.08	13.60
46.23	7.50	22.19	40.50	44.85	45.62	44.39	37.12	11.70
51.29	5.44	20.19	34.92	32.64	38.73	37.25	33.32	9.52
56.35	3.41	18.24	33.08	35.00	34.64	35.42	28.68	6.16
61.40	1.66	16.89	33.48	35.67	37.00	34.44	28.72	3.35
66.46	0.43	15.87	32.60	35.61	37.42	35.67	29.50	4.71
71.52	0.02	15.34	33.15	36.33	37.07	35.79	29.08	3.78
76.58	-1.13	14.03	32.04	35.69	36.50	35.45	28.29	2.73
81.64	-2.38	12.38	30.55	34.69	35.54	34.44	27.19	1.70
86.70	-3.43	11.26	29.58	33.93	34.69	33.47	26.20	0.77

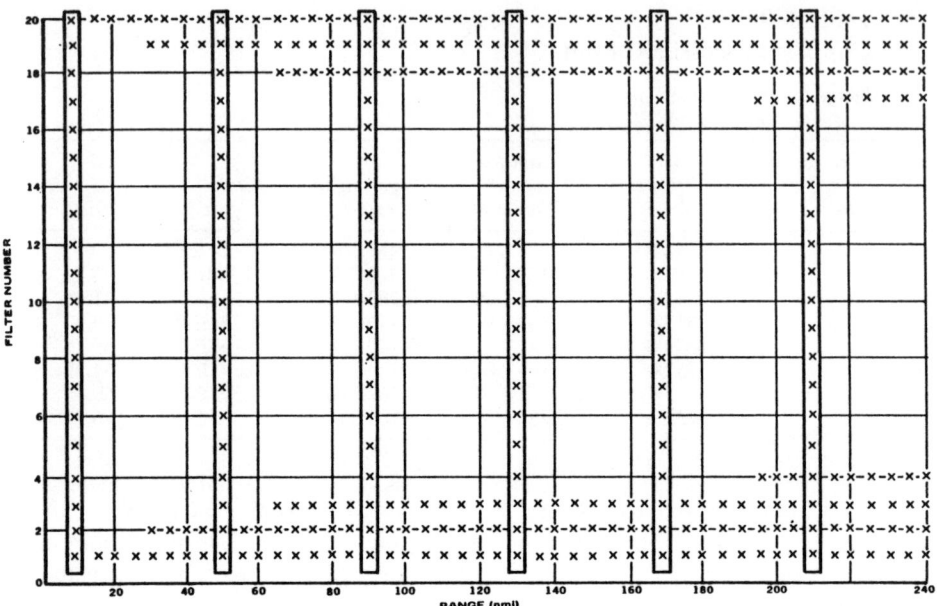

Figure 3–54. Target Visibility—Antenna Average Sidelobes = 100 dB

$R_m/8$, from $H + R_m/8$ to $H + R_m$ where H is the altitude of the platform (5.75 nmi) and R_m is the unambiguous range interval (40.4 nmi for PRF of 2,000 pps). $S/C + N$ data at other ranges were obtained by extrapolating the values of the first eight ranges using the R^4 effect. The particular system considered employs an effective waveform of eight pulses; hence only eight filters were assumed. The $S/C + N$ values for each filter are given in columns 2–9. For strictly MTI systems, the outputs consist only of the table of parameters and the first two tables of outputs. Neither $S/C + N$ versus Doppler frequency nor a summary matrix are given.

In another example, the values in the matrix were extracted to make the plot shown in figure 3–54 of overall visibility at one PRF. The Xs are for all filters whose $S/C + N$ was less than 12 dB. By the use of two or more slightly different PRFs, the overall visibility approaches 100 percent.

4 Radar Performance in an ECM Environment

Of the many electronic countermeasure (ECM) techniques that have been developed for use against radars, only wideband noise, sometimes called *barrage jamming*, is considered quantitatively here. Analysis of the effects of more sophisticated ECM methods is not appropriate to a generalized study of methodology because such ECM is usually optimized against specific radars or classes of radars. In many cases it is necessary to use computer simulations or laboratory representations of actual equipment in order to evaluate performance; only a general discussion of pulse jamming is included here. The effects of wideband noise jamming, however, are easily computed when the jamming waveform has (1) a wider bandwidth than the radar receiver, and (2) a uniform power density over its bandwidth.

Performance in the Presence of Barrage Jamming

The simplest and most general means for evaluating the effect of noise jamming consists of computing an effective input-noise temperature due to jamming (T_{NIJ}), which is added to the input-noise temperature due to receiver noise (T_{NI}). Radar performance is then evaluated in the manner normally used for a benign environment. Effective input jamming-noise temperature is defined as follows:

$$T_{NIJ} = \frac{P_{JR}}{B_J k}, \qquad (4.1)$$

where P_{JR} = jamming power received by the radar antenna;

B_J = jammer bandwidth;

k = Boltzman's constant;

Based on a modification of the radar range equation, T_{NIJ} is computed as follows:

$$T_{NIJ} = \frac{P_J G_J G_R \lambda^2 F^2}{B_J (4\pi R_J)^2 L_J k}, \qquad (4.2)$$

where P_J = jammer power (watts);

G_J = jammer-antenna gain;

G_R = radar-antenna gain in the direction of the jammer;

R_J = jammer range (meters);

L_J = jammer loss factor;

λ, F, F_j, and k as previously defined.

Equation 4.2 is quite general because it can be used to evaluate a jammer in the radar main lobe, in its peak sidelobe, or in an average sidelobe. By establishing T_{NIJ} at the antenna terminals, one avoids the added complexity of receiver microwave losses, receiver characteristics, and signal-processing factors that would be involved in evaluating ECM effects at a point further downstream.

Targets that carry their own ECM equipment, called *self-screening jammers*, represent a special case that allows a simplified analytical approach. Because the radar always sees the target and the ECM with identical antenna gain, at the same range, and with the same propagation factor, it is possible simply to establish a signal-to-jamming (S/J) ratio at the target rather than at the radar. This approach assumes that T_{NIJ} is much larger than T_{NI} (allowing T_{NI} to be neglected), which is nearly always true for jammers in the radar main lobe. A one-way version of the radar range equation is used as follows:

$$R = \left[\frac{E G_T \sigma_T B_J F^2}{4\pi (S/N) P_J G_J L_1} \right]^{1/2}. \quad (4.3)$$

All parameters have been previously defined, with the exception of L_1, which is the one-way radar-loss factor. L_1 is equal to L, with the exception that the dB values of the following loss components are divided by two:

beam shape (pattern) loss (L_p);

atmospheric loss (L_a);

precipitation loss (L_w);

off-broadside loss (L_b).

Performance in the Presence of Pulsed Repeater Jammers

Radar discrimination against repeater jammers is inclined to be a go, no-go situation for unsophisticated radars. That is, the false signal will either be detected and displayed in the same manner as a real target, or it will be completely rejected. This is true for two primary reasons. First, the amount of average power required by a pulsed jammer is trivial compared with that required of CW jammers. It is standard practice simply to assume that pulsed jamming signals will exceed the radar threshold or that they will have sufficient power to saturate a display. Second, their limited extent in time does not allow the radar to use adaptive thresholding techniques, as may be done with CW jamming, to "ride over" the jamming signal.

Nevertheless, several successful techniques for coping with pulsed repeater jammers have been developed. They fall into two general categories; *spatial discrimination* and *parameter discrimination*.

Spatial Discrimination

Spatial discrimination is based on establishing whether or not a signal originated from a point in space covered by the radar main lobe. This is the familiar *sidelobe blanking* approach, often called SLS. Its operation is based on the incorporation of an auxiliary antenna and receiver into the radar. The auxiliary antenna has a broad pattern with a gain factor that exceeds that of the main antenna sidelobes. Thus any signals that originate outside the main antenna's primary lobe will be received at a higher amplitude by the auxiliary antenna and receiver. A simple comparator circuit is used to blank the main receiver channel whenever the auxiliary channel has a greater signal magnitude.

This discrimination technique also provides protection against accidental interference from nearby friendly radars and is widely used for the purpose alone.

Parameter Discrimination

Parameter discrimination takes many forms and degrees of complexity. Basically, the effectiveness of parameter discrimination is related to the ability of a radar receiver and signal-processing system to recognize its own pulse. The primary discriminants are frequency, PRF, pulse width, and pulse code (if used).

All radars incorporate a degree of frequency discrimination as a consequence of the noninfinite bandwidth of their receivers. This capability is sometimes augmented by changing the radar frequency over some wide range in a pseudorandom manner. This technique, called *frequency hopping* or *frequency diversity*, can defeat jammers that are unable to sense and tune to the radar frequency in less than a PRF interval. Even if rapid tuning is incorporated, the jammer cannot create false targets at ranges less than his own; this knowledge is in itself a discrimination factor.

Table 4-1
ECCM against Repeaters

Technique	Evaluation
Sidelobe cancellation: Omnidirectional antenna pattern sets a lower threshold on radar-antenna output to reject all echoes lying outside of main beam.	Demonstrated technique. May result in missed detections if overloaded. Should be incorporated into radar if radar-antenna sidelobes are not sufficiently low to prevent repeater entering via the sidelobes.
Angle sorting: Advantage is taken of the fact that all echoes from the same repeater lie in the same direction, once sidelobe responses are eliminated.	A definite characteristic of repeaters that should be incorporated into the data processing.
Nonuniform search interval: Eliminates second-time-around responses that could appear at ranges closer than that of the repeater vehicle. Permits designating the leading false echo as that associated with the the jamming vehicle.	Readily incorporated into array radar.
Polarization diversity: Tests to determine whether received polarization is consistent with the expected from a real target.	Another reason for wanting dual polarization in the receiving array.
Logical trajectory: The apparent trajectory from a constant-delay repeater might not appear threatening.	A function of the data processor. Might require too much time and computer capacity to utilize fully.
Space diversity: A repeater echo may not appear the same to two or more spaced radars as does the echo from a real target.	Requires netting of spaced radars to be effective, as well as increased computer capacity.

Radar Performance in an ECM Environment 115

Frequency diversity: Use of redundant radars of widely spaced frequencies facilitates target discrimination.	Expensive ECCM tactic, but effective.
Linearity of echo: Utilizes the fact that real-target echoes are linear over a wide dynamic range, whereas those from a repeater are not.	A target identification technique that is not foolproof but is relatively easy to implement.
Doppler and range race: Compares the Doppler velocity and the rate of change of range.	Effectiveness depends on the accuracy of the radar measurements and the degree of sophistication of the simulation. Requires a computer capacity.
Coded waveforms: A simple transponder will not be able to reproduce coded radar signals faithfully.	Most modern radars will have coded waveforms (pulse compression) for reasons other than ECCM.
Guard-band transmission: A transmission at an offset frequency captures the repeater so that the main radar transmission is not affected.	Doubtful effectiveness.
Pulse characteristics: Signal shape examined to test for differences introduced by repeater.	Certain tests (pulse width) relatively easy to implement. Effective primarily against transponders.
Skin echo: Because of the inherent delays in the repeated signal, the skin echo may be detected ahead of the jamming signal.	A target identification technique useful if the jamming vehicle is to be attacked.
Identification pulse: A combination of ECCM technique used to verify detection and discriminate against repeaters.	Desirable ECCM.
Large capacity: Data processing must be sufficient to handle a large number of repeaters.	Necessary if other target identification methods insufficient.

PRF discrimination is useful against nonsynchronous pulse jammers because the jamming pulse does not appear at the same range on successive range sweeps. Either an operator or an electronic sequential detector can fairly easily reject this type of interference, assuming that more than one hit per scan is available on real targets. PRF jitter is sometimes incorporated to enhance the effectiveness of this discrimination technique.

Pulse-width discrimination is sometimes incorporated in radar receivers, although its usefulness is limited by the fact that real targets increase the receiver pulse width by a factor corresponding to their length. Ground clutter and weather returns will almost always exceed any pulse-width discrimination threshold and will therefore be rejected. If it is desired to retain the ability to observe strong targets in the presence of distributed clutter, pulse-width discrimination is usually switched off.

Intrapulse coding is one of the most effective means of defeating pulse jammers. If a degree of randomness can be incorporated in the code, a jammer must literally receive, delay, and retransmit a radar pulse in order to produce false targets. The fidelity required of such jammers is thus related to the code complexity, giving the radar designer an opportunity to exercise ECCM measures by making the code more complex.

Intrapulse coding is most commonly used as a pulse-compression technique rather than a pure ECCM feature. If reasonably large compression ratios are used (greater than 10), it is possible to devise an effective counter to pulse jamming by incorporating a hard limiter ahead of the pulse-compression device. Signals that are not properly coded do not compress and thus always fall below a preset threshold. Binary phase-code schemes normally use a hard limiter because they operate entirely on signal-phase measurements rather than amplitude. FM codes are amplitude sensitive and normally do not use limiters unless the latter are specially incorporated as ECCM devices.

FM coding without limiting results in an improvement in signal-to-jamming ratio for uncoded jamming pulses. Consider what happens to two types of uncoded jamming pulses that might be used. First, consider an uncoded pulse that matches the length of the radar-coded pulse. The jamming pulse will not compress in the radar receiver, giving coded pulse returns from real targets a peak-power advantage equal to the compression ratio for equal pulse energy. Second, consider a short uncoded pulse that matches the FM bandwidth of the coded pulse. The pulse-compression device in the receiver will stretch this pulse to a length approximately equal to that of the coded pulse before compression. Again, coded signal returns enjoy a peak-power advantage approximately equal to the compression ratio for equal pulse energy. In both cases the uncoded jamming pulses leave the receiver with a length equal to the coded pulse before compression, making them vulnerable to pulse-length discrimination.

Table 4–2
ECCM against Noise Jamming

Type of Noise	Protection Methods
Nonsynchronous noise with low repetition frequency.	Pulse delay for repetition period, its comparison with next pulse and blanking.
Wideband noise. Nonsynchronous pulsed noise with high PRF.	Increase observation time and integration. Sidelobe suppression. Retune transmitter.
Unmodulated carrier noise.	Eliminate receive saturation with instantaneous AGC or logarithmic amplifiers, etc.
Multiple-responding pulsed noise.	Repetition-frequency wobble in combination with protection measures against nonsynchronous pulsed noise.
Noise with low-frequency wobble.	Same as nonsynchronous pulsed noise.
Noise with high-frequency wobble.	Same as wideband noise.
Decoys and traps.	Selection based on reflecting properties, mass, and dimensions, with respect to trajectory parameters, degree of deceleration in atmosphere, and ionization of the air.
Dipole reflectors.	Selection based on speed and acceleration.
Diversion of range strobe from true target (RGPO).	Doppler shift of frequency, narrow range of strobe, gating based on acceleration.
Noise from high-altitude nuclear explosions.	Territorial spacing of radars of one group, operation of different frequencies of single-type stations, duplication of objects for the antiaircraft and antimissile defense systems.

Tabulations of ECCM utilization are given in tables 4–1 and 4–2 for repeaters and noise jammers, respectively. Deception repeaters can be countered by techniques such as those shown in table 4–1, which also evaluates the efficacy of these methods. Types of noise jammers and protection methods are listed in table 4–2. These ECCM tabulations represent typical engineering-design approaches for air defense and antiballistic missile radars.

References: Part I

1. Kerr, D. E. *Propagation of Short Radio Waves.* MIT Radiation Laboratory Series, vol. 13. New York: McGraw-Hill, 1951.
2. Blake, L.V. "A Guide to Basic Pulse-Radar Maximum Range Calculation." Naval Research Laboratory Report 5868 (1962).
3. Marcum, J. I., and Swerling, P. "Studies of Target Detection by Pulsed Radar." *IRE Transactions on Information Theory*, Special Monograph Issue (April 1960).
4. Skolnik, M. I. *Introduction to Radar Systems.* New York: McGraw-Hill, 1962.
5. Lawson, J.L., and Uhlenbeck, G.E. *Threshold Signals.* MIT Radiation Laboratory Series, vol. 24. New York: McGraw-Hill, 1950.
6. Mallett, J.D., and Brennan, L.E. "Cumulative Probability of Detection for Targets Approaching a Uniformly Scanning Search Radar." *Proceedings of the IEEE* (April 1963); also "Correction." *Proceedings of the IEEE* (June 1964).
7. Barton, D.K. *Radar System Analysis.* Englewood Cliffs, N.J.: Prentice-Hall, 1964.
8. Bradley, G.A. "How to Determine Antenna Pattern Loss." *Microwaves* (January 1967).
9. Gunn, K.L.S., and East, T.W.R. "The Microwave Properties of Precipitation Particles." *Quarterly Journal of the Royal Meteorological Society* (October 1954).
10. Swerling, P. "Maximum Angular Accuracy of a Pulsed Search Radar." *Proceedings of the IRE* (September 1956).
11. Manasse, R. "Summary of Theoretical Accuracy of Radar Measurements." Mitre Corporation Technical Series Report no. 2, 1 April 1960 (AD 287, 563).
12. Reilly, J.P. "Radar Clutter Reduction by Delay Line Cancellers." Master's thesis, School of Engineering and Applied Science, George Washington University, June 1962.
13. Nathanson, R.E. "Pulse Doppler and Burst Waveforms." Paper written for the Applied Physics Laboratory, Johns Hopkins University.
14. Barlow, E.J. "Doppler Radar." *Proceedings of the IRE* 37 (April 1949): 340–355.
15. Pedgeon, V.W. "Time, Frequency and Spatial Correlation of Radar Sea Return." JHU Applied Physics Laboratory Memo BPD 66U11, July 1966.
16. Pidgeon, V.W. "Radar Land Clutter for Small Grazing Angles at X and L-Band." In American Astronautical Society, Use of Space Systems for Planetary Geology and Geophysics (May 1967).

17. Krason, H., and Randig, G. "Terrain Backscattering Characteristics at Low Grazing Angles for X and S Band." *Proceedings of the IEEE* 54 (December 1966): 1964–1965.
18. Grisetti, R.S.; Santa, M.M.; and Kirkpatrick, G.M. "Effect of Internal Fluctuations and Scanning on Clutter Attenuation in MTI Radar." *IRE Transactions* ANE-2, no. 1 (March 1955): 37–41.
19. Nathanson, F.E., and Reilly, J.P. "Clutter Statistics Which Affect Radar Performance Analysis." *Supplement to IEEE Transactions on Aerospace and Electronic Systems* AES-3, no. 6 (November 1967).
20. Linder, R.A., and Kutz, G.H. "Digital Moving Target Indicators." *Supplement to IEEE Transactions on Electronic Systems* AES-3, no. 6 (November 1967).
21. Wainstein, L.A., and Zubakov, V.D. *Extraction of Signals From Noise*, translated from Russian and edited by R. Silverman. Englewood Cliffs, N.J.: Prentice-Hall, 1962.
22. Resnick, J.B. "High Resolution Waveforms Suitable for a Multiple Target Environment." Master's thesis, Massachusetts Institute of Technology, June 1962.
23. Ares, M. "Some Anticlutter Waveforms Suitable for Phased Array Radars." General Electric (HMED), no. R66EMH8, Syracuse, N.Y., November 1965.
24. Flesher, G.T., and Cohn, G.I. *The General Theory of Comb Filters*. Chicago: Proceedings NEC, 1958, pp. 282–295.
25. Galejs, J. "Enhancement of Pulse Train Signals by Comb Filters." *IRE Transactions of Information Theory* IT-4, no. 3 (September 1958): 114–125.
26. North, D.O. "An Analysis of Factors Which Determine Signal Noise Discrimination in Pulsed Carrier Systems." *Proceedings IEEE* 51, no. 7 (July 1964): 1016–127.
27. Nathanson, F.E. *Radar Design Principles—Signal Processing and the Environment*. New York: McGraw-Hill, 1969.

**Part II
Radar-Target Detection**

5 Radar Cross-Section Models in Detection

Introduction

The approach to radar cross-section (RCS) determination and application is usually determined by the needs of the user. If a specification is to be met, the vehicle designer is primarily interested in whether the measured RCS is below (or above) a certain level, whereas the radar-systems analyst may be interested in the variation of RCS with aspect angles in the azimuthal and/or elevation plane. In the latter case, target motional effects and radar/target geometries can allow evaluation of radar performance in detection and tracking of the vehicle in the presence of natural and man-made interference. If the radar-systems analyst is interested in evaluating the efficacy of the radar in target discrimination and/or target identification, detailed radar cross-section variations as a function of frequencies, bandwidths, and polarizations may be required.

The detection and tracking of low-RCS vehicles in severe-clutter environments is perhaps the area in which statistical detection theory is most applicable in radar-performance assessment. To apply the abundance of these available techniques in a meaningful way requires that the vehicle radar cross-section values be available in a format that allows specification of applicable statistical target models.

Discussion

Determination of the radar cross-section of a target both by analytical means [1,2,3] and by measurement [4,5] has been developed primarily from a diagnostic and estimative point of view. The emphasis has been on identifying sources of contributions to the total signal backscattered when the target is illuminated by an electromagnetic wave, and the variation of this signal-return amplitude as a function of angle (in two orthogonal planes), frequency, and polarizations. A great deal of static-range RCS data were obtained simply because this was less costly than dynamic measurement (where the target is in flight), even though the latter situation may be more realistic.

A compromise radar-analysis approach, made more attractive by the low-cost availability of the digital computer, was to combine static-range RCS data with radar parameters and radar/target geometries to *simulate* the

RCS time history for radar-performance evaluation [6]. The result of such radar-signal simulation was to obtain a measure of the radar signature—the statistical amplitude variation of the radar cross-section as a function of time—as it would affect the radar-system parameters. From the earliest attempts to predict radar performance on a statistical basis [7], a great deal of effort has been expended on definition of useful target models in calculation of detection probabilities, and on accounting for effects of signal processing [8,9,10,20].

The direct passage from measured values of radar cross section into statistical target models for radar system analysis is still not clear, although many attempts have been made to bridge this gap [11,12]. Measurements have been made of cone-cylindrical target geometries, which were not tumbling, but were stable about their pitch and yaw axes, and were viewed with a radar line of sight(s) over a range of aspect angles with respect to the longitudinal (or roll) axes. Both monostatic RCS measurements at a static radar cross-section range [13] and dynamic bistatic radar cross-section measurements [14] of satellite-type targets yielded large data sets. In each case these were reduced for statistical analyses. The conclusions were similar—the so-called log-normal statistical distribution provided the optimum fit to the observed data for both the bistatic and monostatic cases. Satellites in low orbit and cruise missiles both have long cylindrical shapes with vehicle stabilization about the flight-path (longitudinal) axis. Their RCS variations are very similar.

The link to radar-detection-probability prediction for stabilized missiles, which are physically similar to satellites, is first to obtain an "average" radar cross-section value and then to use this value in the radar equation to obtain an "average" signal-to-noise ratio. One can then use available plots of detection probability for fluctuating targets [10,15] with log-normal distributions to obtain detection probabilities for various false-alarm rates and postdetection (pulse-integration) signal processing [16]. Many background-terrain and sea clutter levels can also be described statistically by the log-normal distribution, particularly at low grazing angles. Detection probabilities for log-normal fluctuating targets in log-normal clutter backgrounds for various signal-processing methods have been documented [10,17,18,19].

Some available detection-probability curves are a function of the ratio of range R to a reference range R_0. The range R_0 is obtained from radar parameters as the value of range resulting in a signal-to-noise ratio of unity, or 0 dB. Although this method is of some use in comparisons, most detection-probability curves are now given as a function of number of samples processed and signal-to-noise levels [9,10]. Several available curves display the single-pulse (sample) signal-to-noise ratio with the number of pulses (samples) processed as a parameter [10,16,17]. Chapter 6 summarizes the theory and gives numerical tables in convenient form for applications.

Radar Cross-Section Models in Detection

The choice of target models and their use in computation of detection probabilities, together with a comprehensive set of convenient detectability plots, are available [10] and recommended for use in radar-performance evaluation. Refinements to the knowledge of target models for detection calculation are continuing [20], but such refinements are not expected to improve significantly the variety of methods currently available [10].

The measurement of radar cross section for targets having a large degree of symmetry about their roll axes usually involves recording the backscatter return from nose-on (usually taken as 0°) in the azimuthal plane through broadside (near 90°) to tail-on (180°) at each degree or fraction of a degree. This is repeated at each frequency and transmit/receive polarization of interest and, if the object is not symmetrical, for the full 360° azimuth in each of several roll angles. Clearly, a great deal of RCS data can become available.

The radar-systems analyst is often interested in the RCS values near nose-on, and in some continguous range of aspect angles, depending on anticipated radar/target geometries. Long, cylindrical objects with conical noses have an RCS versus aspect angle characterized by a peak value nose-on (0°) falling off by approximately 10 dB at somewhat less than 10° from nose-on, and remaining at that level to near 60°, where the broadside specular lobe for the cylindrical body is beginning. Therefore, the actual average value of RCS to be used for radar-detection prediction can vary by approximately 10 dB, depending on whether the nose-on value or off-nose-on values are selected.

One method of obtaining an average RCS is to define the value at each of several discrete frequencies as:

$$\sigma_T(f) = \frac{1}{54} \sum_{\theta_R=1}^{3} \sum_{P=1}^{3} \sum_{\theta_A=-3}^{3} \sigma_{10}(f), \qquad (5.1)$$

where $\sigma_T(f)$ = average target RCS at a frequency f in dBm^2;

$\sigma_{10}(f)$ = measured average value of RCS in dBm^2 in a 10° aspect-angle interval at a particular polarization P and roll angle θ_r;

θ_A = aspect-angle intervals (0° is nose-on)

$-30°$ to $-20°$
$-20°$ to $-10°$
$-10°$ to $0°$
$0°$ to $+10°$
$+10°$ to $+20°$
$+20°$ to $+30°$;

θ_R = roll angles: 0°, 60°, and 90°;

P = polarization (transmit/receive) (Vertical, V; Horizontal, H; right circular, RC; left circular, LC) VV, HH, RCLC;

f = 2.0, 2.7, 3.6, 4.9, 6.6, 8.9, 12.0 GHz.

In this method of average RCS computation, numerical values corresponding to the *logarithm* of the radar cross-section are averaged arithmetically, yielding a geometric mean. The definition provided by equation 5.1 is easy to calculate and does yield RCS smoothing, which is appropriate for roll-symmetric target geometries in terms of the roll angle and polarization averaging. However, if the radar-target geometry is such that the nose-on regions of aspect angles are viewed (assuming no effective nose-on RCS reduction), then the resulting value will be weighted in that the "average" RCS value will be less than that which determines radar-detection range. Since detection range varies as the one-fourth root of radar cross-section, all other factors being equal, the detection range could be nearly 1.7 times the values predicted by equation 5.1 if the nose-on value only were viewed by the radar, since it could be a factor of 10 higher.

The aspect-angle range for a radar located on an airborne platform of 10-k altitude, viewing an approaching target at 300-k range might well be limited to a few degrees from nose-on, whereas a ground-based radar viewing the target at 30-k range may have a radar line of sight at several tens of degrees with respect to the vehicle's flight path. The range of aspect angles for each engagement would be quite different, as would the terrain masking and clutter return, which are also dependent on radar-site and target-approach geometries. It is also to be expected that a flight path might be chosen to present the radar with the lowest probability of detecting the vehicle, if hostile radar locations were known.

As will be shown, the concepts of obtaining an "average" RCS can vary. In the interest of providing a numerical comparison, a representative sample data set, as shown in table 5–1, was used to calculate the values for

Table 5–1
RCS Sample Data Set (Amplitude Only)

N	~$\Delta\theta_i$ (Degrees)	~σ_i (m^2)	
1	−30 to −20	0.04	
2	−20 to −10	0.04	
3	−10 to 0	0.4	
4	− 0 to +10	0.4	near nose-on
5	+10 to +10	0.04	
6	+20 to +30	0.04	

Radar Cross-Section Models in Detection 127

arithmetic mean and geometric mean. The difference for this example data set is less than 3 dB and is comparable to variations in RCS due to polarization uncertainties. It is concluded that the decibel-averaging method for defining average RCS is reasonable. The method is convenient, and other comparable unknowns exist in application of the radar equation and detection-probability theory to prediction of radar-system performance and evaluation. RCS modeling for bistatic radar detection, and at higher radar cross-section levels, may yield much larger differences and need yet to be investigated.

Arithmetic-mean calculation:

$$\bar{\sigma}_A = \frac{1}{N} \sum_{i=1}^{N} \sigma_i,$$

where $\bar{\sigma}_A = 0.16$ m^2;
$(\bar{\sigma}_A)$dBm$^2 = -7.96$ dBm^2.

Geometric-mean calculation:

$$\bar{\sigma}_{AF} = \left(\prod_{i=1}^{N} \sigma_i \right)^{1/N}$$

where $\bar{\sigma}_{AF} = 0.086$ m^2;
$\bar{\sigma}_{AF}$ dBm$^2 = -10.6$ dBm^2.

Difference is ≈ 2.64 dB.

RCS-Averaging Concepts

The arithmetic average of values of radar cross-section for targets of rotational symmetry (bodies of revolution), having no variation of RCS in a plane orthogonal to the longitudinal axis, can be formulated for a plane containing the longitudinal axis as

$$\bar{\sigma}_A = \frac{1}{|\theta_2 - \theta_1|} \int_{\theta_1}^{\theta_2} \sigma(\theta) d\theta, \tag{5.2}$$

where $\bar{\sigma}_A$ = arithmetic mean, or average;

$(\theta_2 - \theta_1)$ = aspect-angle interval measurement, $(\theta_2 - \theta_1) \geq 0$;

$\sigma(\theta)$ = variation of RCS as a function of aspect angle.

Equation (5.1) can be written in discrete form as

$$\bar{\sigma}_A = \frac{1}{N} \sum_{i=1}^{N} \sigma_i, \qquad (5.3)$$

where the σ_i are N samples of average RCS values in $\Delta\theta_i$ uniform aspect-angle increments, located within the aspect-angle interval under consideration.

A weighted average may be defined in a manner similar to that of equation 5.2 as [2]

$$\bar{\sigma}_{AW} = \frac{1}{|\theta_2 - \theta_1|} \int_{\theta_1}^{\theta_2} \omega(\theta) \sigma(\theta) d\theta, \qquad (5.4)$$

where the weighting function $\omega(\theta)$ provides a relative weighting over the aspect angles of interest. For example, it might be desirable to weight most heavily the aspect angle subinterval most likely to be observed in a particular radar/target geometry, while placing less emphasis on those aspect angles less likely to be observed. In discrete form, equation 5.4 would become

$$\bar{\sigma}_{AW} = \frac{1}{N} \sum_{i=1}^{N} \omega_i \sigma_i, \qquad (5.5)$$

where the individual weights ω_i might be normalized such that $\omega_i \leq 1$.

Another method of defining an average is to use a functional form such as [2]

$$\bar{\sigma}_{AF} = \frac{1}{|\theta_2 - \theta_1|} \int_{\theta_1}^{\theta_2} F\{\sigma(\theta)\} d\theta. \qquad (5.6)$$

The most convenient functional form has been to use a logarithmic function—the RCS values in decibels referred to a 1-m^2 value. This form is obtained as

$$(\bar{\sigma}_{AF}) \text{dBm}^2 = \frac{1}{|\theta_2 - \theta_1|} \int_{\theta_1}^{\theta_2} 10 \log_{10} \sigma(\theta) d\theta. \qquad (5.7)$$

For convenience, the factor k may be used to relate the decibel logarithmic form (log) to the natural logarithmic form (ln) as

$$K \ln (\,\cdot\,) = 10 \log_{10} (\,\cdot\,). \tag{5.8}$$

Substituting equation 5.7 in equation 5.8 yields

$$K \ln (\bar{\sigma}_{AF}) = \frac{K}{|\theta_2 - \theta_1|} \int_{\theta_1}^{\theta_2} \ln \sigma(\theta) d\theta. \tag{5.9}$$

Equation 5.9 may be written in discrete form as

$$\ln (\bar{\sigma}_{AF}) = \frac{1}{N} \sum_{i=1}^{N} \ln \sigma_i, \tag{5.10}$$

or

$$10 \log_{10} (\bar{\sigma}_{AF}) = \frac{1}{N} \sum_{i=1}^{N} (10 \log_{10} \sigma_i). \tag{5.11}$$

Equation 5.11 affords great computational simplicity, since measured values of RCS are usually available in dBm^2 versus aspect angle. A series of uniform increments of aspect angle, covering the range of interest, yields a series of numerical values of dBm^2 that are simply summed and divided by the number of samples to provide an "average" RCS, also in units of dBm^2.

It is of interest to compare the average RCS values given by application of the definitions of equations 5.3 and 5.11, using 5.10 rewritten as

$$\ln (\bar{\sigma}_{AF}) = \frac{1}{N} \ln \left(\prod_{i=1}^{N} \sigma_i \right), \tag{5.10}$$

which is equivalent to

$$\ln (\bar{\sigma}_{AF}) = \frac{1}{N} \sum_{i=1}^{N} (\ln \sigma_i), \tag{5.12}$$

or

$$\ln (\bar{\sigma}_{AF}) = \ln \left[\left(\prod_{i=1}^{N} \sigma_i \right)^{1/N} \right]. \tag{5.13}$$

Equation 5.13 in numerical form is then

$$(\bar{\sigma}_{AF}) = \left(\prod_{i=1}^{N} \sigma_i \right)^{1/N}, \tag{5.14}$$

and the ratio of the arithmetic "average" to the "logarithmic" average is seen to be the ratio between an arithmetic mean and a geometric mean. Using equations 5.3 and 5.14 yields

$$\frac{\bar{\sigma}^A}{\bar{\sigma}_{AF}} = \frac{\left(\frac{1}{N}\sum_{i=1}^{N}\sigma_i\right)}{\left(\prod_{i=1}^{N}\sigma_i\right)^{1/N}}. \qquad (5.15)$$

For all nonnegative values of $\sigma_i \geq 0$, the geometric mean is equal to or *less* than the arithmetic mean [21], and

$$\left(\prod_{i=1}^{N}\sigma_i\right)^{1/N} \leq \frac{1}{N}\sum_{i=1}^{N}\sigma_i \qquad (5.16)$$

and

$$\bar{\sigma}_{AF} \leq \bar{\sigma}_A.$$

The "average" RCS value obtained by using numerical values in units of dBm² as samples for obtaining an arithmetic mean, as shown in the numerator of equation 5.15 will be *lower* than the value that should be used to calculate average received-to-signal-to-noise levels. Its use in determination of detection probability may yield erroneous results; actual detection ranges may be somewhat greater than those predicted, since the numerical radar cross section, as used in the radar equation, is larger in value than the resulting dBm²-averaged value.

References

1. "Special Issue on Radar Reflectivity," *Proceedings of the IEEE* 53, no. 8 (August 1965): 769–1168.
2. Crispin, J.W., and Seigel, K.M. *Methods of Radar Cross Section Analysis*. New York: Academic Press, 1968.
3. Ruck, G.T., ed. *Radar Cross Section Handbook*. 2 vols. New York: Plenum Press, 1970.
4. *Radar Reflectivity Measurements Symposium*. RADC-TDR-64–25, vols. I and II. AD601364 AD601365 (April 1964).
5. Bachman, C.G.; King, H.E.; and Hansen, R.C. "Techniques for Mea-

surement of Reduced Radar Cross Sections." BSD-TDR-62-300, Contract AF 04(695)-69 (October 1962).
6. Mitchell, R.L. *Radar Signal Simulation*. Dedham, Mass.: Artech House, 1976.
7. Marcum, J.I., and Swerling, P. "Studies of Target Defection by Pulsed Radar." *Transactions of the IRE* Prof. Gp. Information Theory (Special Monograph Issue) IT-6, no. 2 (April 1960).
8. Weinstock, W.W. "Radar Cross Section Target Models," chap. 5, and Illustrative Problems in Radar Detection Analysis," chap. 6. In *Modern Radar: Analysis, Evaluation and System Design*, R.S. Berkowitz, ed. New York: Wiley, 1965.
9. Swerling, P. "Recent Developments in Target Models in Radar Detection Analysis." *AGARD Conference Proceedings*, no. 66, AGARD-CP-66-70; "Advanced Radar Systems." AD 715 485 (November 1970).
10. Meyer, D.P., and Mayer, H.A. *Radar Target Detection: Handbook of Theory and Practice*. New York: Academic Press, 1973.
11. Rheinstein, J. "On the Comparison of Static and Dynamic Radar Cross-Section Measurements." Project report PA-84 (BMRS). Contract AF 19 (628)-500 MIT/Lincoln Laboratory, 28 January 1965.
12. "A Preliminary Investigation of Suitable Techniques for the Expression of Static Radar Cross Section Signatures." Air Force Technical Note no. BSD-TDR-62-381. Chrysler Corporation, Report no. ADB-TN-37-62. Contract AF-(694)-25, 15 December 1962.
13. Kennedy, R.W. "The Spatial and Spectral Characteristics of Satellite-Type Targets." Technical Report AFAL-TR-66-17 (March 1966).
14. Brescia, R.E., and Zirm, R.R. "The Distribution of Reflecting Cross Sections of Satellites." NRL Report 5615, 25 May 1961.
15. Swerling, P. "Probability of Detection for Some Additional Fluctuating Target Cases." Aerospace Corporation Report no. TOR-699 (9990) 14. Contract AF 04 (695)-669 (March 1966).
16. Heidbreder, G.R., and Mitchell, R.L. "Detection Probabilities for Log-Normally Distributed Signals." SSD-TR-66-87, Aerospace Report no. TR-669 (9990)-6 (April 1966); also *IEEE Transactions* PG-AES, vol. AES-3, no. 1 (January 1967): 5–13.
17. Kramer, J.D.R. "The Detection Problem for Log-Normally Distributed Signal and Noise." MITRE Report MTP-136. Contract F 19628-71-C-002 (May 1972).
18. George, S.F. "The Detection of Nonfluctuating Targets in Log-Normal Clutter." NRL Report 6796, 4 October 1968.
19. Trunk, G.V. "Non-Rayleigh Sea Clutter: Properties and Detection of Targets." NRL Report 7986, 25 June 1976.

20. Von Schlacta, K. "Remarks on Target Models for the Design of Radar Systems." *Proceedings of the IEEE* (1975): 4440–4445. International Radar Conference.
21. Rektorys, K., ed. *Survey of Applicable Mathematics*. Cambridge, Mass.: MIT Press, 1969, p. 47.

6

Required S/N Per Pulse *versus* Detection Probability

It is well known that the range of a radar set for a point target is inversely proportional to the fourth root of the required signal-to-noise ratio per pulse. That is,

$$R = R_0 \, x^{-1/4}$$

where R = radar range;

R_0 = constant of proportionality;

x = required signal-to-noise ratio per pulse.

The value of x is specified as that required to produce a specified detection probability P_d when the false-alarm probability P_{fa} (or false-alarm rate f_{fa}) is set to a certain level. The problem of determining x for a square-law detector followed by an unweighted adder has been solved by Marcum [1,2] for nonfluctuating targets and by Swerling [3] for four types of target fluctuation.

Unfortunately, the data in these references are presented in an inconvenient form for use in the radar range equation. The inconvenience is caused by (1) a graphical format in which it is difficult to interpolate to obtain values not on the plotted curves; (2) plotting detection probability versus R/R_0 instead of x; (3) using false-alarm number n as a parameter rather than a factor more simply related to false-alarm probability P_{fa}. As will be shown,

$$n = \frac{N \ln 0.5}{\ln(1 - P_{fa})} \approx \frac{N \ln 2}{P_{fa}} \,.$$

Using an electronic computer, Fehlner [4,5] reproduced Marcum and Swerling's results and presented tables and graphs of more extensive and more accurate computed data. Fehlner plotted his graphs versus x, an improvement; but the presentation was still inconvenient because (1) on the graphs, x is plotted as power ratio instead of in decibels; (2) the parameter n is replaced by n' where $n' = n/N$; (3) on the tables the independent variable is x, with P_d as the dependent variable. Since P_d is usually the specified number, it would be more convenient to make x the dependent variable.

The change of the false-alarm number from n to n' did not simplify anything. Indeed, it still causes some confusion in comparing Fehlner's results with those of Marcum and Swerling. As a result of this change, Fehlner himself mistakenly reported that Swerling's data did not agree with his own, especially for large N, the number of pulses integrated.

Therefore, let us briefly review the false-alarm-number definitions and explore their physical significance.

Most radar sets are used in systems that control the false-alarm rate, f_{fa}. These systems are called CFAR (constant false-alarm rate) systems. The relationship between the false-alarm rate f_{fa} and the false-alarm probability P_{fa} is

$$f_{fa} = f_{dec} \, P_{fa}, \qquad (6.1)$$

where f_{dec} is the decision rate—that is, the number of times per second that the decision as to whether or not a target is present is made. If pulses are not added prior to the decision, then in each range-Doppler cell the decision rate is at the radar pulse-repetition frequency f_r. Thus the overall decision rate is

$$f_{dec} = f_r \, \eta b \qquad (6.2)$$

where $\eta =$ number of range gates per Doppler bin;

$b =$ number of Doppler bins per range gate.

Now suppose that, prior to each decision, m pulses are integrated coherently, and then N of the resulting signals are integrated incoherently. The decision rate will then be slowed down to

$$f_{dec} = \frac{f_r \, \eta b}{mN}. \qquad (6.3)$$

It is seen that equation 6.2 is a special case of equation 6.3 for $mN = 1$. Putting equation 6.3 into equation 6.1, we find that

$$f_{fa} = \frac{f_r \, \eta b}{m} \, \frac{P_{fa}}{N}. \qquad (6.4)$$

Now suppose that the factor

$$\frac{f_r \, \eta b}{m}$$

S/N per Pulse *versus* Detection Probability

is held constant, while N is varied. If a constant false-alarm rate is to be maintained, we should keep the ration P_{fa}/N constant. We can define a false-alarm number to be

$$n_{fa} = \frac{N}{P_{fa}}. \tag{6.5}$$

Then equation 6.4 would become

$$f_{fa} = \frac{f_r\, \eta b}{m\, n_{fa}}. \tag{6.6}$$

Unfortunately, Marcum did not use as simple a definition of false-alarm number as that in equation 6.5. He defined two false-alarm numbers, n and n'.

Marcum visualized the false alarm as the result of a number of independent trials. In order to not get a false alarm, the false alarm must not occur on a certain number of successive independent opportunities. At each opportunity, the probability of getting a false alarm is P_{fa}. Marcum defined the false-alarm number n' as the number of independent opportunities to get (or not get) a false alarm that are necessary to make the multiple-trial probability of not getting a false alarm equal to one-half. That is,

$$(1 - P_{fa})^{n'} = 0.5. \tag{6.7}$$

Solving equation 6.7 for n', we have

$$n' = \frac{\ln 0.5}{\ln (1 - P_{fa})}. \tag{6.8}$$

It is evident that there is a one-to-one correspondence between P_{fa} and n'. If n' is kept constant as N is varied, then P_{fa} will be constant also. Now n' is the parameter that Fehlner chose to keep constant. Hence his curves, plotted with n' as a parameter, are equivalent to having a constant P_{fa} given by

$$P_{fa} = 1 - 0.5^{1/n}, \tag{6.9}$$

The number of opportunities to not get a false alarm, n', is related to the false alarm time, τ, by

$$n' = f_{dec}\, \tau. \tag{6.10}$$

Putting equation 6.3 into equation 6.10 yields

$$n' = \frac{f_r \eta b \tau}{mN}.\qquad(6.11)$$

Marcum's other false-alarm number is defined as

$$n = n'N.\qquad(6.12)$$

Eliminating n' between equations 6.11 and 6.12, we find

$$n = \frac{f_r \eta b}{m}\tau.\qquad(6.13)$$

Thus if we once more assume that the factor

$$\frac{f_r \eta b}{m}$$

is held constant, while N is varied, then a constant value of n implies a constant false-alarm time τ in equation 6.13.

In summary, we may characterize these three false-alarm numbers by the physical entity they represent as constant while N is varied. This is done as follows:

False-Alarm Number	Item Held Constant as N Is Varied
n_{fa}	f_{fa}
n'	P_{fa}
n	τ

It is furthermore assumed that $f_r \eta b/m$ is constant as N is varied for constant n_{fa} and m/n.

If equation 6.1 is divided by equation 6.10, we obtain

$$\frac{f_{fa}}{n'} = \frac{P_{fa}}{\tau}.\qquad(6.14)$$

Using equation 6.8 in equation 6.14, we find

$$f_{fa} = \frac{\ln 0.5}{\ln(1 - P_{fa})\tau} P_{fa}.\qquad(6.15)$$

S/N per Pulse *versus* Detection Probability

Now since (See Appendix A)

$$n' = \frac{\ln 0.5}{\ln(1 - P_{fa})} \approx \frac{\ln 2}{P_{fa}}; \quad (6.16)$$

$$f_{fa} \approx \frac{\ln 2}{\tau}. \quad (6.17)$$

The fact that f_{fa} is not exactly the reciprocal of τ and furthermore is only *approximately* inversely proportional to τ is a source of some confusion. The reason is simply that f_{fa} and $1/\tau$ were independently and differently defined. The average time between false alarms can be defined directly in terms of f_{fa} as

$$\tau_{fa} = \frac{1}{f_{fa}}. \quad (6.18)$$

Now recognizing that radar systems are specified as having either a certain false-alarm rate f_{fa} or a certain false-alarm probability P_{fa}, Morris [6] prepared two tabulations, one with P_{fa}/N held constant and the other with P_{fa} held constant, as N is varied. The tabulation listed S/N in dB; but, like Fehlner, Morris used S/N as the independent variable. This makes the determination of fluctuation loss at a given P_d difficult to ascertain.

It is hoped that the present tabulation overcomes the difficulties of the previous cited works. It uses essentially the same computer program employed by Morris, but employs the Newton-Raphson iteration to solve for the value of x required for a given P_d. Thus P_d is made to be the independent variable. Values of x are computed and tabulated for each P_d. At any P_d the fluctuation loss is simply the difference in the values of x for the fluctuating case and the nonfluctuating case. Relative fluctuation loss between different fluctuation cases can similarly be obtained.

If we define a fourth false-alarm number n'_{fa} as

$$n'_{fa} = \frac{n_{fa}}{N}, \quad (6.19)$$

then we find that, from equation 6.5,

$$n'_{fa} = \frac{1}{P_{fa}}. \quad (6.20)$$

Clearly, this false-alarm number, like n', also keeps P_{fa} constant as N is varied.

It can be seen from the foregoing discussion that the use of any of the four false-alarm numbers is likely to cause confusion, since they are similar, yet somewhat different from each other. They have been discussed at length only so that the relationship between the values presented in this chapter can be properly compared with other sources that use these parameters. The use of n, n', n_{fa}, or n'_{fa} is discouraged. In the tabulation we use the parameter P_{fa}/N for the constant false-alarm-rate case, and the parameter P_{fa} for the constant P_{fa} case.

The tables are arranged so that relative-fluctuation loss and noncoherent-integration loss can be readily determined. For example, suppose we require a 70-percent detection probability at a constant false-alarm rate with $P_{fa} = N \times 10^{-3}$. We would then refer to page 10 of enclosure 10. For $N = 10$ we find that the required signal-to-noise ratio (SNR) for a nonfluctuating target is 0.85 dB, whereas for a Swerling 3 target it is 2.47 dB, so that the fluctuation loss for a Swerling 3 target is 1.62 dB. For $N = 1$, the Swerling 3 target requires 10.94 dB, so that if we had a coherent-integration gain (that is, $10 \log_{10} N$) of 10 dB, we would require only 0.94 dB. Since the noncoherent integration of 10 pulses required 2.47 dB, the noncoherent-integration loss is $2.47 - 0.94 = 1.53$ dB. This can be obtained directly from the tabulated values, as follows:

$$2.47 + 10 - 10.94 = 1.53 \text{ dB}.$$

In formula,

$$L_{int} = (XN)_{dB} + N_{dB} - (X1)_{dB},$$

where L_{int} = noncoherent-integration loss;

$(XN)_{dB}$ = required SNR in dB for N pulses integrated noncoherently;

N_{dB} = coherent-integration gain in dB (second column in table),

$(X1)_{dB}$ = required SNR in dB for 1 pulse.

References

1. Marcum, J.I. "A Statical Theory of Target Detection by Pulsed Radar." RAND Corporation Research Memo Rm-754, 1 December 1947. Reprinted in *Trans*. IT-6, no. 2 (April 1960): 59–144.
2. Marcum, J.I. "A Statistical Theory of Target Detection by Pulsed Radar (Mathematical Appendix)." RAND Corporation Research

Memo RM-753, 1 July 1948. Reprinted in *Trans.* IT-6, no. 2 (April 1960): 145–268.
3. Swerling, P. "Probability of Detection for Fluctuating Targets." RAND Corporation Research Memo RM-1217, 17 March 1954. Reprinted in *Trans.* IT-6, no. 2 (April 1960): 269–308.
4. Fehlner, L.F. "Marcum's and Swerling's Data on Target Detection by a Pulsed Radar." The Johns Hopkins University, Applied Physics Laboratory, Report TG 451 (AD-602121), 2 July 1962.
8. Fehlner, L.F. "Supplement to Marcum's and Swerling's Data on Target Detection by a Pulsed Radar." The Johns Hopkins University, Applied Physics Laboratory, Report TG 451A (AD-608354) (September 1964).
6. Morris, J.D. "Probability of Detection Tables versus Probability of False Alarm for Five Fluctuation Models." Raytheon Company memo RS-71-255, of October 1971.

Appendix 6A: Approximate Formula for n'

We start with equation 6.9, which is rewritten as equation 6A.1:

$$P_{fa} = 1 - 0.5^{1/n}, \qquad (6A.1)$$

Now, employing the identity $x \equiv \exp(\ln x)$

$$P_{fa} = 1 - \exp\left(\frac{\ln 0.5}{n'}\right). \qquad (6A.2)$$

Using the expansion

$$\exp(y) = \sum_{a=0}^{\infty} \frac{y^a}{a!},$$

$$P_{fa} = 1 - \sum_{a=0}^{\infty} \frac{1}{a!} \left(\frac{\ln 0.5}{n'}\right)^a. \qquad (6A.3)$$

For large n' we can neglect all terms beyond the second in the expansion, yielding

$$P_{fa} \approx \frac{\ln 2}{n'}. \qquad (6A.4)$$

Appendix 6B:
No Target Signal:
False-Alarm Probability

Characteristic Function:

$$C_Y(\nu) = (1 + j2\pi\nu)^{-N}. \tag{6B.1}$$

Probability Density

$$p(Y) = \frac{e^{-Y} Y^{N-1}}{(N-1)!} \quad (Y \geq 0). \tag{6B.2}$$

Note that this is the gamma density with parameter $\lambda = 1$. It is also the chi-square density with parameter $\sigma = 1/\sqrt{2}$ and $2N$ degrees of freedom. See Parzen, *Modern Probability Theory and Its Applications* (New York: Wiley, 1960), pp. 180–181.

False-Alarm Probability

$$P_{fa} = e^{-Y_b} \sum_{k=0}^{N-1} \frac{Y_b^b}{k!}. \tag{6B.3}$$

Appendix 6C: Nonfluctuating Target Signal

Characteristic Function

$$C_Y(v) = \frac{e^{-Nx}}{(1 + j2\pi v)^N} e^{Nx/(1+j2\pi v)}. \qquad (6\text{C}.1)$$

Probability Density

$$p(Y) = e^{-Nx-Y} \sum_{k=0}^{\infty} \frac{(Nx)^k}{k!} \frac{Y^{N+k-1}}{(N+k-1)!} \quad (Y \geq 0). \qquad (6\text{C}.2)$$

Detection Probability

$$P_d = e^{-(Y_b + Nx)} \sum_{k=0}^{\infty} \left[\frac{(Nx)^k}{k!} \sum_{a=0}^{N-1+k} \frac{Y_b^a}{a!} \right]. \qquad (6\text{C}.3)$$

Appendix 6D: Swerling Case 1

Characteristic Function

$$C_Y(\nu) = (1 + j2\pi\nu)^{1-N}(1 + j2\pi\nu\,[1 + N\bar{x}])^{-1}. \qquad (6\text{D}.1)$$

Probability Density

For $N = 1$,

$$p(Y) = \frac{1}{1 + \bar{x}}\, e^{-Y/(1+\bar{x})} \qquad (Y \geq 0). \qquad (6\text{D}.2)$$

For $N > 1$,

$$p(Y) = \frac{1}{N\bar{x}}\left(1 + \frac{1}{N\bar{x}}\right)^{N-2} e^{-Y/(1+N\bar{x})}$$

$$\left\{1 - e^{-Y/(1+N\bar{x})} \sum_{k=0}^{N-2} \frac{1}{k!}\left(\frac{Y}{1 + N\bar{x}}\right)^k\right\} \qquad (Y \geq 0). \qquad (6\text{D}.3)$$

Detection Probability

For $N = 1$,

$$P_d = e^{-Y_b/(1+\bar{x})} \qquad (Y \geq 0). \qquad (6\text{D}.4)$$

For $N > 1$,

$$P_d = e^{-Y_b} \sum_{k=0}^{N-2} \frac{Y_b^k}{k!} + P_{da}\left[1 - e^{-Y_b/(1 + 1/N\bar{x})} \sum_{k=0}^{n-2} \frac{1}{k!}\left(\frac{Y_b}{1 + \frac{1}{N\bar{x}}}\right)^k\right], \qquad (6\text{D}.5)$$

where

$$P_{da} = \left(1 + \frac{1}{N\bar{x}}\right)^{N-1} e^{-Y_b/(1+N\bar{x})} = \text{approximate value of } P_d,$$

$(P_d \approx P_{da} \text{ for } N\bar{x} \gg \text{ and } P_{fa} \ll 1).$ \hfill (6D.6)

Appendix 6E:
Swerling Case 2

Characteristic Function

$$C_Y(v) = (1 + j2\pi v[1 + \bar{x}])^{-N}. \qquad (6\text{E}.1)$$

Probability Density

$$p(Y) = \frac{1}{(1 + \bar{x})^N (N-1)!} Y^{N-1} e^{-Y/(1+\bar{x})} \qquad (Y \geq 0). \quad (6\text{E}.2)$$

Detection Probability

$$P_d = e^{-Y_b/(1+\bar{x})} \sum_{k=0}^{N-1} \frac{1}{k!} \left(\frac{Y_b}{1 + \bar{x}}\right)^k. \qquad (6\text{E}.3)$$

Appendix 6F: Swerling Case 3

Characteristic Function

$$C_Y(\nu) = (1 + j2\pi\nu)^{2-N}\left(1 + j2\pi\nu\left[1 + \frac{N\bar{x}}{2}\right]\right)^{-2}. \quad (6\text{F}.1)$$

Probability Density

For $N = 1$,

$$p(Y) = \frac{1}{\left(1 + \frac{\bar{x}}{2}\right)^2}\left(1 + \frac{Y}{1 + \frac{2}{\bar{x}}}\right) e^{-Y/(1+\bar{x}/2)} \quad (Y \geq 0). \quad (6\text{F}.2)$$

for $N = 2$,

$$p(Y) = \frac{Y}{(1 + \bar{x})^2} e^{-Y/(1+\bar{x})} \quad (Y \geq 0). \quad (6\text{F}.3)$$

For $N > 2$,

$$p(Y) = -\frac{(N+2)A^2 e^{-AY}}{(1-A)^{N-1}} + \frac{A^2 Y e^{-AY}}{(1-A)^{N-2}}$$

$$+ A^2 e^{-Y} \sum_{k=0}^{N-3} \frac{(k+1) Y^{N-3-k}}{(N-3-k)!\,(1-A)^{k+2}} \quad (Y \geq 0). \quad (6\text{F}.4)$$

where

$$A = \frac{1}{1 + \frac{N\bar{x}}{2}} \quad (6\text{F}.5)$$

Appendix 6F: Swerling Case 3

Detection Probability

For $N = 1$,

$$P_d = \left[1 + \frac{Y_b}{\left(1 + \dfrac{\bar{x}}{2}\right)\left(1 + \dfrac{2}{\bar{x}}\right)} \right] e^{-Y_b(1+\bar{x}/2)}. \tag{6F.6}$$

For $N = 2$,

$$P_d = \left(1 + \frac{Y_b}{1 + \bar{x}}\right) e^{-Y_b/(1+\bar{x})}. \tag{6F.7}$$

For $N > 2$,

$$P_d = \frac{Y_b^{N-1} e^{-Y_b} A}{(N-2)!} + \sum_{k=0}^{N-2} \frac{e^{-Y_b} Y_b^k}{k!}$$

$$+ \frac{e^{-AY_b}}{(1-A)^{N-2}} \left[1 - \frac{(N-2)A}{1-A} + AY_b \right]$$

$$\left[1 - \sum_{j=0}^{N-2} \frac{e^{-Y_b(1-A)} [Y_b(1-A)]^j}{j!} \right]. \tag{6F.8}$$

Appendix 6G: Swerling Case 4

Characteristic Function

$$C(\nu) = (1 + j2\pi\nu)^N \left(1 + j2\pi\nu\left[1 + \frac{\bar{x}}{2}\right]\right)^{-2N} \quad (6G.1)$$

Probability Density

$$P(Y) = e^{-BY} B^{2N} \sum_{k=0}^{2N-1} \binom{N}{k} (1-B)^{N-k} \frac{Y^{2N-1-k}}{(2N-1-k)!} \quad (Y \geq 0), \quad (6G.2)$$

where

$$B = \frac{1}{1 + \bar{x}/2}. \quad (6G.3)$$

Detection Probability

$$P_d = B^N \sum_{k=0}^{N} \left[\binom{N}{k} \left(\frac{1+B}{B}\right)^{N-k} \sum_{j=0}^{2N-1-k} \frac{e^{-BY_b}(BY_b)^j}{j!} \right]. \quad (6G.4)$$

Appendix 6H: Chi-Square Family

The four Swerling target-fluctuation cases are members of the chi-square family. If x_i is the signal-to-noise ratio for the ith pulse of a scan, we define

$$X = \sum_{i=1}^{N} x_i \qquad (6\text{H}.1)$$

as the integrated signal-to-noise ratio on a given scan. X is said to have a chi-square probability law if it is also equal to the sum of $2K$ squared Gaussian variates—that is, if

$$X = \sum_{i=1}^{2K} \alpha_i^2, \qquad (6\text{H}.2)$$

where the α_i are Gaussian random variates. The number $2K$ is known as the number of degrees of freedom. It can be shown that

$$K = \frac{(\bar{X})^2}{\operatorname{Var} X}. \qquad (6\text{H}.3)$$

The value of K for the four Swerling cases is as follows:

Swerling Case	K
1	1
2	N
3	2
4	$2N$

Characteristic Function

$$C_Y(\nu) = (1 + j2\pi\nu)^{-N}\left(1 + j2\pi \frac{N\bar{x}}{K} \frac{\nu}{1+j2\pi\nu}\right)^{-K}. \qquad (6\text{H}.4)$$

Probability Density

$$p(Y) = \frac{Y^{N-1}e^{-Y}}{(N-1)!}\left(1 + \frac{N\bar{x}}{K}\right)^{-K} M\left(K, N, \frac{Y}{1 + k/N\bar{x}}\right) \quad (Y \geq 0),$$
$$(6\text{H}.5)$$

where M is the confluent hypergeometric function (see appendix 6I).

Appendix 6I: Confluent Hypergeometric Functions

$$M(N,N,Z) = e^Z, \tag{6I.1}$$

$$M(1,N,Z) = Z^{1-N}(n-1)!\left(e^Z - \sum_{k=0}^{N-2}\frac{z^k}{k!}\right) \quad (N>1) \tag{6I.2}$$

$$M(2,1,Z) = (Z+1)e^Z, \tag{6I.3}$$

$$M(2,N,Z) = (Z+2-N)M(1,N,Z) + N - 1. \tag{6I.4}$$

**Appendix:
Radar Detection
Tables**

Appendix: Radar Detection Tables 155

THRESHOLD/NOISE AND REQUIRED SIGNAL/NOISE
==

PROBABILITY OF DETECTION= .5000
CONSTANT FALSE ALARM RATE
LOG10(PFA/N)= 7

N	N DB	THRESH DB	NON-FLUC DB	SWER 1 DB	SWER 2 DB	SWER 3 DB	SWER 4 DB
1	0.00	6.63	6.12	7.52	7.52	6.76	6.76
2	3.01	7.66	3.33	4.67	3.94	3.94	3.61
3	4.77	8.44	1.71	3.00	2.07	2.29	1.88
4	6.02	9.08	.55	1.80	.80	1.10	.66
5	6.99	9.62	-.37	.85	-.18	.17	-.28
6	7.78	10.08	-1.12	.05	-.98	-.61	-1.06
7	8.45	10.50	-1.78	-.64	-1.66	-1.28	-1.72
8	9.03	10.87	-2.35	-1.26	-2.26	-1.88	-2.31
9	9.54	11.21	-2.88	-1.82	-2.80	-2.42	-2.84
10	10.00	11.52	-3.35	-2.33	-3.29	-2.91	-3.32
11	10.41	11.81	-3.80	-2.80	-3.74	-3.37	-3.77
12	10.79	12.08	-4.21	-3.25	-4.16	-3.80	-4.19
13	11.14	12.33	-4.60	-3.68	-4.56	-4.21	-4.58
14	11.46	12.57	-4.98	-4.09	-4.94	-4.60	-4.96
15	11.76	12.79	-5.34	-4.48	-5.30	-4.98	-5.32
16	12.04	13.00	-5.68	-4.85	-5.65	-5.34	-5.67
17	12.30	13.19	-6.02	-5.22	-5.99	-5.68	-6.00
18	12.55	13.38	-6.34	-5.58	-6.32	-6.02	-6.33
19	12.79	13.56	-6.66	-5.93	-6.64	-6.36	-6.65
20	13.01	13.74	-6.97	-6.27	-6.95	-6.68	-6.96
21	13.22	13.90	-7.28	-6.61	-7.26	-7.00	-7.27
22	13.42	14.06	-7.59	-6.94	-7.57	-7.32	-7.58
23	13.62	14.21	-7.89	-7.27	-7.88	-7.64	-7.88
24	13.80	14.36	-8.19	-7.61	-8.18	-7.95	-8.19
25	13.98	14.50	-8.49	-7.94	-8.48	-8.27	-8.49
26	14.15	14.63	-8.80	-8.27	-8.79	-8.58	-8.79
27	14.31	14.76	-9.10	-8.60	-9.09	-8.90	-9.10
28	14.47	14.89	-9.41	-8.94	-9.40	-9.22	-9.40
29	14.62	15.01	-9.72	-9.28	-9.71	-9.54	-9.72
30	14.77	15.13	-10.04	-9.62	-10.03	-9.87	-10.03
31	14.91	15.25	-10.36	-9.97	-10.36	-10.20	-10.36
32	15.05	15.36	-10.69	-10.33	-10.69	-10.55	-10.69
33	15.19	15.47	-11.04	-10.70	-11.03	-10.90	-11.03
34	15.31	15.58	-11.39	-11.08	-11.38	-11.26	-11.39
35	15.44	15.68	-11.76	-11.47	-11.75	-11.64	-11.75
36	15.56	15.78	-12.14	-11.88	-12.13	-12.03	-12.14
37	15.68	15.88	-12.54	-12.30	-12.54	-12.44	-12.54
38	15.80	15.97	-12.96	-12.75	-12.96	-12.88	-12.96
39	15.91	16.07	-13.42	-13.23	-13.41	-13.34	-13.42
40	16.02	16.16	-13.90	-13.73	-13.90	-13.83	-13.90
41	16.13	16.25	-14.43	-14.28	-14.43	-14.37	-14.43
42	16.23	16.33	-15.01	-14.88	-15.01	-14.95	-15.01
43	16.33	16.42	-15.65	-15.55	-15.65	-15.61	-15.65
44	16.43	16.50	-16.38	-16.30	-16.38	-16.35	-16.39
45	16.53	16.58	-17.23	-17.16	-17.23	-17.20	-17.23

THRESHOLD/NOISE AND REQUIRED SIGNAL/NOISE

PROBABILITY OF DETECTION= .6000
CONSTANT FALSE ALARM RATE
LOG10(PFA/N)= -2

N	N DB	THRESH DB	NON-FLUC DB	SWER 1 DB	SWER 2 DB	SWER 3 DB	SWER 4 DB
1	0.00	6.63	6.88	9.04	9.04	7.91	7.91
2	3.01	7.66	4.11	6.21	5.10	5.10	4.59
3	4.77	8.44	2.50	4.56	3.13	3.47	2.81
4	6.02	9.08	1.36	3.37	1.81	2.30	1.58
5	6.99	9.62	.47	2.43	.82	1.39	.64
6	7.78	10.08	-.26	1.65	.01	.63	-.13
7	8.45	10.50	-.89	.98	-.67	-.02	-.78
8	9.03	10.87	-1.45	.38	-1.26	-.60	-1.35
9	9.54	11.21	-1.94	-.16	-1.78	-1.12	-1.86
10	10.00	11.52	-2.39	-.65	-2.25	-1.59	-2.32
11	10.41	11.81	-2.81	-1.10	-2.69	-2.03	-2.75
12	10.79	12.08	-3.20	-1.53	-3.09	-2.44	-3.14
13	11.14	12.33	-3.56	-1.93	-3.46	-2.82	-3.51
14	11.46	12.57	-3.90	-2.32	-3.82	-3.18	-3.86
15	11.76	12.79	-4.23	-2.68	-4.15	-3.53	-4.19
16	12.04	13.00	-4.54	-3.03	-4.47	-3.86	-4.51
17	12.30	13.19	-4.85	-3.37	-4.78	-4.18	-4.81
18	12.55	13.38	-5.14	-3.70	-5.08	-4.49	-5.11
19	12.79	13.56	-5.42	-4.02	-5.36	-4.79	-5.39
20	13.01	13.74	-5.69	-4.33	-5.64	-5.08	-5.67
21	13.22	13.90	-5.96	-4.63	-5.91	-5.36	-5.93
22	13.42	14.06	-6.22	-4.93	-6.18	-5.64	-6.20
23	13.62	14.21	-6.48	-5.23	-6.44	-5.92	-6.46
24	13.80	14.36	-6.73	-5.52	-6.69	-6.19	-6.71
25	13.98	14.50	-6.98	-5.80	-6.94	-6.45	-6.96
26	14.15	14.63	-7.23	-6.09	-7.19	-6.72	-7.21
27	14.31	14.76	-7.47	-6.37	-7.44	-6.98	-7.46
28	14.47	14.89	-7.72	-6.65	-7.69	-7.24	-7.70
29	14.62	15.01	-7.96	-6.93	-7.94	-7.50	-7.95
30	14.77	15.13	-8.21	-7.21	-8.18	-7.76	-8.19
31	14.91	15.25	-8.45	-7.49	-8.43	-8.03	-8.44
32	15.05	15.36	-8.70	-7.77	-8.67	-8.29	-8.69
33	15.19	15.47	-8.94	-8.06	-8.92	-8.55	-8.93
34	15.31	15.58	-9.19	-8.34	-9.18	-8.82	-9.18
35	15.44	15.68	-9.45	-8.63	-9.43	-9.09	-9.44
36	15.56	15.78	-9.70	-8.92	-9.69	-9.36	-9.69
37	15.68	15.88	-9.96	-9.22	-9.95	-9.64	-9.96
38	15.80	15.97	-10.23	-9.52	-10.22	-9.92	-10.22
39	15.91	16.07	-10.50	-9.83	-10.49	-10.21	-10.5
40	16.02	16.16	-10.78	-10.15	-10.77	-10.50	-10.78
41	16.13	16.25	-11.07	-10.47	-11.06	-10.81	-11.07
42	16.23	16.33	-11.37	-10.80	-11.36	-11.12	-11.3
43	16.33	16.42	-11.68	-11.15	-11.67	-11.44	-11.6
44	16.43	16.50	-12.00	-11.50	-11.99	-11.78	-12.0
45	16.53	16.58	-12.34	-11.87	-12.33	-12.13	-12.3

Appendix: Radar Detection Tables

```
         THRESHOLD/NOISE AND REQUIRED SIGNAL/NOISE
         ==============================================

PROBABILITY OF DETECTION= .7000
CONSTANT FALSE ALARM RATE
LOG10(PFA/N)=- 2
```

N	N DB	THRESH DB	NON-FLUC DB	SWER 1 DB	SWER 2 DB	SWER 3 DB	SWER 4 DB
1	0.00	6.63	7.62	10.76	10.76	9.14	9.14
2	3.01	7.66	4.86	7.94	6.35	6.35	5.59
3	4.77	8.44	3.27	6.30	4.23	4.73	3.74
4	6.02	9.08	2.15	5.12	2.85	3.58	2.49
5	6.99	9.62	1.27	4.20	1.82	2.68	1.54
6	7.78	10.08	.55	3.43	.99	1.93	.77
7	8.45	10.50	-.06	2.77	.31	1.29	.12
8	9.03	10.87	-.60	2.19	-.28	.73	-.44
9	9.54	11.21	-1.07	1.66	-.80	.23	-.94
10	10.00	11.52	-1.51	1.18	-1.27	-.23	-1.39
11	10.41	11.81	-1.90	.74	-1.69	-.65	-1.80
12	10.79	12.08	-2.27	.33	-2.08	-1.04	-2.18
13	11.14	12.33	-2.62	-.06	-2.45	-1.41	-2.53
14	11.46	12.57	-2.94	-.43	-2.78	-1.76	-2.86
15	11.76	12.79	-3.25	-.78	-3.10	-2.08	-3.18
16	12.04	13.00	-3.54	-1.11	-3.41	-2.40	-3.47
17	12.30	13.19	-3.81	-1.43	-3.69	-2.70	-3.75
18	12.55	13.38	-4.08	-1.74	-3.97	-2.99	-4.03
19	12.79	13.56	-4.34	-2.04	-4.24	-3.26	-4.29
20	13.01	13.74	-4.59	-2.33	-4.49	-3.53	-4.54
21	13.22	13.90	-4.83	-2.62	-4.74	-3.80	-4.78
22	13.42	14.06	-5.06	-2.89	-4.98	-4.05	-5.02
23	13.62	14.21	-5.29	-3.16	-5.21	-4.30	-5.25
24	13.80	14.36	-5.51	-3.43	-5.44	-4.55	-5.48
25	13.98	14.50	-5.73	-3.69	-5.67	-4.79	-5.70
26	14.15	14.63	-5.95	-3.95	-5.89	-5.02	-5.92
27	14.31	14.76	-6.16	-4.21	-6.10	-5.25	-6.13
28	14.47	14.89	-6.37	-4.46	-6.31	-5.49	-6.34
29	14.62	15.01	-6.58	-4.71	-6.53	-5.71	-6.55
30	14.77	15.13	-6.78	-4.95	-6.73	-5.94	-6.76
31	14.91	15.25	-6.99	-5.20	-6.94	-6.16	-6.96
32	15.05	15.36	-7.19	-5.45	-7.14	-6.39	-7.17
33	15.19	15.47	-7.39	-5.69	-7.35	-6.61	-7.37
34	15.31	15.58	-7.59	-5.94	-7.55	-6.83	-7.57
35	15.44	15.68	-7.79	-6.18	-7.75	-7.05	-7.77
36	15.56	15.78	-7.99	-6.43	-7.96	-7.27	-7.98
37	15.68	15.88	-8.20	-6.67	-8.16	-7.50	-8.18
38	15.80	15.97	-8.40	-6.92	-8.37	-7.72	-8.38
39	15.91	16.07	-8.60	-7.17	-8.57	-7.95	-8.59
40	16.02	16.16	-8.81	-7.42	-8.78	-8.17	-8.80
41	16.13	16.25	-9.02	-7.67	-8.99	-8.40	-9.01
42	16.23	16.33	-9.23	-7.93	-9.21	-8.63	-9.22
43	16.33	16.42	-9.44	-8.19	-9.42	-8.87	-9.43
44	16.43	16.50	-9.66	-8.45	-9.64	-9.11	-9.65
45	16.53	16.58	-9.89	-8.72	-9.86	-9.35	-9.88

THRESHOLD/NOISE AND REQUIRED SIGNAL/NOISE

PROBABILITY OF DETECTION= .8000
CONSTANT FALSE ALARM RATE
$LOG_{10}(PFA/N) = -7$

N	N DB	THRESH DB	NON-FLUC DB	SWER 1 DB	SWER 2 DB	SWER 3 DB	SWER 4 DB
1	0.00	6.63	8.41	12.93	12.93	10.62	10.62
2	3.01	7.66	5.66	10.12	7.84	7.84	6.74
3	4.77	8.44	4.08	8.49	5.50	6.23	4.79
4	6.02	9.08	2.97	7.32	4.01	5.08	3.49
5	6.99	9.62	2.11	6.40	2.93	4.19	2.52
6	7.78	10.08	1.40	5.64	2.07	3.46	1.74
7	8.45	10.50	.80	4.99	1.37	2.83	1.08
8	9.03	10.87	.28	4.42	.77	2.28	.52
9	9.54	11.21	-.19	3.91	.24	1.79	.03
10	10.00	11.52	-.61	3.44	-.23	1.34	-.42
11	10.41	11.81	-.99	3.00	-.65	.93	-.82
12	10.79	12.08	-1.34	2.60	-1.04	.55	-1.19
13	11.14	12.33	-1.67	2.22	-1.39	.20	-1.53
14	11.46	12.57	-1.98	1.87	-1.72	-.13	-1.85
15	11.76	12.79	-2.27	1.53	-2.03	-.45	-2.15
16	12.04	13.00	-2.54	1.21	-2.33	-.75	-2.43
17	12.30	13.19	-2.80	.90	-2.60	-1.03	-2.70
18	12.55	13.38	-3.05	.60	-2.87	-1.31	-2.96
19	12.79	13.56	-3.29	.31	-3.12	-1.57	-3.20
20	13.01	13.74	-3.52	.04	-3.36	-1.82	-3.44
21	13.22	13.90	-3.74	-.23	-3.59	-2.07	-3.67
22	13.42	14.06	-3.96	-.50	-3.82	-2.31	-3.89
23	13.62	14.21	-4.16	-.75	-4.03	-2.54	-4.10
24	13.80	14.36	-4.37	-1.00	-4.24	-2.77	-4.31
25	13.98	14.50	-4.56	-1.25	-4.45	-2.99	-4.51
26	14.15	14.63	-4.76	-1.49	-4.65	-3.21	-4.70
27	14.31	14.76	-4.95	-1.73	-4.84	-3.42	-4.90
28	14.47	14.89	-5.13	-1.96	-5.03	-3.63	-5.08
29	14.62	15.01	-5.32	-2.19	-5.22	-3.84	-5.27
30	14.77	15.13	-5.49	-2.42	-5.41	-4.04	-5.45
31	14.91	15.25	-5.67	-2.65	-5.59	-4.24	-5.63
32	15.05	15.36	-5.85	-2.87	-5.77	-4.44	-5.81
33	15.19	15.47	-6.02	-3.09	-5.94	-4.64	-5.98
34	15.31	15.58	-6.19	-3.32	-6.12	-4.84	-6.16
35	15.44	15.68	-6.36	-3.54	-6.29	-5.03	-6.33
36	15.56	15.78	-6.53	-3.76	-6.46	-5.22	-6.50
37	15.68	15.88	-6.70	-3.97	-6.63	-5.42	-6.67
38	15.80	15.97	-6.86	-4.19	-6.80	-5.61	-6.83
39	15.91	16.07	-7.03	-4.41	-6.97	-5.80	-7.00
40	16.02	16.16	-7.20	-4.62	-7.14	-5.99	-7.17
41	16.13	16.25	-7.37	-4.84	-7.31	-6.18	-7.34
42	16.23	16.33	-7.53	-5.06	-7.48	-6.38	-7.51
43	16.33	16.42	-7.70	-5.28	-7.65	-6.57	-7.68
44	16.43	16.50	-7.87	-5.51	-7.82	-6.76	-7.85
45	16.53	16.58	-8.04	-5.73	-8.00	-6.96	-8.02

Appendix: Radar Detection Tables

THRESHOLD/NOISE AND REQUIRED SIGNAL/NOISE

PROBABILITY OF DETECTION= .9000
CONSTANT FALSE ALARM RATE
$\text{LOG}_{10}(\text{PFA/N}) = -2$

N	N DB	THRESH DB	NON-FLUC DB	SWER 1 DB	SWER 2 DB	SWER 3 DB	SWER 4 DB
1	0.00	6.63	9.40	16.30	16.30	12.76	12.76
2	3.01	7.66	6.65	13.50	9.99	9.99	8.31
3	4.77	8.44	5.08	11.87	7.27	8.39	6.18
4	6.02	9.08	3.98	10.71	5.60	7.25	4.79
5	6.99	9.62	3.12	9.81	4.41	6.37	3.77
6	7.78	10.08	2.43	9.05	3.49	5.64	2.96
7	8.45	10.50	1.84	8.41	2.74	5.03	2.29
8	9.03	10.87	1.33	7.84	2.11	4.49	1.72
9	9.54	11.21	.88	7.33	1.57	4.00	1.22
10	10.00	11.52	.47	6.87	1.08	3.57	.78
11	10.41	11.81	.10	6.45	.65	3.17	.38
12	10.79	12.08	-.24	6.05	.26	2.80	.01
13	11.14	12.33	-.55	5.69	-.10	2.45	-.32
14	11.46	12.57	-.85	5.34	-.43	2.14	-.64
15	11.76	12.79	-1.12	5.01	-.74	1.83	-.93
16	12.04	13.00	-1.38	4.69	-1.02	1.54	-1.20
17	12.30	13.19	-1.63	4.39	-1.29	1.27	-1.46
18	12.55	13.38	-1.86	4.11	-1.55	1.01	-1.71
19	12.79	13.56	-2.08	3.83	-1.79	.76	-1.94
20	13.01	13.74	-2.30	3.56	-2.03	.51	-2.16
21	13.22	13.90	-2.51	3.30	-2.25	.28	-2.38
22	13.42	14.06	-2.70	3.05	-2.46	.05	-2.58
23	13.62	14.21	-2.90	2.80	-2.67	-.17	-2.78
24	13.80	14.36	-3.08	2.56	-2.86	-.38	-2.97
25	13.98	14.50	-3.26	2.33	-3.06	-.59	-3.16
26	14.15	14.63	-3.44	2.10	-3.24	-.79	-3.34
27	14.31	14.76	-3.61	1.87	-3.42	-.99	-3.52
28	14.47	14.89	-3.78	1.65	-3.60	-1.18	-3.69
29	14.62	15.01	-3.94	1.43	-3.77	-1.38	-3.85
30	14.77	15.13	-4.10	1.22	-3.94	-1.56	-4.02
31	14.91	15.25	-4.25	1.00	-4.10	-1.75	-4.18
32	15.05	15.36	-4.41	.80	-4.26	-1.93	-4.33
33	15.19	15.47	-4.56	.59	-4.42	-2.11	-4.49
34	15.31	15.58	-4.71	.38	-4.57	-2.29	-4.64
35	15.44	15.68	-4.85	.18	-4.73	-2.46	-4.79
36	15.56	15.78	-5.00	-.02	-4.88	-2.63	-4.94
37	15.68	15.88	-5.14	-.23	-5.02	-2.81	-5.08
38	15.80	15.97	-5.29	-.42	-5.17	-2.98	-5.23
39	15.91	16.07	-5.43	-.62	-5.32	-3.15	-5.37
40	16.02	16.16	-5.57	-.82	-5.46	-3.32	-5.51
41	16.13	16.25	-5.70	-1.02	-5.60	-3.48	-5.65
42	16.23	16.33	-5.84	-1.22	-5.74	-3.65	-5.79
43	16.33	16.42	-5.98	-1.42	-5.88	-3.82	-5.93
44	16.43	16.50	-6.11	-1.62	-6.02	-3.98	-6.07
45	16.53	16.58	-6.25	-1.82	-6.16	-4.15	-6.21

THRESHOLD/NOISE AND REQUIRED SIGNAL/NOISE

PROBABILITY OF DETECTION= .9500
CONSTANT FALSE ALARM RATE
LOG10(PFA/N)=- 2

N	N DB	THRESH DB	NON-FLUC DB	SWER 1 DB	SWER 2 DB	SWER 3 DB	SWER 4 DB
1	0.00	6.63	10.14	19.48	19.48	14.65	14.65
2	3.01	7.66	7.39	16.68	11.88	11.88	9.62
3	4.77	8.44	5.82	15.06	8.78	10.28	7.30
4	6.02	9.08	4.72	13.91	6.92	9.15	5.82
5	6.99	9.62	3.87	13.00	5.62	8.27	4.75
6	7.78	10.08	3.18	12.25	4.63	7.55	3.91
7	8.45	10.50	2.60	11.61	3.83	6.94	3.22
8	9.03	10.87	2.09	11.04	3.16	6.41	2.63
9	9.54	11.21	1.65	10.54	2.59	5.93	2.12
10	10.00	11.52	1.25	10.08	2.09	5.50	1.67
11	10.41	11.81	.88	9.66	1.65	5.11	1.26
12	10.79	12.08	.55	9.26	1.24	4.74	.90
13	11.14	12.33	.24	8.90	.88	4.40	.56
14	11.46	12.57	-.04	8.55	.54	4.09	.25
15	11.76	12.79	-.31	8.22	.23	3.79	-.04
16	12.04	13.00	-.56	7.91	-.06	3.51	-.31
17	12.30	13.19	-.80	7.62	-.33	3.24	-.57
18	12.55	13.38	-1.03	7.34	-.59	2.98	-.81
19	12.79	13.56	-1.24	7.07	-.83	2.74	-1.04
20	13.01	13.74	-1.45	6.81	-1.06	2.50	-1.26
21	13.22	13.90	-1.65	6.55	-1.28	2.27	-1.46
22	13.42	14.06	-1.84	6.30	-1.49	2.06	-1.66
23	13.62	14.21	-2.02	6.06	-1.69	1.84	-1.86
24	13.80	14.36	-2.20	5.82	-1.88	1.63	-2.04
25	13.98	14.50	-2.37	5.59	-2.07	1.43	-2.22
26	14.15	14.63	-2.54	5.36	-2.25	1.23	-2.39
27	14.31	14.76	-2.70	5.14	-2.43	1.05	-2.56
28	14.47	14.89	-2.85	4.93	-2.60	.85	-2.73
29	14.62	15.01	-3.01	4.71	-2.76	.67	-2.88
30	14.77	15.13	-3.16	4.50	-2.92	.49	-3.04
31	14.91	15.25	-3.30	4.29	-3.08	.31	-3.19
32	15.05	15.36	-3.45	4.09	-3.23	.13	-3.34
33	15.19	15.47	-3.59	3.89	-3.38	-.03	-3.48
34	15.31	15.58	-3.73	3.69	-3.53	-.20	-3.63
35	15.44	15.68	-3.86	3.49	-3.67	-.37	-3.76
36	15.56	15.78	-4.00	3.29	-3.81	-.53	-3.91
37	15.68	15.88	-4.13	3.10	-3.95	-.70	-4.04
38	15.80	15.97	-4.26	2.91	-4.08	-.86	-4.17
39	15.91	16.07	-4.39	2.72	-4.22	-1.02	-4.30
40	16.02	16.16	-4.51	2.53	-4.35	-1.18	-4.43
41	16.13	16.25	-4.64	2.34	-4.48	-1.33	-4.56
42	16.23	16.33	-4.76	2.15	-4.61	-1.49	-4.68
43	16.33	16.42	-4.88	1.95	-4.74	-1.65	-4.81
44	16.43	16.50	-5.01	1.76	-4.87	-1.80	-4.93
45	16.53	16.58	-5.13	1.57	-4.99	-1.96	-5.06

Appendix: Radar Detection Tables

THRESHOLD/NOISE AND REQUIRED SIGNAL/NOISE

PROBABILITY OF DETECTION= .9900
CONSTANT FALSE ALARM RATE
LOG10(PFA/N)=- 2

N	N DB	THRESH DB	NON-FLUC DB	SWER 1 DB	SWER 2 DB	SWER 3 DB	SWER 4 DB
1	0.00	6.63	11.37	26.60	26.60	18.61	18.61
2	3.01	7.66	8.61	23.80	15.84	15.83	12.14
3	4.77	8.44	7.04	22.17	11.77	14.22	9.38
4	6.02	9.08	5.95	21.01	9.46	13.09	7.70
5	6.99	9.62	5.10	20.10	7.90	12.24	6.50
6	7.78	10.08	4.41	19.38	6.73	11.52	5.57
7	8.45	10.50	3.84	18.74	5.80	10.92	4.83
8	9.03	10.87	3.34	18.18	5.06	10.38	4.21
9	9.54	11.21	2.90	17.67	4.43	9.91	3.67
10	10.00	11.52	2.51	17.22	3.87	9.48	3.19
11	10.41	11.81	2.15	16.82	3.39	9.12	2.77
12	10.79	12.08	1.82	16.41	2.96	8.76	2.4
13	11.14	12.33	1.52	16.07	2.57	8.42	2.05
14	11.46	12.57	1.24	15.74	2.21	8.11	1.73
15	11.76	12.79	.98	15.38	1.88	7.81	1.43
16	12.04	13.00	.74	15.07	1.58	7.53	1.16
17	12.30	13.19	.51	14.77	1.29	7.27	.9
18	12.55	13.38	.30	14.49	1.03	7.01	.66
19	12.79	13.56	.09	14.22	.78	6.77	.43
20	13.01	13.74	-.11	13.95	.54	6.54	.22
21	13.22	13.90	-.30	13.70	.32	6.34	.01
22	13.42	14.06	-.48	13.45	.11	6.11	-.19
23	13.62	14.21	-.66	13.21	-.10	5.90	-.38
24	13.80	14.36	-.82	13.00	-.29	5.69	-.56
25	13.98	14.50	-.98	12.78	-.48	5.50	-.73
26	14.15	14.63	-1.13	12.55	-.65	5.30	-.90
27	14.31	14.76	-1.29	12.33	-.83	5.12	-1.06
28	14.47	14.89	-1.43	12.12	-.99	4.93	-1.22
29	14.62	15.01	-1.58	11.91	-1.15	4.75	-1.37
30	14.77	15.13	-1.72	11.70	-1.31	4.58	-1.51
31	14.91	15.25	-1.85	11.52	-1.46	4.43	-1.66
32	15.05	15.36	-1.99	11.30	-1.61	4.26	-1.80
33	15.19	15.47	-2.11	11.11	-1.75	4.10	-1.93
34	15.31	15.58	-2.24	10.93	-1.89	3.94	-2.06
35	15.44	15.68	-2.36	10.71	-2.02	3.78	-2.19
36	15.56	15.78	-2.48	10.51	-2.15	3.62	-2.32
37	15.68	15.88	-2.60	10.32	-2.28	3.47	-2.44
38	15.80	15.97	-2.72	10.13	-2.41	3.32	-2.56
39	15.91	16.07	-2.84	9.94	-2.54	3.16	-2.68
40	16.02	16.16	-2.95	9.75	-2.66	3.04	-2.80
41	16.13	16.25	-3.06	9.56	-2.78	2.89	-2.92
42	16.23	16.33	-3.17	9.38	-2.90	2.73	-3.03
43	16.33	16.42	-3.27	9.19	-3.02	2.59	-3.14
44	16.43	16.50	-3.38	9.03	-3.13	2.44	-3.26
45	16.53	16.58	-3.48	8.84	-3.25	2.29	-3.36

THRESHOLD/NOISE AND REQUIRED SIGNAL/NOISE

PROBABILITY OF DETECTION= .5000
CONSTANT FALSE ALARM RATE
LOG10(PFA/N)=- 3

N	N DB	THRESH DB	NON-FLUC DB	SWER 1 DB	SWER 2 DB	SWER 3 DB	SWER 4 DB
1	0.00	8.39	8.06	9.53	9.53	8.73	8.73
2	3.01	9.27	5.41	6.85	6.07	6.07	5.71
3	4.77	9.96	3.91	5.34	4.32	4.56	4.10
4	6.02	10.52	2.87	4.29	3.16	3.51	3.00
5	6.99	11.00	2.07	3.48	2.29	2.71	2.17
6	7.78	11.42	1.43	2.83	1.61	2.06	1.51
7	8.45	11.80	.89	2.28	1.04	1.51	.96
8	9.03	12.14	.42	1.81	.55	1.04	.48
9	9.54	12.45	.01	1.39	.12	.62	.06
10	10.00	12.74	-.35	1.02	-.26	.25	-.31
11	10.41	13.00	-.68	.68	-.60	-.08	-.64
12	10.79	13.25	-.98	.37	-.91	-.39	-.95
13	11.14	13.48	-1.26	.09	-1.19	-.67	-1.22
14	11.46	13.70	-1.51	-.18	-1.45	-.93	-1.48
15	11.76	13.91	-1.75	-.42	-1.69	-1.17	-1.72
16	12.04	14.11	-1.97	-.65	-1.92	-1.39	-1.95
17	12.30	14.29	-2.18	-.87	-2.13	-1.61	-2.16
18	12.55	14.47	-2.38	-1.07	-2.33	-1.81	-2.36
19	12.79	14.64	-2.56	-1.26	-2.52	-2.00	-2.55
20	13.01	14.80	-2.74	-1.45	-2.70	-2.18	-2.72
21	13.22	14.96	-2.91	-1.62	-2.88	-2.35	-2.89
22	13.42	15.11	-3.07	-1.79	-3.04	-2.52	-3.06
23	13.62	15.25	-3.23	-1.95	-3.20	-2.68	-3.21
24	13.80	15.39	-3.38	-2.11	-3.35	-2.83	-3.36
25	13.98	15.53	-3.52	-2.25	-3.49	-2.98	-3.51
26	14.15	15.66	-3.65	-2.40	-3.63	-3.12	-3.64
27	14.31	15.78	-3.79	-2.53	-3.76	-3.25	-3.78
28	14.47	15.91	-3.91	-2.67	-3.89	-3.38	-3.92
29	14.62	16.02	-4.04	-2.80	-4.02	-3.51	-4.03
30	14.77	16.14	-4.16	-2.92	-4.14	-3.63	-4.15
31	14.91	16.25	-4.27	-3.04	-4.25	-3.75	-4.26
32	15.05	16.36	-4.39	-3.16	-4.37	-3.87	-4.38
33	15.19	16.47	-4.49	-3.27	-4.48	-3.98	-4.49
34	15.31	16.57	-4.60	-3.39	-4.58	-4.09	-4.59
35	15.44	16.67	-4.70	-3.49	-4.69	-4.19	-4.70
36	15.56	16.77	-4.80	-3.60	-4.79	-4.30	-4.80
37	15.68	16.86	-4.90	-3.70	-4.89	-4.40	-4.90
38	15.80	16.96	-5.00	-3.80	-4.98	-4.50	-4.99
39	15.91	17.05	-5.09	-3.90	-5.08	-4.59	-5.09
40	16.02	17.14	-5.18	-4.00	-5.17	-4.69	-5.18
41	16.13	17.22	-5.27	-4.09	-5.26	-4.78	-5.27
42	16.23	17.31	-5.36	-4.19	-5.35	-4.87	-5.35
43	16.33	17.39	-5.45	-4.28	-5.43	-4.96	-5.44
44	16.43	17.48	-5.53	-4.36	-5.52	-5.04	-5.52
45	16.53	17.56	-5.61	-4.45	-5.60	-5.13	-5.61

Appendix: Radar Detection Tables

THRESHOLD/NOISE AND REQUIRED SIGNAL/NOISE
===

PROBABILITY OF DETECTION= .6000
CONSTANT FALSE ALARM RATE
LOG10(PFA/N)=- 3

N	N DB	THRESH DB	NON-FLUC DB	SWER 1 DB	SWER 2 DB	SWER 3 DB	SWER 4 DB
1	0.00	8.39	8.67	10.98	10.98	9.79	9.79
2	3.01	9.27	6.01	8.30	7.12	7.12	6.55
3	4.77	9.96	4.51	6.79	5.23	5.61	4.86
4	6.02	10.52	3.46	5.74	4.00	4.56	3.72
5	6.99	11.00	2.67	4.94	3.09	3.76	2.87
6	7.78	11.42	2.03	4.28	2.37	3.11	2.26
7	8.45	11.80	1.49	3.74	1.78	2.56	1.63
8	9.03	12.14	1.02	3.26	1.28	2.10	1.15
9	9.54	12.45	.62	2.85	.84	1.68	.73
10	10.00	12.74	.25	2.48	.45	1.31	.35
11	10.41	13.00	-.07	2.14	.10	.98	.01
12	10.79	13.25	-.37	1.83	-.21	.68	-.29
13	11.14	13.48	-.64	1.55	-.50	.40	-.57
14	11.46	13.70	-.90	1.29	-.76	.14	-.83
15	11.76	13.91	-1.13	1.05	-1.01	-.10	-1.07
16	12.04	14.11	-1.35	.82	-1.24	-.32	-1.3
17	12.30	14.29	-1.56	.61	-1.45	-.53	-1.51
18	12.55	14.47	-1.75	.40	-1.65	-.73	-1.7
19	12.79	14.64	-1.94	.21	-1.84	-.92	-1.89
20	13.01	14.80	-2.11	.03	-2.02	-1.10	-2.07
21	13.22	14.96	-2.28	-.14	-2.20	-1.27	-2.24
22	13.42	15.11	-2.44	-.31	-2.36	-1.44	-2.4
23	13.62	15.25	-2.59	-.47	-2.51	-1.59	-2.55
24	13.80	15.39	-2.73	-.62	-2.66	-1.74	-2.71
25	13.98	15.53	-2.87	-.77	-2.81	-1.89	-2.84
26	14.15	15.66	-3.01	-.91	-2.94	-2.02	-2.98
27	14.31	15.78	-3.14	-1.04	-3.08	-2.16	-3.11
28	14.47	15.91	-3.26	-1.17	-3.20	-2.29	-3.23
29	14.62	16.02	-3.38	-1.30	-3.33	-2.41	-3.36
30	14.77	16.14	-3.50	-1.43	-3.45	-2.53	-3.47
31	14.91	16.25	-3.61	-1.54	-3.56	-2.65	-3.59
32	15.05	16.36	-3.72	-1.66	-3.67	-2.76	-3.70
33	15.19	16.47	-3.83	-1.77	-3.78	-2.87	-3.81
34	15.31	16.57	-3.93	-1.88	-3.89	-2.98	-3.91
35	15.44	16.67	-4.03	-1.99	-3.99	-3.08	-4.01
36	15.56	16.77	-4.13	-2.09	-4.09	-3.18	-4.11
37	15.68	16.86	-4.23	-2.20	-4.19	-3.28	-4.21
38	15.80	16.96	-4.32	-2.29	-4.28	-3.38	-4.30
39	15.91	17.05	-4.41	-2.39	-4.37	-3.47	-4.39
40	16.02	17.14	-4.50	-2.49	-4.46	-3.57	-4.48
41	16.13	17.22	-4.59	-2.58	-4.55	-3.66	-4.57
42	16.23	17.31	-4.67	-2.67	-4.64	-3.74	-4.65
43	16.33	17.39	-4.75	-2.76	-4.72	-3.83	-4.74
44	16.43	17.48	-4.84	-2.84	-4.80	-3.91	-4.82
45	16.53	17.56	-4.91	-2.93	-4.88	-4.00	-4.90

THRESHOLD/NOISE AND REQUIRED SIGNAL/NOISE

PROBABILITY OF DETECTION= .7000
CONSTANT FALSE ALARM RATE
LOG10(PFA/N) = -3

N	N DB	THRESH DB	NON-FLUC DB	SWER 1 DB	SWER 2 DB	SWER 3 DB	SWER 4 DB
1	0.00	8.39	9.27	12.64	12.64	10.94	10.94
2	3.01	9.27	6.60	9.97	8.27	8.27	7.44
3	4.77	9.96	5.10	8.46	6.21	6.76	5.65
4	6.02	10.52	4.06	7.41	4.88	5.71	4.47
5	6.99	11.00	3.26	6.60	3.92	4.91	3.59
6	7.78	11.42	2.62	5.95	3.17	4.26	2.89
7	8.45	11.80	2.08	5.40	2.55	3.72	2.31
8	9.03	12.14	1.62	4.93	2.02	3.25	1.82
9	9.54	12.45	1.21	4.52	1.57	2.84	1.39
10	10.00	12.74	.85	4.15	1.17	2.47	1.01
11	10.41	13.00	.52	3.81	.81	2.14	.67
12	10.79	13.25	.23	3.51	.49	1.84	.36
13	11.14	13.48	-.04	3.23	.20	1.56	.08
14	11.46	13.70	-.30	2.97	-.07	1.31	-.18
15	11.76	13.91	-.53	2.73	-.32	1.07	-.43
16	12.04	14.11	-.75	2.50	-.55	.84	-.65
17	12.30	14.29	-.95	2.29	-.77	.63	-.86
18	12.55	14.47	-1.15	2.09	-.97	.44	-1.06
19	12.79	14.64	-1.33	1.89	-1.17	.25	-1.25
20	13.01	14.80	-1.50	1.71	-1.35	.07	-1.43
21	13.22	14.96	-1.67	1.54	-1.52	-.10	-1.59
22	13.42	15.11	-1.82	1.38	-1.69	-.26	-1.75
23	13.62	15.25	-1.97	1.22	-1.84	-.41	-1.91
24	13.80	15.39	-2.12	1.07	-1.99	-.56	-2.05
25	13.98	15.53	-2.25	.92	-2.13	-.71	-2.19
26	14.15	15.66	-2.39	.78	-2.27	-.84	-2.33
27	14.31	15.78	-2.51	.65	-2.40	-.97	-2.46
28	14.47	15.91	-2.64	.52	-2.53	-1.10	-2.58
29	14.62	16.02	-2.75	.39	-2.65	-1.22	-2.70
30	14.77	16.14	-2.87	.27	-2.77	-1.34	-2.82
31	14.91	16.25	-2.98	.15	-2.89	-1.46	-2.93
32	15.05	16.36	-3.09	.04	-3.00	-1.57	-3.04
33	15.19	16.47	-3.19	-.07	-3.11	-1.68	-3.15
34	15.31	16.57	-3.29	-.18	-3.21	-1.78	-3.25
35	15.44	16.67	-3.39	-.29	-3.31	-1.89	-3.35
36	15.56	16.77	-3.49	-.39	-3.41	-1.99	-3.45
37	15.68	16.86	-3.58	-.49	-3.51	-2.08	-3.54
38	15.80	16.96	-3.67	-.59	-3.60	-2.18	-3.64
39	15.91	17.05	-3.76	-.68	-3.69	-2.27	-3.73
40	16.02	17.14	-3.85	-.78	-3.78	-2.36	-3.81
41	16.13	17.22	-3.93	-.87	-3.87	-2.45	-3.90
42	16.23	17.31	-4.02	-.96	-3.95	-2.54	-3.98
43	16.33	17.39	-4.10	-1.05	-4.03	-2.62	-4.06
44	16.43	17.48	-4.18	-1.13	-4.11	-2.71	-4.14
45	16.53	17.56	-4.25	-1.21	-4.19	-2.79	-4.22

Appendix: Radar Detection Tables

THRESHOLD/NOISE AND REQUIRED SIGNAL/NOISE

PROBABILITY OF DETECTION= .8000
CONSTANT FALSE ALARM RATE
LOG10(PFA/N)=- 3

N	N DB	THRESH DB	NON-FLUC DB	SWER 1 DB	SWER 2 DB	SWER 3 DB	SWER 4 DB
1	0.00	8.39	9.93	14.76	14.76	12.34	12.34
2	3.01	9.27	7.25	12.09	9.67	9.67	8.48
3	4.77	9.96	5.74	10.58	7.36	8.16	6.56
4	6.02	10.52	4.70	9.53	5.92	7.11	5.32
5	6.99	11.00	3.90	8.73	4.88	6.31	4.40
6	7.78	11.42	3.25	8.08	4.07	5.66	3.67
7	8.45	11.80	2.72	7.53	3.42	5.12	3.07
8	9.03	12.14	2.25	7.06	2.87	4.65	2.56
9	9.54	12.45	1.85	6.64	2.39	4.24	2.12
10	10.00	12.74	1.49	6.27	1.98	3.88	1.73
11	10.41	13.00	1.16	5.94	1.61	3.55	1.39
12	10.79	13.25	.87	5.64	1.27	3.25	1.07
13	11.14	13.48	.59	5.36	.97	2.97	.78
14	11.46	13.70	.34	5.10	.69	2.71	.52
15	11.76	13.91	.11	4.86	.44	2.48	.28
16	12.04	14.11	-.10	4.63	.20	2.25	.05
17	12.30	14.29	-.31	4.42	-.02	2.05	-.16
18	12.55	14.47	-.50	4.22	-.23	1.85	-.36
19	12.79	14.64	-.68	4.03	-.43	1.66	-.55
20	13.01	14.80	-.85	3.85	-.61	1.49	-.73
21	13.22	14.96	-1.02	3.68	-.79	1.32	-.90
22	13.42	15.11	-1.17	3.52	-.95	1.16	-1.06
23	13.62	15.25	-1.32	3.36	-1.11	1.00	-1.22
24	13.80	15.39	-1.46	3.21	-1.26	.86	-1.36
25	13.98	15.53	-1.60	3.07	-1.41	.72	-1.50
26	14.15	15.66	-1.73	2.93	-1.55	.58	-1.64
27	14.31	15.78	-1.85	2.79	-1.68	.45	-1.77
28	14.47	15.91	-1.98	2.67	-1.81	.32	-1.89
29	14.62	16.02	-2.09	2.54	-1.93	.20	-2.01
30	14.77	16.14	-2.21	2.42	-2.05	.08	-2.13
31	14.91	16.25	-2.31	2.30	-2.17	-.03	-2.24
32	15.05	16.36	-2.42	2.19	-2.28	-.14	-2.35
33	15.19	16.47	-2.52	2.08	-2.38	-.25	-2.45
34	15.31	16.57	-2.62	1.97	-2.49	-.35	-2.56
35	15.44	16.67	-2.72	1.87	-2.59	-.45	-2.65
36	15.56	16.77	-2.81	1.76	-2.69	-.55	-2.75
37	15.68	16.86	-2.91	1.66	-2.78	-.65	-2.84
38	15.80	16.96	-3.00	1.57	-2.88	-.74	-2.94
39	15.91	17.05	-3.08	1.47	-2.97	-.83	-3.02
40	16.02	17.14	-3.17	1.38	-3.05	-.92	-3.11
41	16.13	17.22	-3.25	1.29	-3.14	-1.01	-3.20
42	16.23	17.31	-3.33	1.20	-3.22	-1.10	-3.28
43	16.33	17.39	-3.41	1.12	-3.31	-1.18	-3.36
44	16.43	17.48	-3.49	1.03	-3.39	-1.26	-3.44
45	16.53	17.56	-3.56	.95	-3.46	-1.34	-3.51

THRESHOLD/NOISE AND REQUIRED SIGNAL/NOISE

PROBABILITY OF DETECTION= .9000
CONSTANT FALSE ALARM RATE
LOG10(PFA/N)=- 3

N	N DB	THRESH DB	NON-FLUC DB	SWER 1 DB	SWER 2 DB	SWER 3 DB	SWER 4 DB
1	0.00	8.39	10.76	18.10	18.10	14.41	14.41
2	3.01	9.27	8.07	15.43	11.74	11.73	9.93
3	4.77	9.96	6.55	13.92	9.02	10.22	7.81
4	6.02	10.52	5.50	12.87	7.37	9.18	6.46
5	6.99	11.00	4.70	12.06	6.21	8.38	5.47
6	7.78	11.42	4.06	11.41	5.32	7.73	4.73
7	8.45	11.80	3.52	10.86	4.60	7.19	4.07
8	9.03	12.14	3.05	10.39	4.01	6.72	3.54
9	9.54	12.45	2.65	9.98	3.50	6.31	3.08
10	10.00	12.74	2.29	9.62	3.05	5.94	2.67
11	10.41	13.00	1.96	9.28	2.66	5.61	2.31
12	10.79	13.25	1.67	8.98	2.30	5.31	1.99
13	11.14	13.48	1.39	8.70	1.98	5.04	1.69
14	11.46	13.70	1.15	8.44	1.69	4.79	1.42
15	11.76	13.91	.91	8.20	1.42	4.55	1.17
16	12.04	14.11	.70	7.97	1.18	4.33	.94
17	12.30	14.29	.49	7.76	.94	4.12	.72
18	12.55	14.47	.30	7.56	.73	3.92	.52
19	12.79	14.64	.12	7.38	.53	3.74	.33
20	13.01	14.80	-.05	7.20	.34	3.56	.15
21	13.22	14.96	-.21	7.03	.15	3.40	-.03
22	13.42	15.11	-.36	6.86	-.02	3.24	-.19
23	13.62	15.25	-.51	6.71	-.18	3.08	-.34
24	13.80	15.39	-.65	6.56	-.33	2.94	-.49
25	13.98	15.53	-.79	6.41	-.48	2.80	-.63
26	14.15	15.66	-.92	6.28	-.62	2.66	-.77
27	14.31	15.78	-1.04	6.14	-.76	2.54	-.90
28	14.47	15.91	-1.16	6.01	-.89	2.41	-1.02
29	14.62	16.02	-1.28	5.89	-1.02	2.29	-1.14
30	14.77	16.14	-1.39	5.77	-1.14	2.17	-1.26
31	14.91	16.25	-1.50	5.65	-1.25	2.06	-1.37
32	15.05	16.36	-1.60	5.54	-1.37	1.95	-1.48
33	15.19	16.47	-1.70	5.43	-1.47	1.85	-1.59
34	15.31	16.57	-1.80	5.32	-1.58	1.74	-1.69
35	15.44	16.67	-1.90	5.22	-1.68	1.64	-1.79
36	15.56	16.77	-1.99	5.12	-1.78	1.54	-1.88
37	15.68	16.86	-2.08	5.02	-1.88	1.45	-1.98
38	15.80	16.96	-2.17	4.92	-1.97	1.36	-2.07
39	15.91	17.05	-2.25	4.83	-2.06	1.27	-2.16
40	16.02	17.14	-2.34	4.74	-2.15	1.18	-2.24
41	16.13	17.22	-2.42	4.65	-2.24	1.09	-2.33
42	16.23	17.31	-2.50	4.56	-2.32	1.01	-2.41
43	16.33	17.39	-2.57	4.48	-2.40	.93	-2.49
44	16.43	17.48	-2.65	4.39	-2.48	.85	-2.57
45	16.53	17.56	-2.72	4.31	-2.56	.77	-2.64

Appendix: Radar Detection Tables

THRESHOLD/NOISE AND REQUIRED SIGNAL/NOISE

PROBABILITY OF DETECTION= .9500
CONSTANT FALSE ALARM RATE
LOG10(PFA/N)=- 3

N	N DB	THRESH DB	NON-FLUC DB	SWER 1 DB	SWER 2 DB	SWER 3 DB	SWER 4 DB
1	0.00	8.39	11.39	21.25	21.25	16.26	16.26
2	3.01	9.27	8.69	18.59	13.58	13.58	11.17
3	4.77	9.96	7.17	17.08	10.46	12.07	8.84
4	6.02	10.52	6.11	16.03	8.60	11.02	7.38
5	6.99	11.00	5.31	15.22	7.32	10.22	6.33
6	7.78	11.42	4.66	14.57	6.35	9.57	5.52
7	8.45	11.80	4.12	14.03	5.57	9.03	4.86
8	9.03	12.14	3.65	13.56	4.93	8.57	4.3
9	9.54	12.45	3.25	13.14	4.39	8.16	3.83
10	10.00	12.74	2.89	12.77	3.91	7.79	3.41
11	10.41	13.00	2.56	12.44	3.50	7.46	3.04
12	10.79	13.25	2.26	12.14	3.12	7.16	2.7
13	11.14	13.48	1.99	11.86	2.79	6.89	2.4
14	11.46	13.70	1.74	11.60	2.48	6.64	2.12
15	11.76	13.91	1.51	11.36	2.20	6.40	1.86
16	12.04	14.11	1.30	11.14	1.94	6.18	1.62
17	12.30	14.29	1.09	10.94	1.70	5.97	1.40
18	12.55	14.47	.90	10.73	1.48	5.78	1.20
19	12.79	14.64	.72	10.54	1.27	5.59	1.00
20	13.01	14.80	.55	10.36	1.07	5.42	.82
21	13.22	14.96	.39	10.20	.89	5.25	.64
22	13.42	15.11	.24	10.04	.71	5.09	.48
23	13.62	15.25	.09	9.87	.54	4.94	.32
24	13.80	15.39	-.05	9.72	.38	4.80	.17
25	13.98	15.53	-.18	9.58	.23	4.66	.03
26	14.15	15.66	-.31	9.44	.08	4.52	-.11
27	14.31	15.78	-.44	9.31	-.05	4.39	-.24
28	14.47	15.91	-.56	9.18	-.19	4.27	-.37
29	14.62	16.02	-.67	9.06	-.32	4.15	-.49
30	14.77	16.14	-.78	8.94	-.44	4.03	-.61
31	14.91	16.25	-.89	8.82	-.56	3.92	-.72
32	15.05	16.36	-.99	8.71	-.67	3.81	-.83
33	15.19	16.47	-1.10	8.60	-.78	3.71	-.94
34	15.31	16.57	-1.19	8.49	-.89	3.60	-1.04
35	15.44	16.67	-1.29	8.39	-.99	3.50	-1.14
36	15.56	16.77	-1.38	8.29	-1.09	3.41	-1.24
37	15.68	16.86	-1.47	8.19	-1.19	3.31	-1.33
38	15.80	16.96	-1.56	8.10	-1.29	3.22	-1.42
39	15.91	17.05	-1.64	8.00	-1.38	3.13	-1.51
40	16.02	17.14	-1.72	7.91	-1.47	3.04	-1.60
41	16.13	17.22	-1.80	7.82	-1.56	2.96	-1.68
42	16.23	17.31	-1.88	7.74	-1.64	2.87	-1.76
43	16.33	17.39	-1.96	7.65	-1.72	2.79	-1.84
44	16.43	17.48	-2.04	7.57	-1.80	2.71	-1.92
45	16.53	17.56	-2.11	7.49	-1.88	2.63	-1.99

THRESHOLD/NOISE AND REQUIRED SIGNAL/NOISE

PROBABILITY OF DETECTION= .9900
CONSTANT FALSE ALARM RATE
LOG10(PFA/N) = -3

N	N DB	THRESH DB	NON-FLUC DB	SWER 1 DB	SWER 2 DB	SWER 3 DB	SWER 4 DB
1	0.00	8.39	12.46	28.36	28.36	20.15	20.15
2	3.01	9.27	9.75	25.68	17.48	17.48	13.61
3	4.77	9.96	8.21	24.16	13.36	15.97	10.81
4	6.02	10.52	7.15	23.14	11.04	14.92	9.12
5	6.99	11.00	6.34	22.33	9.46	14.11	7.93
6	7.78	11.42	5.68	21.68	8.31	13.47	7.02
7	8.45	11.80	5.14	21.14	7.40	12.95	6.28
8	9.03	12.14	4.67	20.67	6.66	12.48	5.68
9	9.54	12.45	4.26	20.25	6.04	12.07	5.16
10	10.00	12.74	3.89	19.88	5.51	11.70	4.71
11	10.41	13.00	3.57	19.55	5.03	11.38	4.31
12	10.79	13.25	3.27	19.25	4.61	11.08	3.96
13	11.14	13.48	3.00	18.97	4.24	10.80	3.64
14	11.46	13.70	2.75	18.71	3.90	10.55	3.34
15	11.76	13.91	2.52	18.47	3.59	10.31	3.07
16	12.04	14.11	2.30	18.25	3.31	10.09	2.82
17	12.30	14.29	2.09	18.04	3.05	9.88	2.58
18	12.55	14.47	1.90	17.84	2.81	9.69	2.37
19	12.79	14.64	1.72	17.65	2.58	9.50	2.16
20	13.01	14.80	1.55	17.47	2.37	9.33	1.97
21	13.22	14.96	1.40	17.30	2.17	9.16	1.79
22	13.42	15.11	1.24	17.14	1.99	9.00	1.62
23	13.62	15.25	1.09	16.98	1.81	8.85	1.46
24	13.80	15.39	.95	16.84	1.64	8.71	1.30
25	13.98	15.53	.81	16.69	1.48	8.57	1.15
26	14.15	15.66	.69	16.55	1.33	8.44	1.01
27	14.31	15.78	.56	16.42	1.18	8.31	.88
28	14.47	15.91	.45	16.29	1.04	8.18	.75
29	14.62	16.02	.33	16.21	.90	8.06	.62
30	14.77	16.14	.22	16.09	.77	7.95	.50
31	14.91	16.25	.12	15.97	.65	7.84	.38
32	15.05	16.36	.01	15.86	.53	7.73	.27
33	15.19	16.47	-.09	15.75	.41	7.62	.16
34	15.31	16.57	-.19	15.64	.30	7.52	.06
35	15.44	16.67	-.28	15.53	.19	7.42	-.04
36	15.56	16.77	-.37	15.43	.09	7.32	-.14
37	15.68	16.86	-.46	15.33	-.01	7.23	-.24
38	15.80	16.96	-.55	15.23	-.11	7.14	-.33
39	15.91	17.05	-.63	15.14	-.21	7.05	-.42
40	16.02	17.14	-.71	15.05	-.30	6.96	-.51
41	16.13	17.22	-.79	14.95	-.39	6.88	-.59
42	16.23	17.31	-.87	14.87	-.48	6.79	-.68
43	16.33	17.39	-.95	14.78	-.56	6.71	-.76
44	16.43	17.48	-1.02	14.69	-.65	6.63	-.83
45	16.53	17.56	-1.09	14.61	-.73	6.55	-.91

Appendix: Radar Detection Tables

THRESHOLD/NOISE AND REQUIRED SIGNAL/NOISE

PROBABILITY OF DETECTION= .5000
CONSTANT FALSE ALARM RATE
$LOG_{10}(PFA/N) = -4$

N	N DB	THRESH DB	NON-FLUC DB	SWER 1 DB	SWER 2 DB	SWER 3 DB	SWER 4 DB
1	0.00	9.64	9.40	10.89	10.89	10.09	10.09
2	3.01	10.41	6.76	8.25	7.45	7.45	7.08
3	4.77	11.02	5.29	6.77	5.72	5.97	5.49
4	6.02	11.53	4.27	5.75	4.58	4.95	4.41
5	6.99	11.96	3.50	4.97	3.74	4.17	3.61
6	7.78	12.35	2.87	4.35	3.07	3.54	2.96
7	8.45	12.69	2.35	3.83	2.52	3.02	2.43
8	9.03	13.00	1.91	3.38	2.05	2.58	1.97
9	9.54	13.29	1.52	2.99	1.64	2.18	1.58
10	10.00	13.55	1.18	2.64	1.28	1.84	1.23
11	10.41	13.80	.87	2.33	.96	1.53	.91
12	10.79	14.03	.59	2.04	.67	1.24	.62
13	11.14	14.24	.33	1.78	.40	.98	.36
14	11.46	14.45	.09	1.54	.16	.74	.12
15	11.76	14.64	-.13	1.32	-.07	.52	-.10
16	12.04	14.82	-.33	1.11	-.28	.31	-.31
17	12.30	15.00	-.53	.92	-.47	.12	-.53
18	12.55	15.17	-.71	.73	-.66	-.06	-.68
19	12.79	15.33	-.88	.56	-.83	-.23	-.86
20	13.01	15.48	-1.04	.40	-.99	-.40	-1.02
21	13.22	15.63	-1.19	.24	-1.15	-.55	-1.17
22	13.42	15.77	-1.34	.10	-1.30	-.70	-1.32
23	13.62	15.90	-1.48	-.05	-1.44	-.84	-1.46
24	13.80	16.04	-1.61	-.18	-1.57	-.97	-1.59
25	13.98	16.16	-1.74	-.31	-1.70	-1.10	-1.72
26	14.15	16.29	-1.86	-.43	-1.83	-1.23	-1.84
27	14.31	16.41	-1.98	-.55	-1.95	-1.34	-1.96
28	14.47	16.52	-2.09	-.67	-2.06	-1.46	-2.07
29	14.62	16.63	-2.20	-.78	-2.17	-1.57	-2.18
30	14.77	16.74	-2.30	-.88	-2.28	-1.67	-2.29
31	14.91	16.85	-2.40	-.99	-2.38	-1.78	-2.39
32	15.05	16.95	-2.50	-1.09	-2.48	-1.88	-2.49
33	15.19	17.05	-2.60	-1.18	-2.57	-1.97	-2.59
34	15.31	17.15	-2.69	-1.28	-2.67	-2.07	-2.68
35	15.44	17.25	-2.78	-1.37	-2.76	-2.16	-2.77
36	15.56	17.34	-2.86	-1.46	-2.84	-2.24	-2.86
37	15.68	17.43	-2.95	-1.54	-2.93	-2.33	-2.94
38	15.80	17.52	-3.03	-1.63	-3.01	-2.41	-3.02
39	15.91	17.61	-3.11	-1.71	-3.09	-2.49	-3.10
40	16.02	17.69	-3.19	-1.79	-3.17	-2.57	-3.18
41	16.13	17.78	-3.26	-1.86	-3.25	-2.65	-3.26
42	16.23	17.86	-3.34	-1.94	-3.32	-2.72	-3.33
43	16.33	17.94	-3.41	-2.01	-3.39	-2.80	-3.40
44	16.43	18.02	-3.48	-2.09	-3.47	-2.87	-3.47
45	16.53	18.09	-3.55	-2.16	-3.54	-2.94	-3.54

THRESHOLD/NOISE AND REQUIRED SIGNAL/NOISE

PROBABILITY OF DETECTION= .6000
CONSTANT FALSE ALARM RATE
LOG10(PFA/N)=- 4

N	N DB	THRESH DB	NON-FLUC DB	SWER 1 DB	SWER 2 DB	SWER 3 DB	SWER 4 DB
1	0.00	9.64	9.92	12.31	12.31	11.09	11.09
2	3.01	10.41	7.27	9.67	8.45	8.45	7.85
3	4.77	11.02	5.79	8.19	6.57	6.96	6.17
4	6.02	11.53	4.77	7.16	5.35	5.94	5.05
5	6.99	11.96	3.99	6.38	4.45	5.16	4.22
6	7.78	12.35	3.37	5.76	3.75	4.54	3.56
7	8.45	12.69	2.85	5.24	3.17	4.02	3.01
8	9.03	13.00	2.40	4.79	2.68	3.57	2.54
9	9.54	13.29	2.01	4.40	2.26	3.18	2.13
10	10.00	13.55	1.66	4.05	1.89	2.83	1.77
11	10.41	13.80	1.35	3.73	1.56	2.52	1.45
12	10.79	14.03	1.07	3.45	1.26	2.23	1.16
13	11.14	14.24	.81	3.19	.98	1.97	.9
14	11.46	14.45	.57	2.95	.73	1.73	.65
15	11.76	14.64	.35	2.73	.50	1.51	.43
16	12.04	14.82	.15	2.52	.28	1.30	.22
17	12.30	15.00	-.04	2.33	.08	1.11	.02
18	12.55	15.17	-.23	2.14	-.10	.93	-.17
19	12.79	15.33	-.40	1.97	-.28	.76	-.34
20	13.01	15.48	-.56	1.81	-.45	.59	-.50
21	13.22	15.63	-.71	1.65	-.61	.44	-.66
22	13.42	15.77	-.86	1.51	-.76	.29	-.81
23	13.62	15.90	-.99	1.36	-.90	.15	-.95
24	13.80	16.04	-1.13	1.23	-1.04	.02	-1.08
25	13.98	16.16	-1.25	1.10	-1.17	-.11	-1.21
26	14.15	16.29	-1.38	.98	-1.29	-.23	-1.34
27	14.31	16.41	-1.49	.86	-1.41	-.35	-1.45
28	14.47	16.52	-1.61	.74	-1.53	-.46	-1.57
29	14.62	16.63	-1.71	.63	-1.64	-.57	-1.68
30	14.77	16.74	-1.82	.53	-1.75	-.68	-1.78
31	14.91	16.85	-1.92	.43	-1.85	-.78	-1.89
32	15.05	16.95	-2.02	.33	-1.95	-.88	-1.99
33	15.19	17.05	-2.11	.23	-2.05	-.98	-2.08
34	15.31	17.15	-2.20	.14	-2.14	-1.07	-2.17
35	15.44	17.25	-2.29	.04	-2.23	-1.16	-2.26
36	15.56	17.34	-2.38	-.04	-2.32	-1.25	-2.35
37	15.68	17.43	-2.46	-.13	-2.41	-1.33	-2.44
38	15.80	17.52	-2.54	-.21	-2.49	-1.42	-2.52
39	15.91	17.61	-2.62	-.29	-2.57	-1.50	-2.60
40	16.02	17.69	-2.70	-.37	-2.65	-1.58	-2.68
41	16.13	17.78	-2.78	-.45	-2.73	-1.65	-2.75
42	16.23	17.86	-2.85	-.52	-2.80	-1.73	-2.83
43	16.33	17.94	-2.92	-.60	-2.88	-1.80	-2.90
44	16.43	18.02	-2.99	-.67	-2.95	-1.87	-2.97
45	16.53	18.09	-3.06	-.74	-3.02	-1.94	-3.04

Appendix: Radar Detection Tables

THRESHOLD/NOISE AND REQUIRED SIGNAL/NOISE
==

PROBABILITY OF DETECTION= .7000
CONSTANT FALSE ALARM RATE
LOG10(PFA/N)=- 4

N	N DB	THRESH DB	NON-FLUC DB	SWER 1 DB	SWER 2 DB	SWER 3 DB	SWER 4 DB
1	0.00	9.64	10.44	13.95	13.95	12.20	12.20
2	3.01	10.41	7.78	11.30	9.56	9.56	8.69
3	4.77	11.02	6.29	9.82	7.49	8.07	6.90
4	6.02	11.53	5.27	8.79	6.17	7.04	5.73
5	6.99	11.96	4.49	8.01	5.22	6.27	4.86
6	7.78	12.35	3.86	7.39	4.47	5.64	4.17
7	8.45	12.69	3.34	6.86	3.86	5.12	3.6
8	9.03	13.00	2.89	6.42	3.35	4.67	3.12
9	9.54	13.29	2.50	6.02	2.91	4.28	2.7
10	10.00	13.55	2.15	5.68	2.52	3.93	2.34
11	10.41	13.80	1.84	5.36	2.17	3.62	2.01
12	10.79	14.03	1.55	5.08	1.86	3.33	1.71
13	11.14	14.24	1.30	4.82	1.58	3.07	1.44
14	11.46	14.45	1.06	4.58	1.32	2.83	1.19
15	11.76	14.64	.84	4.36	1.08	2.61	.96
16	12.04	14.82	.63	4.15	.86	2.40	.75
17	12.30	15.00	.44	3.95	.65	2.21	.55
18	12.55	15.17	.26	3.77	.46	2.03	.36
19	12.79	15.33	.09	3.60	.28	1.86	.18
20	13.01	15.48	-.08	3.44	.11	1.69	.02
21	13.22	15.63	-.23	3.28	-.05	1.54	-.14
22	13.42	15.77	-.38	3.13	-.21	1.39	-.29
23	13.62	15.90	-.51	2.99	-.35	1.25	-.43
24	13.80	16.04	-.65	2.86	-.49	1.12	-.57
25	13.98	16.16	-.77	2.73	-.63	.99	-.70
26	14.15	16.29	-.90	2.61	-.76	.87	-.83
27	14.31	16.41	-1.01	2.49	-.88	.75	-.95
28	14.47	16.52	-1.12	2.37	-1.00	.63	-1.06
29	14.62	16.63	-1.23	2.26	-1.11	.53	-1.17
30	14.77	16.74	-1.34	2.16	-1.22	.42	-1.28
31	14.91	16.85	-1.44	2.06	-1.32	.32	-1.38
32	15.05	16.95	-1.54	1.96	-1.42	.22	-1.48
33	15.19	17.05	-1.63	1.86	-1.52	.12	-1.58
34	15.31	17.15	-1.72	1.77	-1.62	.03	-1.67
35	15.44	17.25	-1.81	1.68	-1.71	-.06	-1.76
36	15.56	17.34	-1.90	1.59	-1.80	-.15	-1.85
37	15.68	17.43	-1.98	1.50	-1.88	-.23	-1.93
38	15.80	17.52	-2.06	1.42	-1.97	-.31	-2.02
39	15.91	17.61	-2.14	1.34	-2.05	-.40	-2.10
40	16.02	17.69	-2.22	1.26	-2.13	-.47	-2.18
41	16.13	17.78	-2.30	1.18	-2.21	-.55	-2.25
42	16.23	17.86	-2.37	1.11	-2.28	-.62	-2.33
43	16.33	17.94	-2.44	1.03	-2.36	-.70	-2.40
44	16.43	18.02	-2.51	.96	-2.43	-.77	-2.47
45	16.53	18.09	-2.58	.89	-2.50	-.84	-2.54

THRESHOLD/NOISE AND REQUIRED SIGNAL/NOISE

PROBABILITY OF DETECTION= .8000
CONSTANT FALSE ALARM RATE
LOG10(PFA/N)=- 4

N	N DB	THRESH DB	NON-FLUC DB	SWER 1 DB	SWER 2 DB	SWER 3 DB	SWER 4 DB
1	0.00	9.64	11.01	16.05	16.05	13.57	13.57
2	3.01	10.41	8.35	13.40	10.92	10.92	9.67
3	4.77	11.02	6.85	11.92	8.60	9.43	7.75
4	6.02	11.53	5.81	10.89	7.15	8.40	6.50
5	6.99	11.96	5.03	10.11	6.11	7.62	5.59
6	7.78	12.35	4.40	9.48	5.31	6.99	4.87
7	8.45	12.69	3.87	8.96	4.66	6.47	4.28
8	9.03	13.00	3.42	8.51	4.12	6.02	3.78
9	9.54	13.29	3.03	8.12	3.65	5.63	3.34
10	10.00	13.55	2.68	7.77	3.24	5.28	2.96
11	10.41	13.80	2.37	7.46	2.88	4.97	2.63
12	10.79	14.03	2.08	7.17	2.55	4.68	2.32
13	11.14	14.24	1.82	6.91	2.26	4.42	2.04
14	11.46	14.45	1.58	6.67	1.99	4.18	1.79
15	11.76	14.64	1.36	6.45	1.74	3.96	1.55
16	12.04	14.82	1.15	6.24	1.51	3.75	1.33
17	12.30	15.00	.96	6.05	1.29	3.56	1.13
18	12.55	15.17	.78	5.87	1.09	3.38	.94
19	12.79	15.33	.61	5.70	.91	3.21	.76
20	13.01	15.48	.45	5.53	.73	3.04	.59
21	13.22	15.63	.29	5.38	.56	2.89	.43
22	13.42	15.77	.15	5.23	.41	2.74	.28
23	13.62	15.90	.01	5.09	.26	2.60	.13
24	13.80	16.04	-.13	4.96	.11	2.47	-.01
25	13.98	16.16	-.25	4.83	-.02	2.34	-.14
26	14.15	16.29	-.37	4.70	-.16	2.22	-.26
27	14.31	16.41	-.49	4.59	-.28	2.10	-.39
28	14.47	16.52	-.60	4.47	-.40	1.99	-.50
29	14.62	16.63	-.71	4.36	-.52	1.88	-.61
30	14.77	16.74	-.82	4.26	-.63	1.77	-.72
31	14.91	16.85	-.92	4.15	-.74	1.67	-.83
32	15.05	16.95	-1.02	4.05	-.84	1.57	-.93
33	15.19	17.05	-1.11	3.96	-.94	1.47	-1.02
34	15.31	17.15	-1.20	3.86	-1.04	1.38	-1.12
35	15.44	17.25	-1.29	3.77	-1.13	1.29	-1.21
36	15.56	17.34	-1.38	3.69	-1.22	1.20	-1.30
37	15.68	17.43	-1.46	3.60	-1.31	1.12	-1.38
38	15.80	17.52	-1.54	3.52	-1.39	1.04	-1.47
39	15.91	17.61	-1.62	3.44	-1.48	.96	-1.55
40	16.02	17.69	-1.70	3.36	-1.56	.88	-1.63
41	16.13	17.78	-1.77	3.28	-1.64	.80	-1.70
42	16.23	17.86	-1.85	3.21	-1.71	.73	-1.78
43	16.33	17.94	-1.92	3.13	-1.79	.66	-1.85
44	16.43	18.02	-1.99	3.06	-1.86	.58	-1.92
45	16.53	18.09	-2.06	2.99	-1.93	.52	-1.99

Appendix: Radar Detection Tables

THRESHOLD/NOISE AND REQUIRED SIGNAL/NOISE

PROBABILITY OF DETECTION= .9000
CONSTANT FALSE ALARM RATE
$LOG_{10}(PFA/N) = -4$

N	N DB	THRESH DB	NON-FLUC DB	SWER 1 DB	SWER 2 DB	SWER 3 DB	SWER 4 DB
1	0.00	9.64	11.75	19.37	19.37	15.60	15.6
2	3.01	10.41	9.07	16.71	12.94	12.94	11.07
3	4.77	11.02	7.56	15.23	10.20	11.45	8.93
4	6.02	11.53	6.52	14.20	8.54	10.42	7.57
5	6.99	11.96	5.73	13.42	7.37	9.64	6.58
6	7.78	12.35	5.09	12.80	6.48	9.01	5.81
7	8.45	12.69	4.56	12.27	5.76	8.49	5.18
8	9.03	13.00	4.11	11.82	5.17	8.04	4.65
9	9.54	13.29	3.71	11.43	4.66	7.65	4.20
10	10.00	13.55	3.36	11.08	4.22	7.30	3.80
11	10.41	13.80	3.04	10.77	3.83	6.98	3.45
12	10.79	14.03	2.76	10.48	3.48	6.70	3.13
13	11.14	14.24	2.50	10.22	3.17	6.44	2.84
14	11.46	14.45	2.25	9.98	2.88	6.20	2.57
15	11.76	14.64	2.03	9.76	2.62	5.98	2.33
16	12.04	14.82	1.82	9.55	2.37	5.77	2.10
17	12.30	15.00	1.63	9.36	2.15	5.58	1.89
18	12.55	15.17	1.45	9.18	1.94	5.39	1.70
19	12.79	15.33	1.27	9.01	1.74	5.22	1.51
20	13.01	15.48	1.11	8.84	1.56	5.06	1.34
21	13.22	15.63	.96	8.69	1.38	4.90	1.17
22	13.42	15.77	.81	8.54	1.22	4.76	1.02
23	13.62	15.90	.67	8.40	1.06	4.62	.87
24	13.80	16.04	.54	8.27	.91	4.48	.73
25	13.98	16.16	.41	8.14	.77	4.36	.59
26	14.15	16.29	.29	8.01	.63	4.23	.46
27	14.31	16.41	.17	7.90	.50	4.12	.34
28	14.47	16.52	.06	7.78	.38	4.00	.22
29	14.62	16.63	-.05	7.67	.26	3.89	.10
30	14.77	16.74	-.16	7.57	.14	3.79	-.01
31	14.91	16.85	-.26	7.46	.03	3.69	-.11
32	15.05	16.95	-.36	7.36	-.07	3.59	-.21
33	15.19	17.05	-.45	7.27	-.18	3.49	-.31
34	15.31	17.15	-.54	7.18	-.28	3.40	-.41
35	15.44	17.25	-.63	7.09	-.37	3.31	-.50
36	15.56	17.34	-.72	7.00	-.47	3.22	-.59
37	15.68	17.43	-.80	6.91	-.56	3.14	-.68
38	15.80	17.52	-.88	6.83	-.65	3.06	-.76
39	15.91	17.61	-.96	6.75	-.73	2.97	-.85
40	16.02	17.69	-1.04	6.67	-.81	2.90	-.93
41	16.13	17.78	-1.12	6.59	-.90	2.82	-1.00
42	16.23	17.86	-1.19	6.52	-.97	2.75	-1.08
43	16.33	17.94	-1.26	6.45	-1.05	2.67	-1.16
44	16.43	18.02	-1.33	6.37	-1.12	2.60	-1.23
45	16.53	18.09	-1.40	6.30	-1.20	2.53	-1.30

THRESHOLD/NOISE AND REQUIRED SIGNAL/NOISE

PROBABILITY OF DETECTION= .9500
CONSTANT FALSE ALARM RATE
LOG10(PFA/N)= -4

N	N DB	THRESH DB	NON-FLUC DB	SWER 1 DB	SWER 2 DB	SWER 3 DB	SWER 4 DB
1	0.00	9.64	12.31	22.52	22.52	17.43	17.43
2	3.01	10.41	9.62	19.86	14.77	14.77	12.26
3	4.77	11.02	8.10	18.38	11.61	13.27	9.92
4	6.02	11.53	7.06	17.35	9.73	12.25	8.44
5	6.99	11.96	6.26	16.57	8.43	11.46	7.39
6	7.78	12.35	5.62	15.94	7.46	10.83	6.57
7	8.45	12.69	5.09	15.41	6.68	10.31	5.91
8	9.03	13.00	4.63	14.96	6.04	9.86	5.35
9	9.54	13.29	4.23	14.57	5.49	9.46	4.88
10	10.00	13.55	3.88	14.23	5.02	9.11	4.46
11	10.41	13.80	3.56	13.92	4.60	8.80	4.10
12	10.79	14.03	3.27	13.63	4.23	8.52	3.77
13	11.14	14.24	3.01	13.37	3.90	8.25	3.47
14	11.46	14.45	2.77	13.13	3.60	8.02	3.19
15	11.76	14.64	2.54	12.91	3.32	7.79	2.94
16	12.04	14.82	2.33	12.70	3.07	7.59	2.71
17	12.30	15.00	2.14	12.51	2.83	7.39	2.49
18	12.55	15.17	1.95	12.33	2.61	7.21	2.29
19	12.79	15.33	1.78	12.16	2.40	7.04	2.10
20	13.01	15.48	1.62	11.99	2.21	6.88	1.92
21	13.22	15.63	1.46	11.84	2.03	6.72	1.75
22	13.42	15.77	1.31	11.69	1.86	6.57	1.59
23	13.62	15.90	1.17	11.55	1.69	6.43	1.44
24	13.80	16.04	1.04	11.42	1.54	6.30	1.29
25	13.98	16.16	.91	11.29	1.39	6.17	1.16
26	14.15	16.29	.79	11.16	1.25	6.05	1.02
27	14.31	16.41	.67	11.04	1.12	5.93	.90
28	14.47	16.52	.56	10.92	.99	5.82	.78
29	14.62	16.63	.45	10.81	.86	5.71	.66
30	14.77	16.74	.34	10.71	.74	5.60	.55
31	14.91	16.85	.24	10.61	.63	5.50	.44
32	15.05	16.95	.14	10.51	.52	5.40	.33
33	15.19	17.05	.05	10.41	.41	5.31	.23
34	15.31	17.15	-.04	10.32	.31	5.22	.14
35	15.44	17.25	-.13	10.23	.21	5.13	.04
36	15.56	17.34	-.22	10.14	.12	5.04	-.05
37	15.68	17.43	-.30	10.05	.02	4.95	-.14
38	15.80	17.52	-.39	9.97	-.07	4.87	-.22
39	15.91	17.61	-.47	9.89	-.15	4.79	-.31
40	16.02	17.69	-.54	9.81	-.24	4.71	-.39
41	16.13	17.78	-.62	9.74	-.32	4.64	-.47
42	16.23	17.86	-.69	9.66	-.40	4.56	-.55
43	16.33	17.94	-.76	9.59	-.48	4.49	-.62
44	16.43	18.02	-.83	9.52	-.56	4.42	-.69
45	16.53	18.09	-.90	9.45	-.63	4.35	-.76

Appendix: Radar Detection Tables

THRESHOLD/NOISE AND REQUIRED SIGNAL/NOISE

PROBABILITY OF DETECTION= .9900
CONSTANT FALSE ALARM RATE
$LOG_{10}(PFA/N)=-4$

N	N DB	THRESH DB	NON-FLUC DB	SWER 1 DB	SWER 2 DB	SWER 3 DB	SWER 4 DB
1	0.00	9.64	13.28	29.62	29.62	21.31	21.31
2	3.01	10.41	10.57	26.97	18.65	18.64	14.66
3	4.77	11.02	9.04	25.51	14.48	17.14	11.82
4	6.02	11.53	7.98	24.45	12.11	16.10	10.10
5	6.99	11.96	7.17	23.67	10.52	15.31	8.90
6	7.78	12.35	6.53	23.04	9.36	14.68	7.98
7	8.45	12.69	5.99	22.51	8.43	14.15	7.25
8	9.03	13.00	5.53	22.07	7.69	13.70	6.64
9	9.54	13.29	5.13	21.67	7.05	13.31	6.11
10	10.00	13.55	4.76	21.32	6.52	12.96	5.67
11	10.41	13.80	4.45	21.01	6.05	12.64	5.27
12	10.79	14.03	4.15	20.73	5.63	12.36	4.91
13	11.14	14.24	3.89	20.47	5.26	12.10	4.59
14	11.46	14.45	3.64	20.23	4.93	11.86	4.30
15	11.76	14.64	3.41	20.00	4.61	11.63	4.03
16	12.04	14.82	3.20	19.80	4.33	11.43	3.78
17	12.30	15.00	3.00	19.60	4.07	11.23	3.55
18	12.55	15.17	2.82	19.42	3.83	11.05	3.34
19	12.79	15.33	2.64	19.25	3.60	10.88	3.14
20	13.01	15.48	2.48	19.08	3.39	10.72	2.95
21	13.22	15.63	2.32	18.93	3.20	10.56	2.77
22	13.42	15.77	2.17	18.78	3.01	10.42	2.60
23	13.62	15.90	2.03	18.64	2.84	10.28	2.44
24	13.80	16.04	1.90	18.51	2.67	10.14	2.29
25	13.98	16.16	1.77	18.38	2.51	10.01	2.15
26	14.15	16.29	1.64	18.26	2.36	9.89	2.01
27	14.31	16.41	1.53	18.14	2.22	9.77	1.88
28	14.47	16.52	1.41	18.02	2.08	9.66	1.75
29	14.62	16.63	1.30	17.91	1.95	9.55	1.63
30	14.77	16.74	1.19	17.81	1.82	9.45	1.51
31	14.91	16.85	1.09	17.71	1.70	9.34	1.40
32	15.05	16.95	.99	17.61	1.58	9.25	1.29
33	15.19	17.05	.90	17.51	1.47	9.15	1.19
34	15.31	17.15	.80	17.42	1.36	9.06	1.09
35	15.44	17.25	.71	17.33	1.25	8.97	.99
36	15.56	17.34	.63	17.24	1.15	8.88	.89
37	15.68	17.43	.54	17.16	1.05	8.80	.80
38	15.80	17.52	.46	17.07	.96	8.72	.71
39	15.91	17.61	.38	16.99	.87	8.64	.63
40	16.02	17.69	.30	16.91	.78	8.56	.54
41	16.13	17.78	.22	16.84	.69	8.48	.46
42	16.23	17.86	.15	16.76	.61	8.41	.38
43	16.33	17.94	.08	16.69	.52	8.34	.31
44	16.43	18.02	.01	16.62	.44	8.27	.23
45	16.53	18.09	-.06	16.55	.37	8.20	.16

THRESHOLD/NOISE AND REQUIRED SIGNAL/NOISE

PROBABILITY OF DETECTION= .5000
CONSTANT FALSE ALARM RATE
LOG10(PFA/N)= -5

N	N DB	THRESH DB	NON-FLUC DB	SWER 1 DB	SWER 2 DB	SWER 3 DB	SWER 4 DB
1	0.00	10.61	10.42	11.93	11.93	11.12	11.12
2	3.01	11.30	7.78	9.29	8.48	8.48	8.10
3	4.77	11.85	6.30	7.81	6.74	7.00	6.51
4	6.02	12.31	5.28	6.79	5.60	5.98	5.43
5	6.99	12.71	4.51	6.02	4.76	5.21	4.63
6	7.78	13.06	3.89	5.40	4.10	4.59	3.99
7	8.45	13.38	3.38	4.88	3.55	4.07	3.46
8	9.03	13.67	2.94	4.44	3.09	3.63	3.01
9	9.54	13.94	2.56	4.06	2.68	3.24	2.61
10	10.00	14.18	2.22	3.72	2.33	2.90	2.27
11	10.41	14.41	1.91	3.41	2.01	2.60	1.96
12	10.79	14.63	1.63	3.13	1.72	2.32	1.68
13	11.14	14.83	1.38	2.88	1.46	2.07	1.42
14	11.46	15.02	1.15	2.64	1.22	1.83	1.18
15	11.76	15.20	.93	2.43	1.00	1.62	.96
16	12.04	15.38	.73	2.23	.80	1.41	.76
17	12.30	15.54	.54	2.04	.60	1.23	.57
18	12.55	15.70	.37	1.86	.42	1.05	.39
19	12.79	15.85	.20	1.69	.25	.88	.23
20	13.01	16.00	.05	1.54	.09	.72	.07
21	13.22	16.13	-.10	1.39	-.06	.57	-.08
22	13.42	16.27	-.24	1.24	-.20	.43	-.22
23	13.62	16.40	-.38	1.11	-.34	.30	-.36
24	13.80	16.52	-.51	.98	-.47	.17	-.49
25	13.98	16.65	-.63	.86	-.59	.04	-.61
26	14.15	16.76	-.75	.74	-.71	-.07	-.73
27	14.31	16.88	-.86	.62	-.83	-.19	-.85
28	14.47	16.99	-.97	.51	-.94	-.30	-.96
29	14.62	17.09	-1.07	.41	-1.04	-.40	-1.06
30	14.77	17.20	-1.18	.31	-1.15	-.50	-1.16
31	14.91	17.30	-1.27	.21	-1.24	-.60	-1.26
32	15.05	17.40	-1.37	.11	-1.34	-.70	-1.35
33	15.19	17.49	-1.46	.02	-1.43	-.79	-1.45
34	15.31	17.59	-1.55	-.07	-1.52	-.88	-1.53
35	15.44	17.68	-1.63	-.15	-1.61	-.96	-1.62
36	15.56	17.77	-1.72	-.24	-1.69	-1.05	-1.70
37	15.68	17.86	-1.80	-.32	-1.77	-1.13	-1.79
38	15.80	17.94	-1.87	-.40	-1.85	-1.21	-1.86
39	15.91	18.03	-1.95	-.48	-1.93	-1.28	-1.94
40	16.02	18.11	-2.02	-.55	-2.00	-1.36	-2.02
41	16.13	18.19	-2.10	-.62	-2.08	-1.43	-2.09
42	16.23	18.27	-2.17	-.69	-2.15	-1.50	-2.16
43	16.33	18.34	-2.24	-.76	-2.22	-1.57	-2.23
44	16.43	18.42	-2.30	-.83	-2.29	-1.64	-2.30
45	16.53	18.49	-2.37	-.90	-2.35	-1.71	-2.36

Appendix: Radar Detection Tables

THRESHOLD/NOISE AND REQUIRED SIGNAL/NOISE
===

PROBABILITY OF DETECTION= .6000
CONSTANT FALSE ALARM RATE
LOG$_{10}$(PFA/N)= -5

N	N DB	THRESH DB	NON-FLUC DB	SWER 1 DB	SWER 2 DB	SWER 3 DB	SWER 4 DB
1	0.00	10.61	10.88	13.33	13.33	12.10	12.10
2	3.01	11.30	8.23	10.68	9.45	9.45	8.84
3	4.77	11.85	6.74	9.20	7.56	7.96	7.15
4	6.02	12.31	5.72	8.18	6.33	6.94	6.03
5	6.99	12.71	4.95	7.41	5.44	6.17	5.19
6	7.78	13.06	4.33	6.79	4.73	5.55	4.53
7	8.45	13.38	3.81	6.27	4.16	5.03	3.98
8	9.03	13.67	3.36	5.83	3.67	4.59	3.52
9	9.54	13.94	2.98	5.45	3.25	4.20	3.11
10	10.00	14.18	2.64	5.10	2.88	3.86	2.76
11	10.41	14.41	2.33	4.80	2.55	3.55	2.44
12	10.79	14.63	2.05	4.52	2.25	3.28	2.15
13	11.14	14.83	1.80	4.26	1.98	3.02	1.89
14	11.46	15.02	1.56	4.03	1.74	2.79	1.65
15	11.76	15.20	1.35	3.81	1.51	2.57	1.43
16	12.04	15.38	1.15	3.61	1.30	2.37	1.22
17	12.30	15.54	.96	3.42	1.10	2.18	1.03
18	12.55	15.70	.78	3.25	.91	2.00	.85
19	12.79	15.85	.61	3.08	.74	1.84	.68
20	13.01	16.00	.46	2.92	.58	1.68	.52
21	13.22	16.13	.31	2.77	.42	1.53	.36
22	13.42	16.27	.16	2.63	.27	1.39	.22
23	13.62	16.40	.03	2.49	.13	1.25	.08
24	13.80	16.52	-.10	2.36	.00	1.12	-.05
25	13.98	16.65	-.22	2.24	-.13	1.00	-.18
26	14.15	16.76	-.34	2.12	-.25	.88	-.30
27	14.31	16.88	-.45	2.01	-.37	.77	-.41
28	14.47	16.99	-.56	1.90	-.48	.66	-.52
29	14.62	17.09	-.67	1.79	-.59	.55	-.63
30	14.77	17.20	-.77	1.69	-.69	.45	-.73
31	14.91	17.30	-.87	1.59	-.79	.35	-.83
32	15.05	17.40	-.96	1.50	-.89	.26	-.93
33	15.19	17.49	-1.05	1.41	-.98	.16	-1.02
34	15.31	17.59	-1.14	1.32	-1.07	.08	-1.11
35	15.44	17.68	-1.23	1.23	-1.16	-.01	-1.20
36	15.56	17.77	-1.31	1.15	-1.25	-.09	-1.28
37	15.68	17.86	-1.39	1.06	-1.33	-.18	-1.36
38	15.80	17.94	-1.47	.99	-1.41	-.26	-1.44
39	15.91	18.03	-1.55	.91	-1.49	-.33	-1.52
40	16.02	18.11	-1.62	.83	-1.56	-.41	-1.59
41	16.13	18.19	-1.69	.76	-1.64	-.48	-1.67
42	16.23	18.27	-1.76	.69	-1.71	-.55	-1.74
43	16.33	18.34	-1.83	.62	-1.78	-.62	-1.81
44	16.43	18.42	-1.90	.55	-1.85	-.69	-1.88
45	16.53	18.49	-1.97	.49	-1.92	-.75	-1.94

THRESHOLD/NOISE AND REQUIRED SIGNAL/NOISE

PROBABILITY OF DETECTION= .7000
CONSTANT FALSE ALARM RATE
LOG10(PFA/N)= -5

N	N DB	THRESH DB	NON-FLUC DB	SWER 1 DB	SWER 2 DB	SWER 3 DB	SWER 4 DB
1	0.00	10.61	11.35	14.95	14.95	13.18	13.18
2	3.01	11.30	8.68	12.30	10.53	10.53	9.64
3	4.77	11.85	7.19	10.82	8.45	9.04	7.84
4	6.02	12.31	6.16	9.80	7.12	8.02	6.66
5	6.99	12.71	5.39	9.03	6.16	7.25	5.79
6	7.78	13.06	4.76	8.41	5.42	6.62	5.10
7	8.45	13.38	4.24	7.89	4.81	6.11	4.53
8	9.03	13.67	3.79	7.44	4.29	5.66	4.05
9	9.54	13.94	3.41	7.06	3.85	5.28	3.63
10	10.00	14.18	3.06	6.72	3.46	4.93	3.27
11	10.41	14.41	2.75	6.41	3.12	4.63	2.94
12	10.79	14.63	2.47	6.13	2.81	4.35	2.64
13	11.14	14.83	2.22	5.88	2.53	4.09	2.38
14	11.46	15.02	1.98	5.64	2.27	3.86	2.13
15	11.76	15.20	1.76	5.42	2.04	3.64	1.90
16	12.04	15.38	1.56	5.22	1.82	3.44	1.69
17	12.30	15.54	1.37	5.03	1.61	3.25	1.49
18	12.55	15.70	1.19	4.86	1.42	3.07	1.31
19	12.79	15.85	1.03	4.69	1.24	2.90	1.14
20	13.01	16.00	.87	4.53	1.07	2.75	.97
21	13.22	16.13	.72	4.38	.92	2.60	.82
22	13.42	16.27	.58	4.24	.76	2.45	.67
23	13.62	16.40	.44	4.10	.62	2.32	.53
24	13.80	16.52	.31	3.97	.48	2.19	.40
25	13.98	16.65	.19	3.85	.35	2.07	.27
26	14.15	16.76	.07	3.73	.23	1.95	.15
27	14.31	16.88	-.05	3.62	.11	1.83	.03
28	14.47	16.99	-.16	3.51	-.01	1.72	-.08
29	14.62	17.09	-.26	3.40	-.12	1.62	-.19
30	14.77	17.20	-.36	3.30	-.22	1.52	-.29
31	14.91	17.30	-.46	3.20	-.33	1.42	-.39
32	15.05	17.40	-.56	3.11	-.43	1.32	-.49
33	15.19	17.49	-.65	3.01	-.52	1.23	-.58
34	15.31	17.59	-.74	2.92	-.61	1.14	-.68
35	15.44	17.68	-.82	2.84	-.70	1.05	-.76
36	15.56	17.77	-.91	2.75	-.79	.97	-.85
37	15.68	17.86	-.99	2.67	-.88	.89	-.93
38	15.80	17.94	-1.07	2.59	-.96	.81	-1.01
39	15.91	18.03	-1.14	2.52	-1.04	.73	-1.09
40	16.02	18.11	-1.22	2.44	-1.11	.66	-1.17
41	16.13	18.19	-1.29	2.37	-1.19	.59	-1.24
42	16.23	18.27	-1.36	2.30	-1.26	.52	-1.31
43	16.33	18.34	-1.43	2.23	-1.33	.45	-1.38
44	16.43	18.42	-1.50	2.16	-1.40	.38	-1.45
45	16.53	18.49	-1.56	2.10	-1.47	.31	-1.52

Appendix: Radar Detection Tables

THRESHOLD/NOISE AND REQUIRED SIGNAL/NOISE
===

PROBABILITY OF DETECTION= .8000
CONSTANT FALSE ALARM RATE
$LOG_{10}(PFA/N)=-5$

N	N DB	THRESH DB	NON-FLUC DB	SWER 1 DB	SWER 2 DB	SWER 3 DB	SWER 4 DB
1	0.00	10.61	11.86	17.04	17.04	14.52	14.52
2	3.01	11.30	9.19	14.39	11.87	11.87	10.59
3	4.77	11.85	7.69	12.91	9.53	10.38	8.65
4	6.02	12.31	6.66	11.88	8.07	9.35	7.39
5	6.99	12.71	5.87	11.11	7.02	8.58	6.47
6	7.78	13.06	5.24	10.49	6.22	7.95	5.75
7	8.45	13.38	4.72	9.97	5.56	7.43	5.15
8	9.03	13.67	4.27	9.53	5.02	6.99	4.65
9	9.54	13.94	3.88	9.14	4.55	6.60	4.22
10	10.00	14.18	3.53	8.80	4.14	6.26	3.84
11	10.41	14.41	3.22	8.49	3.78	5.95	3.51
12	10.79	14.63	2.94	8.21	3.45	5.67	3.20
13	11.14	14.83	2.68	7.96	3.16	5.42	2.92
14	11.46	15.02	2.44	7.72	2.89	5.18	2.67
15	11.76	15.20	2.23	7.50	2.64	4.96	2.44
16	12.04	15.38	2.02	7.30	2.41	4.76	2.22
17	12.30	15.54	1.83	7.11	2.20	4.57	2.02
18	12.55	15.70	1.65	6.93	2.00	4.39	1.83
19	12.79	15.85	1.48	6.77	1.82	4.23	1.65
20	13.01	16.00	1.32	6.61	1.64	4.07	1.49
21	13.22	16.13	1.17	6.46	1.48	3.92	1.33
22	13.42	16.27	1.03	6.32	1.32	3.78	1.18
23	13.62	16.40	.89	6.18	1.17	3.64	1.03
24	13.80	16.52	.76	6.05	1.03	3.51	.91
25	13.98	16.65	.64	5.93	.89	3.39	.77
26	14.15	16.76	.52	5.81	.77	3.27	.64
27	14.31	16.88	.40	5.69	.64	3.15	.52
28	14.47	16.99	.29	5.58	.52	3.04	.41
29	14.62	17.09	.19	5.48	.41	2.94	.30
30	14.77	17.20	.09	5.38	.30	2.84	.19
31	14.91	17.30	-.01	5.28	.19	2.74	.09
32	15.05	17.40	-.11	5.18	.09	2.64	-.01
33	15.19	17.49	-.20	5.09	-.00	2.55	-.10
34	15.31	17.59	-.29	5.00	-.10	2.46	-.19
35	15.44	17.68	-.38	4.92	-.19	2.38	-.28
36	15.56	17.77	-.46	4.83	-.28	2.29	-.37
37	15.68	17.86	-.54	4.75	-.37	2.21	-.45
38	15.80	17.94	-.62	4.67	-.45	2.13	-.53
39	15.91	18.03	-.70	4.60	-.53	2.05	-.61
40	16.02	18.11	-.77	4.52	-.61	1.98	-.69
41	16.13	18.19	-.85	4.45	-.69	1.91	-.77
42	16.23	18.27	-.92	4.38	-.76	1.84	-.84
43	16.33	18.34	-.99	4.31	-.84	1.77	-.91
44	16.43	18.42	-1.05	4.24	-.91	1.70	-.98
45	16.53	18.49	-1.12	4.17	-.98	1.63	-1.05

THRESHOLD/NOISE AND REQUIRED SIGNAL/NOISE

PROBABILITY OF DETECTION= .9000
CONSTANT FALSE ALARM RATE
LOG10(PFA/N)=- 5

N	N DB	THRESH DB	NON-FLUC DB	SWER 1 DB	SWER 2 DB	SWER 3 DB	SWER 4 DB
1	0.00	10.61	12.53	20.35	20.35	16.53	16.53
2	3.01	11.30	9.84	17.69	13.87	13.87	11.95
3	4.77	11.85	8.33	16.21	11.10	12.38	9.79
4	6.02	12.31	7.29	15.18	9.42	11.35	8.41
5	6.99	12.71	6.50	14.41	8.24	10.57	7.41
6	7.78	13.06	5.86	13.78	7.34	9.95	6.63
7	8.45	13.38	5.34	13.27	6.62	9.43	6.00
8	9.03	13.67	4.88	12.82	6.02	8.98	5.47
9	9.54	13.94	4.49	12.43	5.51	8.59	5.02
10	10.00	14.18	4.14	12.09	5.07	8.25	4.62
11	10.41	14.41	3.82	11.78	4.67	7.94	4.26
12	10.79	14.63	3.54	11.50	4.32	7.66	3.94
13	11.14	14.83	3.28	11.25	4.01	7.41	3.66
14	11.46	15.02	3.04	11.01	3.72	7.17	3.39
15	11.76	15.20	2.82	10.80	3.46	6.95	3.15
16	12.04	15.38	2.61	10.60	3.22	6.75	2.92
17	12.30	15.54	2.42	10.41	2.99	6.56	2.71
18	12.55	15.70	2.24	10.23	2.78	6.38	2.52
19	12.79	15.85	2.07	10.06	2.58	6.21	2.33
20	13.01	16.00	1.91	9.90	2.40	6.06	2.16
21	13.22	16.13	1.76	9.75	2.23	5.91	2.00
22	13.42	16.27	1.61	9.61	2.06	5.76	1.84
23	13.62	16.40	1.48	9.48	1.91	5.63	1.70
24	13.80	16.52	1.34	9.35	1.76	5.50	1.56
25	13.98	16.65	1.22	9.22	1.62	5.37	1.42
26	14.15	16.76	1.10	9.10	1.48	5.25	1.29
27	14.31	16.88	.98	8.99	1.35	5.14	1.17
28	14.47	16.99	.87	8.88	1.23	5.03	1.05
29	14.62	17.09	.77	8.77	1.11	4.92	.94
30	14.77	17.20	.66	8.67	1.00	4.82	.83
31	14.91	17.30	.56	8.57	.89	4.72	.73
32	15.05	17.40	.47	8.48	.78	4.63	.63
33	15.19	17.49	.37	8.39	.68	4.54	.53
34	15.31	17.59	.28	8.30	.58	4.45	.44
35	15.44	17.68	.20	8.21	.49	4.36	.34
36	15.56	17.77	.11	8.13	.39	4.28	.26
37	15.68	17.86	.03	8.05	.30	4.19	.17
38	15.80	17.94	-.05	7.97	.22	4.12	.09
39	15.91	18.03	-.13	7.89	.13	4.04	.01
40	16.02	18.11	-.20	7.81	.05	3.96	-.07
41	16.13	18.19	-.28	7.74	-.03	3.89	-.15
42	16.23	18.27	-.35	7.67	-.10	3.82	-.22
43	16.33	18.34	-.42	7.60	-.18	3.75	-.30
44	16.43	18.42	-.48	7.53	-.25	3.68	-.37
45	16.53	18.49	-.55	7.47	-.32	3.62	-.44

THRESHOLD/NOISE AND REQUIRED SIGNAL/NOISE

PROBABILITY OF DETECTION= .9500
CONSTANT FALSE ALARM RATE
$LOG_{10}(PFA/N) = -5$

N	N DB	THRESH DB	NON-FLUC DB	SWER 1 DB	SWER 2 DB	SWER 3 DB	SWER 4 DB
1	0.00	10.61	13.05	23.50	23.50	18.34	18.34
2	3.01	11.30	10.35	20.84	15.68	15.68	13.12
3	4.77	11.85	8.83	19.35	12.49	14.19	10.74
4	6.02	12.31	7.78	18.33	10.59	13.16	9.25
5	6.99	12.71	6.98	17.55	9.28	12.38	8.18
6	7.78	13.06	6.34	16.93	8.29	11.75	7.36
7	8.45	13.38	5.81	16.41	7.50	11.23	6.69
8	9.03	13.67	5.36	15.97	6.86	10.79	6.14
9	9.54	13.94	4.96	15.58	6.30	10.40	5.66
10	10.00	14.18	4.61	15.24	5.83	10.05	5.24
11	10.41	14.41	4.29	14.93	5.41	9.74	4.87
12	10.79	14.63	4.00	14.65	5.04	9.46	4.54
13	11.14	14.83	3.74	14.39	4.71	9.21	4.24
14	11.46	15.02	3.50	14.16	4.40	8.97	3.97
15	11.76	15.20	3.28	13.94	4.12	8.75	3.71
16	12.04	15.38	3.07	13.74	3.87	8.55	3.48
17	12.30	15.54	2.88	13.55	3.63	8.36	3.26
18	12.55	15.70	2.69	13.37	3.41	8.18	3.06
19	12.79	15.85	2.52	13.21	3.20	8.01	2.87
20	13.01	16.00	2.36	13.05	3.01	7.86	2.69
21	13.22	16.13	2.21	12.90	2.83	7.70	2.53
22	13.42	16.27	2.06	12.75	2.66	7.56	2.37
23	13.62	16.40	1.92	12.62	2.49	7.43	2.22
24	13.80	16.52	1.79	12.49	2.34	7.30	2.07
25	13.98	16.65	1.67	12.37	2.19	7.17	1.94
26	14.15	16.76	1.54	12.25	2.05	7.05	1.80
27	14.31	16.88	1.43	12.13	1.92	6.94	1.68
28	14.47	16.99	1.32	12.02	1.79	6.83	1.56
29	14.62	17.09	1.21	11.92	1.67	6.72	1.44
30	14.77	17.20	1.10	11.81	1.55	6.62	1.33
31	14.91	17.30	1.00	11.72	1.44	6.52	1.23
32	15.05	17.40	.91	11.63	1.33	6.43	1.12
33	15.19	17.49	.81	11.54	1.22	6.33	1.02
34	15.31	17.59	.72	11.45	1.12	6.24	.93
35	15.44	17.68	.64	11.36	1.02	6.16	.83
36	15.56	17.77	.55	11.27	.93	6.07	.74
37	15.68	17.86	.47	11.19	.83	5.99	.66
38	15.80	17.94	.39	11.11	.75	5.91	.57
39	15.91	18.03	.31	11.03	.66	5.84	.49
40	16.02	18.11	.24	10.96	.58	5.76	.41
41	16.13	18.19	.16	10.89	.49	5.69	.33
42	16.23	18.27	.09	10.82	.41	5.62	.25
43	16.33	18.34	.02	10.75	.34	5.55	.18
44	16.43	18.42	-.05	10.68	.26	5.48	.11
45	16.53	18.49	-.12	10.61	.19	5.41	.04

THRESHOLD/NOISE AND REQUIRED SIGNAL/NOISE

PROBABILITY OF DETECTION= .9900
CONSTANT FALSE ALARM RATE
$LOG_{10}(PFA/N) = -5$

N	N DB	THRESH DB	NON-FLUC DB	SWER 1 DB	SWER 2 DB	SWER 3 DB	SWER 4 DB
1	0.00	10.61	13.94	30.57	30.57	22.20	22.20
2	3.01	11.30	11.22	27.94	19.55	19.55	15.47
3	4.77	11.85	9.69	26.45	15.32	18.04	12.61
4	6.02	12.31	8.63	25.42	12.94	17.01	10.87
5	6.99	12.71	7.82	24.65	11.33	16.25	9.65
6	7.78	13.06	7.17	24.06	10.15	15.61	8.71
7	8.45	13.38	6.64	23.54	9.21	15.09	7.98
8	9.03	13.67	6.17	23.09	8.46	14.65	7.36
9	9.54	13.94	5.77	22.70	7.82	14.27	6.84
10	10.00	14.18	5.41	22.36	7.28	13.90	6.38
11	10.41	14.41	5.09	22.05	6.80	13.59	5.98
12	10.79	14.63	4.80	21.77	6.38	13.31	5.62
13	11.14	14.83	4.53	21.52	6.01	13.05	5.29
14	11.46	15.02	4.29	21.28	5.67	12.82	5.00
15	11.76	15.20	4.06	21.06	5.36	12.60	4.73
16	12.04	15.38	3.85	20.86	5.07	12.42	4.48
17	12.30	15.54	3.66	20.68	4.81	12.23	4.25
18	12.55	15.70	3.47	20.50	4.57	12.05	4.04
19	12.79	15.85	3.30	20.33	4.35	11.88	3.84
20	13.01	16.00	3.14	20.18	4.13	11.71	3.65
21	13.22	16.13	2.98	20.03	3.93	11.56	3.48
22	13.42	16.27	2.83	19.89	3.74	11.42	3.31
23	13.62	16.40	2.69	19.75	3.57	11.28	3.14
24	13.80	16.52	2.56	19.62	3.40	11.15	2.99
25	13.98	16.65	2.43	19.40	3.24	11.02	2.85
26	14.15	16.76	2.31	19.34	3.09	10.90	2.71
27	14.31	16.88	2.19	19.22	2.94	10.78	2.58
28	14.47	16.99	2.08	19.11	2.81	10.68	2.45
29	14.62	17.09	1.97	19.01	2.67	10.57	2.33
30	14.77	17.20	1.87	18.91	2.55	10.48	2.22
31	14.91	17.30	1.76	18.81	2.42	10.38	2.11
32	15.05	17.40	1.67	18.71	2.31	10.29	2.00
33	15.19	17.49	1.57	18.62	2.19	10.20	1.89
34	15.31	17.59	1.48	18.53	2.09	10.09	1.79
35	15.44	17.68	1.39	18.45	1.99	10.00	1.70
36	15.56	17.77	1.31	18.36	1.88	9.92	1.61
37	15.68	17.86	1.22	18.28	1.79	9.84	1.51
38	15.80	17.94	1.14	18.20	1.69	9.76	1.43
39	15.91	18.03	1.06	18.13	1.60	9.68	1.34
40	16.02	18.11	.99	18.05	1.51	9.61	1.26
41	16.13	18.19	.91	17.98	1.42	9.53	1.17
42	16.23	18.27	.84	17.91	1.34	9.46	1.10
43	16.33	18.34	.77	17.84	1.26	9.39	1.02
44	16.43	18.42	.70	17.77	1.18	9.32	.94
45	16.53	18.49	.63	17.70	1.10	9.26	.87

Appendix: Radar Detection Tables

THRESHOLD/NOISE AND REQUIRED SIGNAL/NOISE

PROBABILITY OF DETECTION= .5000
CONSTANT FALSE ALARM RATE
LOG10(PFA/N)=- 6

N	N DB	THRESH DB	NON-FLUC DB	SWER 1 DB	SWER 2 DB	SWER 3 DB	SWER 4 DB
1	0.00	11.40	11.24	12.77	12.77	11.95	11.95
2	3.01	12.03	8.59	10.12	9.30	9.30	8.92
3	4.77	12.53	7.10	8.63	7.56	7.81	7.31
4	6.02	12.95	6.08	7.61	6.41	6.79	6.23
5	6.99	13.32	5.31	6.83	5.57	6.02	5.43
6	7.78	13.65	4.69	6.21	4.90	5.39	4.79
7	8.45	13.95	4.17	5.70	4.35	4.88	4.25
8	9.03	14.22	3.73	5.26	3.88	4.44	3.80
9	9.54	14.47	3.35	4.87	3.48	4.05	3.41
10	10.00	14.70	3.01	4.53	3.12	3.71	3.06
11	10.41	14.92	2.70	4.22	2.81	3.41	2.75
12	10.79	15.12	2.43	3.95	2.52	3.13	2.47
13	11.14	15.31	2.18	3.70	2.26	2.88	2.21
14	11.46	15.50	1.94	3.46	2.02	2.64	1.98
15	11.76	15.67	1.73	3.25	1.80	2.43	1.76
16	12.04	15.83	1.53	3.05	1.60	2.23	1.56
17	12.30	15.99	1.34	2.86	1.41	2.04	1.37
18	12.55	16.14	1.17	2.69	1.23	1.87	1.20
19	12.79	16.28	1.00	2.52	1.06	1.70	1.03
20	13.01	16.42	.85	2.36	.90	1.55	.87
21	13.22	16.55	.70	2.22	.75	1.40	.72
22	13.42	16.68	.56	2.08	.61	1.26	.58
23	13.62	16.81	.43	1.94	.47	1.12	.45
24	13.80	16.93	.30	1.81	.34	1.00	.32
25	13.98	17.04	.18	1.69	.22	.87	.20
26	14.15	17.15	.06	1.58	.10	.76	.08
27	14.31	17.26	-.05	1.46	-.01	.64	-.03
28	14.47	17.37	-.16	1.36	-.12	.54	-.14
29	14.62	17.47	-.26	1.25	-.23	.43	-.25
30	14.77	17.57	-.36	1.15	-.33	.33	-.35
31	14.91	17.67	-.46	1.05	-.43	.24	-.44
32	15.05	17.76	-.55	.96	-.52	.14	-.54
33	15.19	17.86	-.64	.87	-.61	.05	-.63
34	15.31	17.95	-.73	.78	-.70	-.04	-.71
35	15.44	18.03	-.81	.70	-.79	-.12	-.80
36	15.56	18.12	-.89	.62	-.87	-.20	-.88
37	15.68	18.20	-.97	.54	-.95	-.28	-.96
38	15.80	18.29	-1.05	.46	-1.03	-.36	-1.04
39	15.91	18.37	-1.12	.38	-1.10	-.43	-1.11
40	16.02	18.45	-1.20	.31	-1.18	-.51	-1.19
41	16.13	18.52	-1.27	.24	-1.25	-.58	-1.26
42	16.23	18.60	-1.34	.17	-1.32	-.65	-1.33
43	16.33	18.67	-1.41	.10	-1.39	-.72	-1.40
44	16.43	18.75	-1.47	.04	-1.45	-.78	-1.46
45	16.53	18.82	-1.54	-.03	-1.52	-.85	-1.53

THRESHOLD/NOISE AND REQUIRED SIGNAL/NOISE

PROBABILITY OF DETECTION= .6000
CONSTANT FALSE ALARM RATE
LOG10(PFA/N) = -6

N	N DB	THRESH DB	NON-FLUC DB	SWER 1 DB	SWER 2 DB	SWER 3 DB	SWER 4 DB
1	0.00	11.40	11.66	14.16	14.16	12.91	12.91
2	3.01	12.03	9.00	11.50	10.25	10.25	9.63
3	4.77	12.53	7.50	10.01	8.35	8.76	7.93
4	6.02	12.95	6.48	8.99	7.12	7.74	6.80
5	6.99	13.32	5.70	8.21	6.21	6.96	5.96
6	7.78	13.65	5.08	7.59	5.51	6.34	5.29
7	8.45	13.95	4.56	7.07	4.93	5.82	4.74
8	9.03	14.22	4.11	6.63	4.44	5.38	4.28
9	9.54	14.47	3.73	6.25	4.02	4.99	3.87
10	10.00	14.70	3.39	5.91	3.65	4.65	3.52
11	10.41	14.92	3.08	5.60	3.32	4.35	3.20
12	10.79	15.12	2.80	5.32	3.02	4.07	2.91
13	11.14	15.31	2.55	5.07	2.75	3.81	2.65
14	11.46	15.50	2.31	4.84	2.50	3.58	2.41
15	11.76	15.67	2.10	4.62	2.27	3.37	2.19
16	12.04	15.83	1.90	4.42	2.06	3.17	1.98
17	12.30	15.99	1.71	4.23	1.86	2.98	1.79
18	12.55	16.14	1.53	4.06	1.68	2.80	1.61
19	12.79	16.28	1.37	3.89	1.51	2.64	1.44
20	13.01	16.42	1.21	3.74	1.34	2.48	1.28
21	13.22	16.55	1.06	3.59	1.19	2.33	1.13
22	13.42	16.68	.92	3.45	1.04	2.19	.98
23	13.62	16.81	.79	3.31	.90	2.06	.85
24	13.80	16.93	.66	3.19	.77	1.93	.71
25	13.98	17.04	.54	3.06	.64	1.81	.59
26	14.15	17.15	.42	2.95	.52	1.69	.47
27	14.31	17.26	.31	2.83	.41	1.58	.36
28	14.47	17.37	.20	2.73	.29	1.47	.25
29	14.62	17.47	.10	2.62	.19	1.37	.14
30	14.77	17.57	-.00	2.52	.08	1.27	.04
31	14.91	17.67	-.10	2.42	-.02	1.17	-.06
32	15.05	17.76	-.19	2.33	-.11	1.08	-.15
33	15.19	17.86	-.28	2.24	-.21	.98	-.24
34	15.31	17.95	-.37	2.15	-.30	.90	-.33
35	15.44	18.03	-.46	2.07	-.38	.81	-.42
36	15.56	18.12	-.54	1.99	-.47	.73	-.50
37	15.68	18.20	-.62	1.91	-.55	.65	-.58
38	15.80	18.29	-.70	1.83	-.63	.57	-.66
39	15.91	18.37	-.77	1.75	-.71	.50	-.74
40	16.02	18.45	-.84	1.68	-.78	.42	-.81
41	16.13	18.52	-.92	1.61	-.85	.35	-.88
42	16.23	18.60	-.99	1.54	-.92	.28	-.95
43	16.33	18.67	-1.05	1.47	-.99	.21	-1.02
44	16.43	18.75	-1.12	1.40	-1.06	.15	-1.09
45	16.53	18.82	-1.18	1.34	-1.13	.08	-1.16

Appendix: Radar Detection Tables

```
THRESHOLD/NOISE AND REQUIRED SIGNAL/NOISE
=========================================

PROBABILITY OF DETECTION= .7000
CONSTANT FALSE ALARM RATE
LOG10(PFA/N)=- 6
```

N	N DB	THRESH DB	NON-FLUC DB	SWER 1 DB	SWER 2 DB	SWER 3 DB	SWER 4 DB
1	0.00	11.40	12.19	15.77	15.77	13.98	13.98
2	3.01	12.03	9.41	13.11	11.32	11.32	10.41
3	4.77	12.53	7.91	11.62	9.22	9.82	8.59
4	6.02	12.95	6.88	10.59	7.88	8.80	7.40
5	6.99	13.32	6.10	9.82	6.92	8.02	6.52
6	7.78	13.65	5.47	9.19	6.16	7.40	5.83
7	8.45	13.95	4.95	8.68	5.55	6.88	5.26
8	9.03	14.22	4.50	8.23	5.03	6.44	4.78
9	9.54	14.47	4.12	7.85	4.59	6.05	4.36
10	10.00	14.70	3.77	7.51	4.20	5.71	3.99
11	10.41	14.92	3.46	7.20	3.85	5.40	3.66
12	10.79	15.12	3.18	6.92	3.54	5.12	3.37
13	11.14	15.31	2.93	6.67	3.26	4.87	3.10
14	11.46	15.50	2.69	6.44	3.00	4.63	2.85
15	11.76	15.67	2.47	6.22	2.77	4.42	2.62
16	12.04	15.83	2.27	6.02	2.55	4.22	2.41
17	12.30	15.99	2.08	5.83	2.34	4.03	2.22
18	12.55	16.14	1.91	5.66	2.15	3.85	2.03
19	12.79	16.28	1.74	5.49	1.97	3.69	1.86
20	13.01	16.42	1.58	5.33	1.81	3.53	1.70
21	13.22	16.55	1.43	5.19	1.65	3.38	1.54
22	13.42	16.68	1.29	5.05	1.49	3.24	1.39
23	13.62	16.81	1.16	4.91	1.35	3.11	1.26
24	13.80	16.93	1.03	4.78	1.21	2.98	1.12
25	13.98	17.04	.91	4.66	1.08	2.86	1.00
26	14.15	17.15	.79	4.54	.96	2.74	.87
27	14.31	17.26	.67	4.43	.84	2.63	.75
28	14.47	17.37	.57	4.32	.73	2.52	.65
29	14.62	17.47	.46	4.22	.62	2.41	.54
30	14.77	17.57	.36	4.12	.51	2.31	.44
31	14.91	17.67	.26	4.02	.41	2.22	.34
32	15.05	17.76	.17	3.93	.31	2.12	.24
33	15.19	17.86	.08	3.84	.22	2.03	.15
34	15.31	17.95	-.01	3.75	.12	1.95	.06
35	15.44	18.03	-.10	3.67	.03	1.86	-.03
36	15.56	18.12	-.18	3.58	-.05	1.78	-.11
37	15.68	18.20	-.26	3.50	-.14	1.70	-.20
38	15.80	18.29	-.34	3.43	-.22	1.62	-.28
39	15.91	18.37	-.41	3.35	-.30	1.54	-.35
40	16.02	18.45	-.49	3.28	-.37	1.47	-.43
41	16.13	18.52	-.56	3.21	-.45	1.40	-.50
42	16.23	18.60	-.63	3.14	-.52	1.33	-.57
43	16.33	18.67	-.70	3.07	-.59	1.26	-.64
44	16.43	18.75	-.76	3.00	-.66	1.20	-.71
45	16.53	18.82	-.83	2.94	-.73	1.13	-.78

THRESHOLD/NOISE AND REQUIRED SIGNAL/NOISE
===

PROBABILITY OF DETECTION= .8000
CONSTANT FALSE ALARM RATE
LOG10(PFA/N)=- 6

N	N DB	THRESH DB	NON-FLUC DB	SWER 1 DB	SWER 2 DB	SWER 3 DB	SWER 4 DB
1	0.00	11.40	12.57	17.85	17.85	15.31	15.31
2	3.01	12.03	9.88	15.18	12.64	12.64	11.33
3	4.77	12.53	8.37	13.69	10.28	11.14	9.38
4	6.02	12.95	7.33	12.67	8.81	10.11	8.11
5	6.99	13.32	6.54	11.89	7.75	9.34	7.18
6	7.78	13.65	5.91	11.27	6.94	8.71	6.45
7	8.45	13.95	5.39	10.75	6.28	8.19	5.85
8	9.03	14.22	4.94	10.31	5.73	7.75	5.35
9	9.54	14.47	4.55	9.92	5.26	7.36	4.91
10	10.00	14.70	4.20	9.58	4.84	7.02	4.53
11	10.41	14.92	3.89	9.27	4.48	6.71	4.19
12	10.79	15.12	3.61	8.99	4.15	6.43	3.89
13	11.14	15.31	3.35	8.74	3.86	6.18	3.61
14	11.46	15.50	3.11	8.51	3.59	5.94	3.36
15	11.76	15.67	2.89	8.29	3.34	5.73	3.12
16	12.04	15.83	2.69	8.09	3.11	5.53	2.90
17	12.30	15.99	2.50	7.90	2.90	5.34	2.70
18	12.55	16.14	2.32	7.73	2.70	5.16	2.51
19	12.79	16.28	2.15	7.56	2.51	4.99	2.34
20	13.01	16.42	1.99	7.40	2.34	4.84	2.17
21	13.22	16.55	1.84	7.26	2.17	4.69	2.01
22	13.42	16.68	1.70	7.12	2.01	4.55	1.86
23	13.62	16.81	1.56	6.98	1.86	4.41	1.72
24	13.80	16.93	1.43	6.85	1.72	4.28	1.58
25	13.98	17.04	1.31	6.73	1.59	4.16	1.45
26	14.15	17.15	1.19	6.61	1.46	4.04	1.33
27	14.31	17.26	1.08	6.50	1.34	3.93	1.21
28	14.47	17.37	.97	6.39	1.22	3.82	1.10
29	14.62	17.47	.86	6.29	1.10	3.72	.99
30	14.77	17.57	.76	6.19	1.00	3.62	.88
31	14.91	17.67	.66	6.09	.89	3.52	.78
32	15.05	17.76	.57	6.00	.79	3.43	.68
33	15.19	17.86	.48	5.91	.69	3.34	.59
34	15.31	17.95	.39	5.82	.60	3.25	.50
35	15.44	18.03	.30	5.73	.51	3.16	.41
36	15.56	18.12	.22	5.65	.42	3.08	.32
37	15.68	18.20	.14	5.57	.33	3.00	.24
38	15.80	18.29	.06	5.49	.25	2.92	.16
39	15.91	18.37	-.01	5.42	.17	2.85	.08
40	16.02	18.45	-.09	5.35	.09	2.77	.00
41	16.13	18.52	-.16	5.27	.01	2.70	-.07
42	16.23	18.60	-.23	5.20	-.06	2.63	-.15
43	16.33	18.67	-.30	5.14	-.13	2.56	-.22
44	16.43	18.75	-.37	5.07	-.20	2.50	-.28
45	16.53	18.82	-.43	5.00	-.27	2.43	-.35

Appendix: Radar Detection Tables

THRESHOLD/NOISE AND REQUIRED SIGNAL/NOISE

PROBABILITY OF DETECTION= .9000
CONSTANT FALSE ALARM RATE
LOG10(PFA/N)=- 6

N	N DB	THRESH DB	NON-FLUC DB	SWER 1 DB	SWER 2 DB	SWER 3 DB	SWER 4 DB
1	0.00	11.40	13.18	21.14	21.14	17.30	17.3
2	3.01	12.03	10.48	18.48	14.62	14.62	12.67
3	4.77	12.53	8.96	16.99	11.83	13.13	10.48
4	6.02	12.95	7.92	15.96	10.13	12.10	9.09
5	6.99	13.32	7.12	15.18	8.94	11.32	8.08
6	7.78	13.65	6.49	14.56	8.03	10.69	7.30
7	8.45	13.95	5.95	14.04	7.30	10.17	6.66
8	9.03	14.22	5.50	13.60	6.70	9.72	6.13
9	9.54	14.47	5.10	13.21	6.18	9.33	5.67
10	10.00	14.70	4.75	12.87	5.73	8.99	5.27
11	10.41	14.92	4.44	12.56	5.34	8.68	4.91
12	10.79	15.12	4.16	12.28	4.99	8.40	4.59
13	11.14	15.31	3.90	12.03	4.67	8.15	4.30
14	11.46	15.50	3.66	11.80	4.38	7.91	4.03
15	11.76	15.67	3.44	11.58	4.12	7.70	3.79
16	12.04	15.83	3.23	11.38	3.87	7.50	3.56
17	12.30	15.99	3.04	11.19	3.65	7.31	3.35
18	12.55	16.14	2.86	11.02	3.44	7.13	3.16
19	12.79	16.28	2.69	10.85	3.24	6.96	2.97
20	13.01	16.42	2.53	10.69	3.05	6.81	2.80
21	13.22	16.55	2.38	10.54	2.88	6.66	2.63
22	13.42	16.68	2.23	10.40	2.71	6.52	2.48
23	13.62	16.81	2.09	10.27	2.56	6.38	2.33
24	13.80	16.93	1.96	10.14	2.41	6.25	2.19
25	13.98	17.04	1.84	10.02	2.27	6.13	2.06
26	14.15	17.15	1.72	9.90	2.13	6.01	1.93
27	14.31	17.26	1.60	9.79	2.00	5.90	1.81
28	14.47	17.37	1.49	9.68	1.88	5.79	1.69
29	14.62	17.47	1.39	9.58	1.76	5.69	1.58
30	14.77	17.57	1.28	9.48	1.65	5.59	1.47
31	14.91	17.67	1.19	9.38	1.54	5.49	1.37
32	15.05	17.76	1.09	9.28	1.43	5.40	1.26
33	15.19	17.86	1.00	9.19	1.33	5.30	1.17
34	15.31	17.95	.91	9.11	1.23	5.22	1.07
35	15.44	18.03	.82	9.02	1.14	5.13	.98
36	15.56	18.12	.74	8.94	1.04	5.05	.89
37	15.68	18.20	.66	8.86	.95	4.97	.81
38	15.80	18.29	.58	8.78	.87	4.89	.73
39	15.91	18.37	.50	8.71	.78	4.81	.65
40	16.02	18.45	.43	8.63	.70	4.74	.57
41	16.13	18.52	.35	8.56	.62	4.67	.49
42	16.23	18.60	.28	8.49	.55	4.60	.42
43	16.33	18.67	.21	8.42	.47	4.53	.35
44	16.43	18.75	.15	8.36	.40	4.46	.27
45	16.53	18.82	.08	8.29	.33	4.40	.21

THRESHOLD/NOISE AND REQUIRED SIGNAL/NOISE

PROBABILITY OF DETECTION= .9500
CONSTANT FALSE ALARM RATE
LOG10(PFA/N) = -6

N	N DB	THRESH DB	NON-FLUC DB	SWER 1 DB	SWER 2 DB	SWER 3 DB	SWER 4 DB
1	0.00	11.40	13.66	24.29	24.29	19.10	19.10
2	3.01	12.03	10.95	21.63	16.42	16.42	13.83
3	4.77	12.53	9.42	20.13	13.20	14.93	11.42
4	6.02	12.95	8.37	19.10	11.29	13.90	9.91
5	6.99	13.32	7.57	18.33	9.96	13.11	8.83
6	7.78	13.65	6.93	17.70	8.96	12.49	8.00
7	8.45	13.95	6.39	17.18	8.17	11.97	7.33
8	9.03	14.22	5.94	16.74	7.51	11.52	6.76
9	9.54	14.47	5.54	16.35	6.96	11.13	6.28
10	10.00	14.70	5.19	16.01	6.48	10.79	5.86
11	10.41	14.92	4.87	15.70	6.05	10.48	5.49
12	10.79	15.12	4.58	15.42	5.68	10.20	5.15
13	11.14	15.31	4.32	15.17	5.34	9.94	4.85
14	11.46	15.50	4.08	14.94	5.03	9.71	4.58
15	11.76	15.67	3.86	14.72	4.75	9.49	4.32
16	12.04	15.83	3.65	14.52	4.50	9.29	4.09
17	12.30	15.99	3.46	14.33	4.26	9.10	3.87
18	12.55	16.14	3.27	14.16	4.04	8.92	3.67
19	12.79	16.28	3.10	13.99	3.83	8.76	3.48
20	13.01	16.42	2.94	13.83	3.63	8.60	3.30
21	13.22	16.55	2.79	13.68	3.45	8.45	3.13
22	13.42	16.68	2.64	13.54	3.28	8.31	2.97
23	13.62	16.81	2.51	13.41	3.12	8.17	2.82
24	13.80	16.93	2.37	13.28	2.96	8.05	2.68
25	13.98	17.04	2.25	13.16	2.81	7.92	2.54
26	14.15	17.15	2.13	13.04	2.67	7.80	2.41
27	14.31	17.26	2.01	12.93	2.54	7.69	2.28
28	14.47	17.37	1.90	12.82	2.41	7.58	2.16
29	14.62	17.47	1.79	12.72	2.29	7.48	2.05
30	14.77	17.57	1.69	12.62	2.17	7.38	1.93
31	14.91	17.67	1.59	12.52	2.05	7.28	1.83
32	15.05	17.76	1.49	12.43	1.94	7.19	1.72
33	15.19	17.86	1.40	12.33	1.84	7.09	1.63
34	15.31	17.95	1.31	12.25	1.74	7.01	1.53
35	15.44	18.03	1.22	12.16	1.64	6.92	1.44
36	15.56	18.12	1.14	12.08	1.54	6.84	1.35
37	15.68	18.20	1.06	12.00	1.45	6.76	1.26
38	15.80	18.29	.98	11.92	1.36	6.68	1.17
39	15.91	18.37	.90	11.85	1.28	6.60	1.09
40	16.02	18.45	.82	11.77	1.19	6.53	1.01
41	16.13	18.52	.75	11.70	1.11	6.46	.93
42	16.23	18.60	.68	11.63	1.03	6.39	.86
43	16.33	18.67	.61	11.56	.95	6.32	.78
44	16.43	18.75	.54	11.50	.88	6.25	.71
45	16.53	18.82	.48	11.43	.80	6.19	.64

Appendix: Radar Detection Tables 189

THRESHOLD/NOISE AND REQUIRED SIGNAL/NOISE

PROBABILITY OF DETECTION= .9900
CONSTANT FALSE ALARM RATE
$LOG_{10}(PFA/N) = -6$

N	N DB	THRESH DB	NON-FLUC DB	SWER 1 DB	SWER 2 DB	SWER 3 DB	SWER 4 DB
1	0.00	11.40	14.50	31.37	31.37	22.95	22.95
2	3.01	12.03	11.76	28.70	20.27	20.27	16.16
3	4.77	12.53	10.23	27.23	16.03	18.79	13.27
4	6.02	12.95	9.16	26.20	13.61	17.75	11.50
5	6.99	13.32	8.35	25.42	11.98	16.96	10.26
6	7.78	13.65	7.70	24.80	10.79	16.33	9.32
7	8.45	13.95	7.16	24.28	9.86	15.81	8.56
8	9.03	14.22	6.70	23.83	9.08	15.36	7.95
9	9.54	14.47	6.29	23.45	8.44	14.97	7.42
10	10.00	14.70	5.94	23.10	7.89	14.63	6.96
11	10.41	14.92	5.61	22.80	7.40	14.32	6.55
12	10.79	15.12	5.32	22.52	6.98	14.04	6.19
13	11.14	15.31	5.06	22.26	6.60	13.78	5.86
14	11.46	15.50	4.81	22.03	6.25	13.55	5.56
15	11.76	15.67	4.59	21.81	5.95	13.33	5.29
16	12.04	15.83	4.38	21.61	5.66	13.13	5.04
17	12.30	15.99	4.18	21.42	5.40	12.94	4.81
18	12.55	16.14	4.00	21.25	5.15	12.76	4.60
19	12.79	16.28	3.83	21.08	4.92	12.60	4.39
20	13.01	16.42	3.66	20.93	4.71	12.44	4.20
21	13.22	16.55	3.51	20.78	4.51	12.29	4.03
22	13.42	16.68	3.36	20.64	4.32	12.15	3.86
23	13.62	16.81	3.22	20.50	4.15	12.01	3.70
24	13.80	16.93	3.09	20.37	3.98	11.89	3.54
25	13.98	17.04	2.96	20.25	3.82	11.76	3.40
26	14.15	17.15	2.84	20.13	3.66	11.64	3.26
27	14.31	17.26	2.72	20.02	3.52	11.53	3.13
28	14.47	17.37	2.61	19.91	3.38	11.42	3.00
29	14.62	17.47	2.50	19.81	3.24	11.32	2.88
30	14.77	17.57	2.39	19.71	3.12	11.22	2.77
31	14.91	17.67	2.29	19.61	2.99	11.12	2.65
32	15.05	17.76	2.20	19.52	2.88	11.03	2.55
33	15.19	17.86	2.10	19.43	2.76	10.94	2.44
34	15.31	17.95	2.01	19.34	2.66	10.85	2.34
35	15.44	18.03	1.92	19.26	2.55	10.76	2.24
36	15.56	18.12	1.84	19.17	2.45	10.68	2.15
37	15.68	18.20	1.75	19.09	2.35	10.60	2.06
38	15.80	18.29	1.67	19.02	2.25	10.52	1.97
39	15.91	18.37	1.59	18.94	2.16	10.45	1.88
40	16.02	18.45	1.52	18.87	2.07	10.38	1.80
41	16.13	18.52	1.44	18.80	1.99	10.30	1.72
42	16.23	18.60	1.37	18.73	1.90	10.23	1.64
43	16.33	18.67	1.30	18.66	1.82	10.17	1.57
44	16.43	18.75	1.23	18.59	1.74	10.10	1.49
45	16.53	18.82	1.16	18.53	1.66	10.03	1.41

THRESHOLD/NOISE AND REQUIRED SIGNAL/NOISE
==

PROBABILITY OF DETECTION= .5000
CONSTANT FALSE ALARM RATE
LOG10(PFA/N)= -7

N	N DB	THRESH DB	NON-FLUC DB	SWER 1 DB	SWER 2 DB	SWER 3 DB	SWER 4 DB
1	0.00	12.07	11.94	13.47	13.47	12.65	12.65
2	3.01	12.65	9.26	10.80	9.98	9.98	9.60
3	4.77	13.11	7.77	9.31	8.23	8.49	7.98
4	6.02	13.50	6.74	8.28	7.08	7.46	6.90
5	6.99	13.85	5.96	7.50	6.23	6.68	6.09
6	7.78	14.16	5.34	6.88	5.55	6.06	5.44
7	8.45	14.44	4.82	6.36	5.00	5.54	4.91
8	9.03	14.69	4.38	5.92	4.53	5.09	4.45
9	9.54	14.93	3.99	5.53	4.13	4.71	4.06
10	10.00	15.15	3.65	5.19	3.77	4.36	3.71
11	10.41	15.35	3.35	4.88	3.45	4.06	3.40
12	10.79	15.55	3.07	4.60	3.17	3.78	3.11
13	11.14	15.73	2.82	4.35	2.90	3.53	2.86
14	11.46	15.90	2.58	4.12	2.66	3.30	2.62
15	11.76	16.07	2.37	3.90	2.44	3.08	2.40
16	12.04	16.22	2.17	3.70	2.24	2.88	2.20
17	12.30	16.37	1.98	3.51	2.05	2.69	2.01
18	12.55	16.52	1.81	3.34	1.87	2.52	1.83
19	12.79	16.65	1.64	3.17	1.70	2.35	1.67
20	13.01	16.79	1.49	3.02	1.54	2.20	1.51
21	13.22	16.91	1.34	2.87	1.39	2.05	1.36
22	13.42	17.04	1.20	2.73	1.25	1.91	1.22
23	13.62	17.16	1.07	2.60	1.11	1.77	1.09
24	13.80	17.27	.94	2.47	.98	1.65	.96
25	13.98	17.38	.82	2.35	.86	1.53	.84
26	14.15	17.49	.70	2.23	.74	1.41	.72
27	14.31	17.59	.59	2.12	.63	1.30	.61
28	14.47	17.70	.48	2.01	.52	1.19	.50
29	14.62	17.80	.38	1.91	.41	1.09	.39
30	14.77	17.89	.28	1.81	.31	.99	.29
31	14.91	17.99	.18	1.71	.21	.89	.20
32	15.05	18.08	.09	1.62	.12	.80	.10
33	15.19	18.17	.00	1.53	.03	.71	.01
34	15.31	18.25	-.09	1.44	-.06	.62	-.07
35	15.44	18.34	-.17	1.36	-.14	.54	-.16
36	15.56	18.42	-.25	1.29	-.22	.46	-.24
37	15.68	18.50	-.33	1.20	-.30	.38	-.32
38	15.80	18.58	-.41	1.12	-.38	.30	-.40
39	15.91	18.66	-.48	1.05	-.46	.22	-.47
40	16.02	18.74	-.55	.97	-.53	.15	-.54
41	16.13	18.81	-.62	.90	-.60	.08	-.61
42	16.23	18.89	-.69	.83	-.67	.01	-.68
43	16.33	18.96	-.76	.77	-.74	-.06	-.75
44	16.43	19.03	-.83	.70	-.81	-.12	-.82
45	16.53	19.10	-.89	.64	-.87	-.19	-.88

Appendix: Radar Detection Tables

THRESHOLD/NOISE AND REQUIRED SIGNAL/NOISE
===

PROBABILITY OF DETECTION= .6000
CONSTANT FALSE ALARM RATE
LOG10(PFA/N)= -7

N	N DB	THRESH DB	NON-FLUC DB	SWER 1 DB	SWER 2 DB	SWER 3 DB	SWER 4 DB
1	0.00	12.07	12.32	14.85	14.85	13.60	13.6
2	3.01	12.65	9.64	12.18	10.92	10.92	10.29
3	4.77	13.11	8.14	10.68	9.01	9.42	8.58
4	6.02	13.50	7.11	9.65	7.76	8.39	7.44
5	6.99	13.85	6.33	8.87	6.86	7.61	6.60
6	7.78	14.16	5.70	8.25	6.14	6.99	5.92
7	8.45	14.44	5.18	7.73	5.56	6.47	5.37
8	9.03	14.69	4.73	7.28	5.07	6.02	4.90
9	9.54	14.93	4.34	6.90	4.64	5.64	4.49
10	10.00	15.15	4.00	6.55	4.27	5.29	4.14
11	10.41	15.35	3.69	6.25	3.94	4.99	3.82
12	10.79	15.55	3.41	5.97	3.64	4.71	3.53
13	11.14	15.73	3.16	5.72	3.37	4.45	3.26
14	11.46	15.90	2.92	5.48	3.12	4.22	3.02
15	11.76	16.07	2.71	5.27	2.89	4.00	2.80
16	12.04	16.22	2.51	5.07	2.68	3.80	2.59
17	12.30	16.37	2.32	4.88	2.48	3.62	2.40
18	12.55	16.52	2.14	4.70	2.29	3.44	2.22
19	12.79	16.65	1.98	4.54	2.12	3.28	2.05
20	13.01	16.79	1.82	4.38	1.96	3.12	1.89
21	13.22	16.91	1.67	4.23	1.80	2.97	1.74
22	13.42	17.04	1.53	4.09	1.66	2.83	1.59
23	13.62	17.16	1.40	3.96	1.52	2.70	1.46
24	13.80	17.27	1.27	3.83	1.38	2.57	1.33
25	13.98	17.38	1.15	3.71	1.26	2.45	1.20
26	14.15	17.49	1.03	3.59	1.14	2.33	1.08
27	14.31	17.59	.92	3.48	1.02	2.22	.97
28	14.47	17.70	.81	3.37	.91	2.11	.86
29	14.62	17.80	.70	3.27	.80	2.01	.75
30	14.77	17.89	.60	3.17	.70	1.91	.65
31	14.91	17.99	.51	3.07	.60	1.81	.55
32	15.05	18.08	.41	2.98	.50	1.72	.46
33	15.19	18.17	.32	2.89	.41	1.63	.37
34	15.31	18.25	.24	2.80	.32	1.54	.28
35	15.44	18.34	.15	2.72	.23	1.46	.19
36	15.56	18.42	.07	2.64	.15	1.37	.11
37	15.68	18.50	-.01	2.56	.07	1.29	.03
38	15.80	18.58	-.09	2.48	-.01	1.22	-.05
39	15.91	18.66	-.16	2.41	-.09	1.14	-.13
40	16.02	18.74	-.23	2.33	-.16	1.07	-.20
41	16.13	18.81	-.30	2.26	-.24	1.00	-.27
42	16.23	18.89	-.37	2.19	-.31	.93	-.34
43	16.33	18.96	-.44	2.13	-.38	.86	-.41
44	16.43	19.03	-.51	2.06	-.44	.80	-.48
45	16.53	19.10	-.57	2.00	-.51	.73	-.54

THRESHOLD/NOISE AND REQUIRED SIGNAL/NOISE

PROBABILITY OF DETECTION= .7000
CONSTANT FALSE ALARM RATE
LOG10(PFA/N)=- 7

N	N DB	THRESH DB	NON-FLUC DB	SWER 1 DB	SWER 2 DB	SWER 3 DB	SWER 4 DB
1	0.00	12.07	12.72	16.45	16.45	14.65	14.65
2	3.01	12.65	10.03	13.78	11.97	11.97	11.05
3	4.77	13.11	8.52	12.28	9.86	10.47	9.23
4	6.02	13.50	7.48	11.25	8.52	9.44	8.02
5	6.99	13.85	6.70	10.47	7.54	8.66	7.14
6	7.78	14.16	6.06	9.84	6.78	8.03	6.44
7	8.45	14.44	5.54	9.32	6.16	7.51	5.86
8	9.03	14.69	5.09	8.88	5.64	7.07	5.38
9	9.54	14.93	4.70	8.49	5.19	6.68	4.96
10	10.00	15.15	4.35	8.15	4.80	6.34	4.59
11	10.41	15.35	4.04	7.84	4.45	6.03	4.26
12	10.79	15.55	3.76	7.56	4.14	5.75	3.96
13	11.14	15.73	3.51	7.31	3.86	5.50	3.69
14	11.46	15.90	3.27	7.08	3.60	5.26	3.44
15	11.76	16.07	3.05	6.86	3.36	5.05	3.21
16	12.04	16.22	2.85	6.66	3.14	4.84	3.00
17	12.30	16.37	2.66	6.47	2.94	4.66	2.80
18	12.55	16.52	2.48	6.30	2.74	4.48	2.62
19	12.79	16.65	2.32	6.13	2.56	4.32	2.44
20	13.01	16.79	2.16	5.97	2.40	4.16	2.28
21	13.22	16.91	2.01	5.83	2.24	4.01	2.13
22	13.42	17.04	1.87	5.69	2.08	3.87	1.98
23	13.62	17.16	1.73	5.55	1.94	3.74	1.84
24	13.80	17.27	1.60	5.42	1.80	3.61	1.71
25	13.98	17.38	1.48	5.30	1.67	3.49	1.58
26	14.15	17.49	1.36	5.18	1.55	3.37	1.46
27	14.31	17.59	1.25	5.07	1.43	3.26	1.34
28	14.47	17.70	1.14	4.97	1.31	3.15	1.23
29	14.62	17.80	1.04	4.86	1.20	3.04	1.12
30	14.77	17.89	.94	4.76	1.10	2.94	1.02
31	14.91	17.99	.84	4.67	1.00	2.85	.92
32	15.05	18.08	.75	4.57	.90	2.75	.82
33	15.19	18.17	.66	4.48	.80	2.66	.73
34	15.31	18.25	.57	4.39	.71	2.58	.64
35	15.44	18.34	.48	4.31	.62	2.49	.55
36	15.56	18.42	.40	4.23	.54	2.41	.47
37	15.68	18.50	.32	4.15	.45	2.33	.39
38	15.80	18.58	.24	4.07	.37	2.25	.31
39	15.91	18.66	.17	4.00	.29	2.18	.23
40	16.02	18.74	.09	3.92	.22	2.11	.15
41	16.13	18.81	.02	3.85	.14	2.03	.08
42	16.23	18.89	-.05	3.78	.07	1.97	.01
43	16.33	18.96	-.12	3.72	-.00	1.90	-.06
44	16.43	19.03	-.18	3.65	-.07	1.83	-.13
45	16.53	19.10	-.25	3.59	-.14	1.77	-.19

Appendix: Radar Detection Tables

THRESHOLD/NOISE AND REQUIRED SIGNAL/NOISE

PROBABILITY OF DETECTION= .8000
CONSTANT FALSE ALARM RATE
LOG10(PFA/N)= -7

N	N DB	THRESH DB	NON-FLUC DB	SWER 1 DB	SWER 2 DB	SWER 3 DB	SWER 4 DB
1	0.00	12.07	13.16	18.53	18.53	15.97	15.97
2	3.01	12.65	10.46	15.85	13.29	13.28	11.96
3	4.77	13.11	8.94	14.35	10.91	11.78	9.99
4	6.02	13.50	7.90	13.32	9.42	10.75	8.71
5	6.99	13.85	7.11	12.54	8.36	9.97	7.77
6	7.78	14.16	6.47	11.91	7.54	9.34	7.03
7	8.45	14.44	5.94	11.39	6.87	8.82	6.43
8	9.03	14.69	5.49	10.95	6.32	8.37	5.92
9	9.54	14.93	5.10	10.56	5.84	7.98	5.49
10	10.00	15.15	4.75	10.21	5.43	7.64	5.10
11	10.41	15.35	4.44	9.91	5.06	7.33	4.76
12	10.79	15.55	4.15	9.63	4.73	7.05	4.45
13	11.14	15.73	3.90	9.37	4.43	6.79	4.17
14	11.46	15.90	3.66	9.14	4.16	6.56	3.92
15	11.76	16.07	3.44	8.93	3.91	6.34	3.68
16	12.04	16.22	3.24	8.72	3.68	6.14	3.46
17	12.30	16.37	3.04	8.54	3.46	5.95	3.26
18	12.55	16.52	2.87	8.36	3.26	5.78	3.07
19	12.79	16.65	2.70	8.19	3.08	5.61	2.89
20	13.01	16.79	2.54	8.04	2.90	5.45	2.72
21	13.22	16.91	2.39	7.89	2.73	5.31	2.57
22	13.42	17.04	2.24	7.75	2.58	5.16	2.42
23	13.62	17.16	2.11	7.62	2.43	5.03	2.27
24	13.80	17.27	1.98	7.49	2.29	4.90	2.14
25	13.98	17.38	1.86	7.37	2.15	4.78	2.01
26	14.15	17.49	1.74	7.25	2.02	4.66	1.88
27	14.31	17.59	1.62	7.14	1.90	4.55	1.76
28	14.47	17.70	1.51	7.03	1.78	4.44	1.65
29	14.62	17.80	1.41	6.92	1.66	4.34	1.54
30	14.77	17.89	1.31	6.82	1.56	4.24	1.43
31	14.91	17.99	1.21	6.73	1.45	4.14	1.33
32	15.05	18.08	1.11	6.64	1.35	4.05	1.23
33	15.19	18.17	1.02	6.55	1.25	3.96	1.14
34	15.31	18.25	.93	6.46	1.16	3.87	1.05
35	15.44	18.34	.85	6.37	1.06	3.79	.96
36	15.56	18.42	.77	6.29	.98	3.70	.87
37	15.68	18.50	.69	6.21	.89	3.62	.79
38	15.80	18.58	.61	6.13	.81	3.55	.71
39	15.91	18.66	.53	6.06	.73	3.47	.63
40	16.02	18.74	.46	5.99	.65	3.40	.55
41	16.13	18.81	.38	5.92	.57	3.33	.48
42	16.23	18.89	.31	5.85	.50	3.26	.41
43	16.33	18.96	.25	5.78	.42	3.19	.34
44	16.43	19.03	.18	5.71	.35	3.12	.27
45	16.53	19.10	.11	5.65	.29	3.05	.20

THRESHOLD/NOISE AND REQUIRED SIGNAL/NOISE
===

PROBABILITY OF DETECTION= .9000
CONSTANT FALSE ALARM RATE
LOG10(PFA/N)=- 7

N	N DB	THRESH DB	NON-FLUC DB	SWER 1 DB	SWER 2 DB	SWER 3 DB	SWER 4 DB
1	0.00	12.07	13.74	21.82	21.82	17.95	17.95
2	3.01	12.65	11.03	19.14	15.26	15.26	13.28
3	4.77	13.11	9.50	17.64	12.45	13.76	11.08
4	6.02	13.50	8.45	16.61	10.73	12.72	9.67
5	6.99	13.85	7.65	15.83	9.53	11.94	8.65
6	7.78	14.16	7.01	15.20	8.61	11.31	7.86
7	8.45	14.44	6.47	14.68	7.87	10.78	7.22
8	9.03	14.69	6.02	14.23	7.26	10.34	6.68
9	9.54	14.93	5.62	13.84	6.74	9.95	6.21
10	10.00	15.15	5.27	13.50	6.29	9.60	5.81
11	10.41	15.35	4.95	13.19	5.89	9.29	5.45
12	10.79	15.55	4.67	12.91	5.54	9.01	5.12
13	11.14	15.73	4.41	12.66	5.22	8.76	4.83
14	11.46	15.90	4.17	12.43	4.93	8.52	4.56
15	11.76	16.07	3.94	12.21	4.66	8.31	4.32
16	12.04	16.22	3.74	12.01	4.41	8.10	4.09
17	12.30	16.37	3.55	11.82	4.19	7.92	3.88
18	12.55	16.52	3.36	11.65	3.97	7.74	3.68
19	12.79	16.65	3.19	11.48	3.77	7.57	3.50
20	13.01	16.79	3.03	11.32	3.59	7.41	3.32
21	13.22	16.91	2.88	11.18	3.41	7.27	3.16
22	13.42	17.04	2.74	11.03	3.25	7.12	3.00
23	13.62	17.16	2.60	10.90	3.09	6.99	2.85
24	13.80	17.27	2.47	10.77	2.94	6.86	2.71
25	13.98	17.38	2.34	10.65	2.80	6.74	2.58
26	14.15	17.49	2.22	10.53	2.66	6.62	2.45
27	14.31	17.59	2.11	10.42	2.53	6.51	2.33
28	14.47	17.70	2.00	10.31	2.41	6.40	2.21
29	14.62	17.80	1.89	10.21	2.29	6.30	2.10
30	14.77	17.89	1.79	10.11	2.17	6.20	1.99
31	14.91	17.99	1.69	10.01	2.06	6.10	1.88
32	15.05	18.08	1.60	9.92	1.96	6.01	1.78
33	15.19	18.17	1.50	9.83	1.86	5.92	1.68
34	15.31	18.25	1.41	9.74	1.76	5.83	1.59
35	15.44	18.34	1.33	9.66	1.66	5.74	1.50
36	15.56	18.42	1.24	9.58	1.57	5.66	1.41
37	15.68	18.50	1.16	9.50	1.48	5.58	1.33
38	15.80	18.58	1.08	9.42	1.39	5.50	1.24
39	15.91	18.66	1.01	9.34	1.31	5.43	1.16
40	16.02	18.74	.93	9.27	1.23	5.36	1.08
41	16.13	18.81	.86	9.20	1.15	5.28	1.01
42	16.23	18.89	.79	9.13	1.07	5.21	.93
43	16.33	18.96	.72	9.06	1.00	5.15	.86
44	16.43	19.03	.65	8.99	.92	5.08	.79
45	16.53	19.10	.59	8.93	.85	5.01	.72

Appendix: Radar Detection Tables

THRESHOLD/NOISE AND REQUIRED SIGNAL/NOISE
==

PROBABILITY OF DETECTION= .9500
CONSTANT FALSE ALARM RATE
LOG10(PFA/N) = - 7

N	N DB	THRESH DB	NON-FLUC DB	SWER 1 DB	SWER 2 DB	SWER 3 DB	SWER 4 DB
1	0.00	12.07	14.19	24.95	24.95	19.74	19.74
2	3.01	12.65	11.46	22.28	17.05	17.05	14.43
3	4.77	13.11	9.93	20.79	13.81	15.55	12.00
4	6.02	13.50	8.87	19.75	11.88	14.51	10.47
5	6.99	13.85	8.07	18.97	10.54	13.73	9.38
6	7.78	14.16	7.42	18.34	9.53	13.10	8.54
7	8.45	14.44	6.89	17.82	8.72	12.57	7.86
8	9.03	14.69	6.43	17.37	8.06	12.12	7.29
9	9.54	14.93	6.03	16.98	7.50	11.73	6.80
10	10.00	15.15	5.67	16.64	7.02	11.39	6.38
11	10.41	15.35	5.35	16.33	6.59	11.08	6.00
12	10.79	15.55	5.07	16.05	6.21	10.80	5.67
13	11.14	15.73	4.80	15.80	5.87	10.54	5.36
14	11.46	15.90	4.56	15.57	5.56	10.31	5.08
15	11.76	16.07	4.34	15.35	5.28	10.09	4.83
16	12.04	16.22	4.13	15.15	5.02	9.89	4.59
17	12.30	16.37	3.94	14.96	4.78	9.70	4.37
18	12.55	16.52	3.75	14.79	4.55	9.52	4.17
19	12.79	16.65	3.58	14.62	4.34	9.36	3.98
20	13.01	16.79	3.42	14.46	4.15	9.20	3.80
21	13.22	16.91	3.27	14.31	3.97	9.05	3.63
22	13.42	17.04	3.12	14.17	3.79	8.91	3.47
23	13.62	17.16	2.98	14.04	3.63	8.77	3.32
24	13.80	17.27	2.85	13.91	3.47	8.65	3.17
25	13.98	17.38	2.72	13.79	3.32	8.52	3.03
26	14.15	17.49	2.60	13.67	3.18	8.41	2.90
27	14.31	17.59	2.49	13.56	3.04	8.29	2.78
28	14.47	17.70	2.38	13.45	2.91	8.18	2.65
29	14.62	17.80	2.27	13.35	2.79	8.08	2.54
30	14.77	17.89	2.17	13.25	2.67	7.98	2.43
31	14.91	17.99	2.07	13.15	2.56	7.88	2.32
32	15.05	18.08	1.97	13.06	2.45	7.79	2.22
33	15.19	18.17	1.88	12.97	2.34	7.70	2.12
34	15.31	18.25	1.79	12.88	2.24	7.61	2.02
35	15.44	18.34	1.70	12.80	2.14	7.53	1.93
36	15.56	18.42	1.62	12.72	2.04	7.44	1.84
37	15.68	18.50	1.53	12.64	1.95	7.36	1.75
38	15.80	18.58	1.45	12.56	1.86	7.29	1.66
39	15.91	18.66	1.38	12.48	1.78	7.21	1.58
40	16.02	18.74	1.30	12.41	1.69	7.14	1.50
41	16.13	18.81	1.23	12.34	1.61	7.07	1.42
42	16.23	18.89	1.16	12.27	1.53	7.00	1.35
43	16.33	18.96	1.09	12.20	1.45	6.93	1.27
44	16.43	19.03	1.02	12.14	1.38	6.86	1.20
45	16.53	19.10	.95	12.07	1.30	6.80	1.13

THRESHOLD/NOISE AND REQUIRED SIGNAL/NOISE

PROBABILITY OF DETECTION= .9900
CONSTANT FALSE ALARM RATE
LOG10(PFA/N)=- 7

N	N DB	THRESH DB	NON-FLUC DB	SWER 1 DB	SWER 2 DB	SWER 3 DB	SWER 4 DB
1	0.00	12.07	14.97	32.05	32.05	23.56	23.56
2	3.01	12.65	12.23	29.36	20.89	20.89	16.75
3	4.77	13.11	10.68	27.85	16.63	19.38	13.82
4	6.02	13.50	9.62	26.82	14.19	18.34	12.03
5	6.99	13.85	8.80	26.06	12.54	17.58	10.79
6	7.78	14.16	8.15	25.43	11.34	16.95	9.83
7	8.45	14.44	7.60	24.91	10.39	16.42	9.08
8	9.03	14.69	7.14	24.47	9.61	15.97	8.44
9	9.54	14.93	6.74	24.08	8.95	15.58	7.91
10	10.00	15.15	6.38	23.73	8.40	15.23	7.44
11	10.41	15.35	6.05	23.43	7.92	14.92	7.03
12	10.79	15.55	5.76	23.15	7.48	14.64	6.67
13	11.14	15.73	5.50	22.89	7.11	14.39	6.34
14	11.46	15.90	5.25	22.66	6.76	14.15	6.04
15	11.76	16.07	5.03	22.44	6.44	13.93	5.76
16	12.04	16.22	4.81	22.24	6.15	13.73	5.51
17	12.30	16.37	4.62	22.05	5.89	13.54	5.27
18	12.55	16.52	4.43	21.88	5.64	13.37	5.06
19	12.79	16.65	4.26	21.71	5.41	13.20	4.86
20	13.01	16.79	4.10	21.55	5.19	13.04	4.67
21	13.22	16.91	3.94	21.41	4.99	12.89	4.49
22	13.42	17.04	3.79	21.27	4.80	12.75	4.32
23	13.62	17.16	3.65	21.13	4.62	12.62	4.15
24	13.80	17.27	3.52	21.00	4.45	12.49	4.00
25	13.98	17.38	3.39	20.88	4.29	12.37	3.85
26	14.15	17.49	3.27	20.76	4.14	12.25	3.72
27	14.31	17.59	3.15	20.65	3.99	12.14	3.58
28	14.47	17.70	3.04	20.54	3.85	12.03	3.46
29	14.62	17.80	2.93	20.44	3.72	11.92	3.34
30	14.77	17.89	2.83	20.34	3.59	11.82	3.22
31	14.91	17.99	2.72	20.24	3.46	11.73	3.11
32	15.05	18.08	2.63	20.15	3.34	11.63	3.00
33	15.19	18.17	2.53	20.06	3.23	11.54	2.89
34	15.31	18.25	2.44	19.95	3.12	11.46	2.79
35	15.44	18.34	2.35	19.86	3.02	11.35	2.69
36	15.56	18.42	2.27	19.78	2.91	11.27	2.60
37	15.68	18.50	2.18	19.70	2.81	11.19	2.51
38	15.80	18.58	2.10	19.62	2.72	11.11	2.42
39	15.91	18.66	2.02	19.55	2.63	11.03	2.33
40	16.02	18.74	1.95	19.48	2.54	10.96	2.25
41	16.13	18.81	1.87	19.41	2.45	10.89	2.16
42	16.23	18.89	1.80	19.34	2.36	10.82	2.09
43	16.33	18.96	1.73	19.27	2.28	10.75	2.01
44	16.43	19.03	1.66	19.20	2.20	10.68	1.94
45	16.53	19.10	1.60	19.14	2.12	10.62	1.86

Appendix: Radar Detection Tables

THRESHOLD/NOISE AND REQUIRED SIGNAL/NOISE

PROBABILITY OF DETECTION= .5000
CONSTANT FALSE ALARM RATE
LOG10(PFA/N)=- 8

N	N DB	THRESH DB	NON-FLUC DB	SWER 1 DB	SWER 2 DB	SWER 3 DB	SWER 4 DB
1	0.00	12.65	12.53	14.08	14.08	13.26	13.26
2	3.01	13.18	9.85	11.39	10.57	10.57	10.19
3	4.77	13.61	8.34	9.89	8.81	9.06	8.56
4	6.02	13.98	7.31	8.85	7.65	8.03	7.46
5	6.99	14.31	6.52	8.07	6.79	7.24	6.65
6	7.78	14.60	5.89	7.44	6.11	6.62	6.00
7	8.45	14.86	5.37	6.92	5.56	6.09	5.46
8	9.03	15.10	4.93	6.47	5.08	5.65	5.00
9	9.54	15.33	4.54	6.08	4.68	5.26	4.60
10	10.00	15.54	4.19	5.74	4.32	4.91	4.25
11	10.41	15.73	3.89	5.43	4.00	4.61	3.94
12	10.79	15.92	3.61	5.15	3.71	4.33	3.65
13	11.14	16.09	3.35	4.90	3.44	4.07	3.40
14	11.46	16.26	3.12	4.66	3.20	3.84	3.16
15	11.76	16.42	2.90	4.45	2.98	3.62	2.94
16	12.04	16.57	2.70	4.25	2.78	3.42	2.74
17	12.30	16.71	2.52	4.06	2.58	3.23	2.55
18	12.55	16.85	2.34	3.88	2.40	3.06	2.37
19	12.79	16.98	2.17	3.72	2.23	2.89	2.20
20	13.01	17.11	2.02	3.56	2.07	2.74	2.04
21	13.22	17.23	1.87	3.41	1.92	2.59	1.89
22	13.42	17.35	1.73	3.27	1.78	2.45	1.75
23	13.62	17.46	1.60	3.14	1.64	2.31	1.62
24	13.80	17.57	1.47	3.01	1.51	2.19	1.49
25	13.98	17.68	1.35	2.89	1.39	2.06	1.37
26	14.15	17.79	1.23	2.77	1.27	1.95	1.25
27	14.31	17.89	1.12	2.66	1.16	1.84	1.14
28	14.47	17.99	1.01	2.55	1.05	1.73	1.03
29	14.62	18.08	.91	2.45	.94	1.62	.92
30	14.77	18.17	.81	2.35	.84	1.53	.82
31	14.91	18.27	.71	2.25	.75	1.43	.73
32	15.05	18.35	.62	2.16	.65	1.34	.63
33	15.19	18.44	.53	2.07	.56	1.25	.54
34	15.31	18.53	.44	1.98	.47	1.16	.46
35	15.44	18.61	.36	1.90	.39	1.08	.37
36	15.56	18.69	.28	1.82	.31	.99	.29
37	15.68	18.77	.20	1.74	.23	.91	.21
38	15.80	18.84	.12	1.66	.15	.84	.13
39	15.91	18.92	.05	1.59	.07	.76	.06
40	16.02	18.99	-.03	1.51	-.00	.69	-.01
41	16.13	19.07	-.10	1.44	-.07	.62	-.09
42	16.23	19.14	-.17	1.37	-.14	.55	-.15
43	16.33	19.21	-.23	1.31	-.21	.48	-.22
44	16.43	19.28	-.30	1.24	-.28	.42	-.29
45	16.53	19.34	-.36	1.18	-.34	.35	-.35

THRESHOLD/NOISE AND REQUIRED SIGNAL/NOISE
==

PROBABILITY OF DETECTION= .6000
CONSTANT FALSE ALARM RATE
LOG10(PFA/N)=- 8

N	N DB	THRESH DB	NON-FLUC DB	SWER 1 DB	SWER 2 DB	SWER 3 DB	SWER 4 DB
1	0.00	12.65	12.90	15.45	15.45	14.19	14.19
2	3.01	13.18	10.20	12.76	11.50	11.50	10.87
3	4.77	13.61	8.69	11.25	9.57	9.99	9.14
4	6.02	13.98	7.65	10.22	8.32	8.95	7.99
5	6.99	14.31	6.86	9.43	7.40	8.17	7.14
6	7.78	14.60	6.23	8.80	6.69	7.54	6.46
7	8.45	14.86	5.70	8.28	6.10	7.01	5.90
8	9.03	15.10	5.25	7.83	5.60	6.57	5.43
9	9.54	15.33	4.86	7.44	5.18	6.18	5.02
10	10.00	15.54	4.52	7.10	4.80	5.83	4.66
11	10.41	15.73	4.21	6.79	4.47	5.52	4.34
12	10.79	15.92	3.93	6.51	4.16	5.25	4.05
13	11.14	16.09	3.67	6.26	3.89	4.99	3.78
14	11.46	16.26	3.44	6.02	3.64	4.76	3.54
15	11.76	16.42	3.22	5.81	3.41	4.54	3.32
16	12.04	16.57	3.02	5.61	3.20	4.34	3.11
17	12.30	16.71	2.83	5.42	3.00	4.15	2.91
18	12.55	16.85	2.65	5.24	2.81	3.97	2.73
19	12.79	16.98	2.48	5.07	2.64	3.81	2.56
20	13.01	17.11	2.33	4.92	2.47	3.65	2.40
21	13.22	17.23	2.18	4.77	2.32	3.50	2.25
22	13.42	17.35	2.04	4.63	2.17	3.36	2.10
23	13.62	17.46	1.90	4.50	2.03	3.23	1.97
24	13.80	17.57	1.77	4.37	1.90	3.10	1.84
25	13.98	17.68	1.65	4.25	1.77	2.98	1.71
26	14.15	17.79	1.53	4.13	1.65	2.86	1.59
27	14.31	17.89	1.42	4.02	1.53	2.75	1.48
28	14.47	17.99	1.31	3.91	1.42	2.64	1.37
29	14.62	18.08	1.21	3.81	1.31	2.54	1.26
30	14.77	18.17	1.11	3.71	1.21	2.44	1.16
31	14.91	18.27	1.01	3.61	1.11	2.34	1.06
32	15.05	18.35	.92	3.52	1.01	2.25	.97
33	15.19	18.44	.83	3.43	.92	2.16	.87
34	15.31	18.53	.74	3.34	.83	2.07	.78
35	15.44	18.61	.66	3.25	.74	1.99	.70
36	15.56	18.69	.58	3.17	.66	1.90	.62
37	15.68	18.77	.50	3.09	.57	1.82	.54
38	15.80	18.84	.42	3.02	.50	1.75	.46
39	15.91	18.92	.34	2.94	.42	1.67	.38
40	16.02	18.99	.27	2.87	.34	1.60	.31
41	16.13	19.07	.20	2.80	.27	1.53	.23
42	16.23	19.14	.13	2.73	.20	1.46	.16
43	16.33	19.21	.06	2.66	.13	1.39	.10
44	16.43	19.28	-.00	2.60	.06	1.33	.03
45	16.53	19.34	-.07	2.53	-.00	1.26	-.04

Appendix: Radar Detection Tables

THRESHOLD/NOISE AND REQUIRED SIGNAL/NOISE
===

PROBABILITY OF DETECTION= .7000
CONSTANT FALSE ALARM RATE
LOG10(PFA/N)=- 8

N	N DB	THRESH DB	NON-FLUC DB	SWER 1 DB	SWER 2 DB	SWER 3 DB	SWER 4 DB
1	0.00	12.65	13.27	17.05	17.05	15.24	15.24
2	3.01	13.18	10.56	14.35	12.54	12.54	11.61
3	4.77	13.61	9.04	12.85	10.42	11.03	9.77
4	6.02	13.98	8.00	11.81	9.06	10.00	8.56
5	6.99	14.31	7.21	11.02	8.08	9.21	7.66
6	7.78	14.60	6.57	10.39	7.31	8.58	6.96
7	8.45	14.86	6.04	9.87	6.69	8.05	6.38
8	9.03	15.10	5.59	9.42	6.16	7.61	5.89
9	9.54	15.33	5.20	9.03	5.71	7.22	5.47
10	10.00	15.54	4.85	8.69	5.32	6.87	5.09
11	10.41	15.73	4.54	8.38	4.97	6.56	4.76
12	10.79	15.92	4.26	8.10	4.65	6.28	4.46
13	11.14	16.09	4.00	7.85	4.37	6.03	4.19
14	11.46	16.26	3.76	7.61	4.10	5.79	3.94
15	11.76	16.42	3.54	7.40	3.87	5.57	3.71
16	12.04	16.57	3.34	7.19	3.64	5.37	3.50
17	12.30	16.71	3.15	7.01	3.44	5.18	3.30
18	12.55	16.85	2.97	6.83	3.24	5.01	3.11
19	12.79	16.98	2.80	6.66	3.06	4.84	2.94
20	13.01	17.11	2.64	6.51	2.89	4.68	2.77
21	13.22	17.23	2.49	6.36	2.73	4.53	2.62
22	13.42	17.35	2.35	6.22	2.58	4.39	2.47
23	13.62	17.46	2.22	6.08	2.43	4.26	2.33
24	13.80	17.57	2.09	5.96	2.30	4.13	2.19
25	13.98	17.68	1.96	5.83	2.17	4.01	2.07
26	14.15	17.79	1.85	5.72	2.04	3.89	1.95
27	14.31	17.89	1.73	5.60	1.92	3.78	1.83
28	14.47	17.99	1.62	5.50	1.81	3.67	1.72
29	14.62	18.08	1.52	5.39	1.69	3.57	1.61
30	14.77	18.17	1.42	5.29	1.59	3.47	1.50
31	14.91	18.27	1.32	5.20	1.49	3.37	1.40
32	15.05	18.35	1.23	5.10	1.39	3.28	1.31
33	15.19	18.44	1.14	5.01	1.29	3.19	1.22
34	15.31	18.53	1.05	4.93	1.20	3.10	1.13
35	15.44	18.61	.96	4.84	1.11	3.01	1.04
36	15.56	18.69	.88	4.76	1.02	2.93	.95
37	15.68	18.77	.80	4.68	.94	2.85	.87
38	15.80	18.84	.72	4.60	.86	2.78	.79
39	15.91	18.92	.65	4.53	.78	2.70	.71
40	16.02	18.99	.57	4.45	.70	2.63	.64
41	16.13	19.07	.50	4.38	.63	2.56	.57
42	16.23	19.14	.43	4.31	.56	2.49	.50
43	16.33	19.21	.36	4.25	.49	2.42	.43
44	16.43	19.28	.30	4.18	.42	2.35	.36
45	16.53	19.34	.23	4.12	.35	2.29	.29

THRESHOLD/NOISE AND REQUIRED SIGNAL/NOISE
===

PROBABILITY OF DETECTION= .8000
CONSTANT FALSE ALARM RATE
LOG10(PFA/N)=- 8

N	N DB	THRESH DB	NON-FLUC DB	SWER 1 DB	SWER 2 DB	SWER 3 DB	SWER 4 DB
1	0.00	12.65	13.68	19.11	19.11	16.54	16.54
2	3.01	13.18	10.97	16.42	13.85	13.85	12.51
3	4.77	13.61	9.44	14.91	11.45	12.33	10.52
4	6.02	13.98	8.39	13.88	9.95	11.29	9.23
5	6.99	14.31	7.59	13.09	8.88	10.51	8.28
6	7.78	14.60	6.95	12.46	8.05	9.87	7.54
7	8.45	14.86	6.42	11.93	7.38	9.35	6.93
8	9.03	15.10	5.97	11.49	6.82	8.90	6.42
9	9.54	15.33	5.57	11.10	6.34	8.51	5.98
10	10.00	15.54	5.22	10.75	5.92	8.16	5.59
11	10.41	15.73	4.91	10.44	5.55	7.85	5.25
12	10.79	15.92	4.62	10.16	5.22	7.57	4.94
13	11.14	16.09	4.36	9.91	4.92	7.32	4.65
14	11.46	16.26	4.12	9.67	4.65	7.08	4.40
15	11.76	16.42	3.90	9.46	4.39	6.86	4.16
16	12.04	16.57	3.70	9.25	4.16	6.66	3.94
17	12.30	16.71	3.51	9.07	3.95	6.47	3.73
18	12.55	16.85	3.33	8.89	3.74	6.30	3.54
19	12.79	16.98	3.16	8.72	3.56	6.13	3.36
20	13.01	17.11	3.00	8.57	3.38	5.97	3.20
21	13.22	17.23	2.85	8.42	3.21	5.82	3.04
22	13.42	17.35	2.70	8.28	3.05	5.68	2.88
23	13.62	17.46	2.57	8.14	2.90	5.55	2.74
24	13.80	17.57	2.44	8.01	2.76	5.42	2.60
25	13.98	17.68	2.31	7.89	2.62	5.30	2.47
26	14.15	17.79	2.19	7.78	2.49	5.18	2.35
27	14.31	17.89	2.08	7.66	2.37	5.07	2.23
28	14.47	17.99	1.97	7.56	2.25	4.96	2.11
29	14.62	18.08	1.86	7.45	2.14	4.85	2.00
30	14.77	18.17	1.76	7.35	2.03	4.75	1.90
31	14.91	18.27	1.67	7.25	1.92	4.66	1.80
32	15.05	18.35	1.57	7.16	1.82	4.56	1.70
33	15.19	18.44	1.48	7.07	1.72	4.47	1.60
34	15.31	18.53	1.39	6.98	1.62	4.38	1.51
35	15.44	18.61	1.30	6.90	1.53	4.30	1.42
36	15.56	18.69	1.22	6.82	1.44	4.22	1.34
37	15.68	18.77	1.14	6.74	1.36	4.14	1.25
38	15.80	18.84	1.06	6.66	1.27	4.06	1.17
39	15.91	18.92	.99	6.59	1.19	3.99	1.09
40	16.02	18.99	.91	6.51	1.11	3.91	1.02
41	16.13	19.07	.84	6.44	1.04	3.84	.94
42	16.23	19.14	.77	6.37	.96	3.77	.87
43	16.33	19.21	.70	6.31	.89	3.70	.80
44	16.43	19.28	.63	6.24	.82	3.64	.73
45	16.53	19.34	.57	6.18	.75	3.57	.66

Appendix: Radar Detection Tables   201

```
            THRESHOLD/NOISE AND REQUIRED SIGNAL/NOISE
            ============================================

PROBABILITY OF DETECTION= .9000
CONSTANT FALSE ALARM RATE
LOG10(PFA/N)=- 8
```

N	N DB	THRESH DB	NON-FLUC DB	SWER 1 DB	SWER 2 DB	SWER 3 DB	SWER 4 DB
1	0.00	12.65	14.23	22.40	22.40	18.51	18.51
2	3.01	13.18	11.50	19.71	15.81	15.81	13.82
3	4.77	13.61	9.97	18.20	12.98	14.30	11.59
4	6.02	13.98	8.91	17.16	11.25	13.26	10.17
5	6.99	14.31	8.10	16.37	10.04	12.47	9.14
6	7.78	14.60	7.46	15.75	9.11	11.84	8.34
7	8.45	14.86	6.92	15.22	8.37	11.31	7.69
8	9.03	15.10	6.46	14.77	7.75	10.86	7.15
9	9.54	15.33	6.06	14.38	7.22	10.47	6.68
10	10.00	15.54	5.71	14.04	6.77	10.12	6.27
11	10.41	15.73	5.39	13.73	6.37	9.81	5.91
12	10.79	15.92	5.10	13.45	6.01	9.53	5.58
13	11.14	16.09	4.84	13.19	5.69	9.27	5.29
14	11.46	16.26	4.60	12.96	5.39	9.04	5.02
15	11.76	16.42	4.38	12.74	5.12	8.82	4.77
16	12.04	16.57	4.17	12.54	4.88	8.62	4.54
17	12.30	16.71	3.98	12.35	4.65	8.43	4.33
18	12.55	16.85	3.80	12.17	4.43	8.25	4.13
19	12.79	16.98	3.63	12.01	4.23	8.08	3.94
20	13.01	17.11	3.46	11.85	4.04	7.93	3.77
21	13.22	17.23	3.31	11.70	3.87	7.78	3.60
22	13.42	17.35	3.17	11.56	3.70	7.63	3.44
23	13.62	17.46	3.03	11.43	3.54	7.50	3.29
24	13.80	17.57	2.90	11.30	3.39	7.37	3.15
25	13.98	17.68	2.77	11.17	3.25	7.25	3.02
26	14.15	17.79	2.65	11.05	3.11	7.13	2.89
27	14.31	17.89	2.54	10.94	2.98	7.02	2.77
28	14.47	17.99	2.43	10.83	2.85	6.91	2.65
29	14.62	18.08	2.32	10.73	2.73	6.80	2.53
30	14.77	18.17	2.22	10.63	2.62	6.70	2.42
31	14.91	18.27	2.12	10.53	2.51	6.61	2.32
32	15.05	18.35	2.02	10.44	2.40	6.51	2.22
33	15.19	18.44	1.93	10.35	2.30	6.42	2.12
34	15.31	18.53	1.84	10.26	2.20	6.33	2.03
35	15.44	18.61	1.75	10.18	2.10	6.25	1.93
36	15.56	18.69	1.67	10.10	2.01	6.17	1.85
37	15.68	18.77	1.59	10.02	1.92	6.09	1.76
38	15.80	18.84	1.51	9.94	1.83	6.01	1.68
39	15.91	18.92	1.43	9.86	1.75	5.93	1.60
40	16.02	18.99	1.36	9.79	1.67	5.86	1.52
41	16.13	19.07	1.28	9.72	1.59	5.79	1.44
42	16.23	19.14	1.21	9.65	1.51	5.72	1.37
43	16.33	19.21	1.14	9.58	1.43	5.65	1.29
44	16.43	19.28	1.08	9.52	1.36	5.59	1.22
45	16.53	19.34	1.01	9.45	1.29	5.52	1.15

THRESHOLD/NOISE AND REQUIRED SIGNAL/NOISE
==

PROBABILITY OF DETECTION= .9500
CONSTANT FALSE ALARM RATE
LOG10(PFA/N)=- 8

N	N DB	THRESH DB	NON-FLUC DB	SWER 1 DB	SWER 2 DB	SWER 3 DB	SWER 4 DB
1	0.00	12.65	14.65	25.54	25.54	20.31	20.31
2	3.01	13.18	11.92	22.85	17.60	17.60	14.96
3	4.77	13.61	10.37	21.34	14.33	16.09	12.50
4	6.02	13.98	9.31	20.31	12.39	15.04	10.96
5	6.99	14.31	8.50	19.52	11.03	14.25	9.86
6	7.78	14.60	7.85	18.89	10.01	13.62	9.01
7	8.45	14.86	7.31	18.36	9.20	13.09	8.32
8	9.03	15.10	6.85	17.91	8.54	12.64	7.75
9	9.54	15.33	6.45	17.52	7.97	12.25	7.26
10	10.00	15.54	6.09	17.18	7.48	11.90	6.83
11	10.41	15.73	5.77	16.87	7.05	11.59	6.45
12	10.79	15.92	5.48	16.59	6.67	11.31	6.11
13	11.14	16.09	5.22	16.33	6.32	11.05	5.80
14	11.46	16.26	4.97	16.10	6.01	10.82	5.52
15	11.76	16.42	4.75	15.88	5.73	10.60	5.26
16	12.04	16.57	4.54	15.68	5.46	10.40	5.03
17	12.30	16.71	4.35	15.49	5.22	10.21	4.81
18	12.55	16.85	4.16	15.31	5.00	10.03	4.60
19	12.79	16.98	3.99	15.15	4.79	9.86	4.41
20	13.01	17.11	3.83	14.99	4.59	9.70	4.23
21	13.22	17.23	3.67	14.84	4.40	9.56	4.05
22	13.42	17.35	3.53	14.70	4.23	9.41	3.89
23	13.62	17.46	3.39	14.56	4.06	9.28	3.74
24	13.80	17.57	3.26	14.44	3.90	9.15	3.59
25	13.98	17.68	3.13	14.31	3.76	9.03	3.45
26	14.15	17.79	3.01	14.20	3.61	8.91	3.32
27	14.31	17.89	2.89	14.08	3.48	8.80	3.20
28	14.47	17.99	2.78	13.98	3.34	8.69	3.07
29	14.62	18.08	2.67	13.87	3.22	8.58	2.96
30	14.77	18.17	2.57	13.77	3.10	8.48	2.84
31	14.91	18.27	2.47	13.68	2.99	8.39	2.74
32	15.05	18.35	2.37	13.58	2.87	8.29	2.63
33	15.19	18.44	2.28	13.49	2.77	8.20	2.53
34	15.31	18.53	2.19	13.41	2.66	8.11	2.44
35	15.44	18.61	2.10	13.32	2.57	8.03	2.34
36	15.56	18.69	2.02	13.24	2.47	7.95	2.25
37	15.68	18.77	1.94	13.16	2.38	7.87	2.16
38	15.80	18.84	1.86	13.08	2.29	7.79	2.08
39	15.91	18.92	1.78	13.00	2.20	7.71	2.00
40	16.02	18.99	1.70	12.93	2.11	7.64	1.91
41	16.13	19.07	1.63	12.85	2.03	7.57	1.84
42	16.23	19.14	1.56	12.79	1.95	7.50	1.76
43	16.33	19.21	1.49	12.72	1.87	7.43	1.69
44	16.43	19.28	1.42	12.65	1.80	7.36	1.61
45	16.53	19.34	1.35	12.59	1.72	7.30	1.54

Appendix: Radar Detection Tables

THRESHOLD/NOISE AND REQUIRED SIGNAL/NOISE

```
PROBABILITY OF DETECTION= .9900
CONSTANT FALSE ALARM RATE
LOG10(PFA/N)=- 8
```

N	N DB	THRESH DB	NON-FLUC DB	SWER 1 DB	SWER 2 DB	SWER 3 DB	SWER 4 DB
1	0.00	12.65	15.39	32.66	32.66	24.12	24.12
2	3.01	13.18	12.65	29.93	21.43	21.43	17.27
3	4.77	13.61	11.09	28.41	17.14	19.92	14.30
4	6.02	13.98	10.02	27.37	14.68	18.87	12.50
5	6.99	14.31	9.20	26.58	13.03	18.08	11.24
6	7.78	14.60	8.54	25.95	11.81	17.44	10.28
7	8.45	14.86	8.00	25.43	10.85	16.94	9.51
8	9.03	15.10	7.53	25.00	10.06	16.49	8.88
9	9.54	15.33	7.12	24.61	9.40	16.10	8.34
10	10.00	15.54	6.76	24.27	8.85	15.75	7.87
11	10.41	15.73	6.44	23.96	8.36	15.44	7.45
12	10.79	15.92	6.14	23.68	7.92	15.15	7.09
13	11.14	16.09	5.88	23.42	7.53	14.90	6.75
14	11.46	16.26	5.63	23.19	7.18	14.66	6.45
15	11.76	16.42	5.40	22.97	6.86	14.44	6.17
16	12.04	16.57	5.19	22.77	6.57	14.24	5.91
17	12.30	16.71	4.99	22.58	6.31	14.05	5.68
18	12.55	16.85	4.81	22.40	6.06	13.87	5.46
19	12.79	16.98	4.63	22.21	5.83	13.71	5.26
20	13.01	17.11	4.47	22.05	5.61	13.55	5.06
21	13.22	17.23	4.31	21.90	5.40	13.40	4.88
22	13.42	17.35	4.16	21.76	5.21	13.26	4.71
23	13.62	17.46	4.02	21.63	5.03	13.12	4.55
24	13.80	17.57	3.89	21.50	4.86	12.97	4.4
25	13.98	17.68	3.76	21.38	4.70	12.85	4.25
26	14.15	17.79	3.64	21.26	4.54	12.73	4.11
27	14.31	17.89	3.52	21.15	4.40	12.61	3.98
28	14.47	17.99	3.41	21.04	4.26	12.51	3.85
29	14.62	18.08	3.30	20.94	4.12	12.40	3.72
30	14.77	18.17	3.19	20.84	3.99	12.30	3.61
31	14.91	18.27	3.09	20.74	3.86	12.20	3.49
32	15.05	18.35	2.99	20.65	3.74	12.11	3.38
33	15.19	18.44	2.90	20.56	3.63	12.02	3.28
34	15.31	18.53	2.81	20.47	3.52	11.93	3.18
35	15.44	18.61	2.72	20.39	3.41	11.85	3.08
36	15.56	18.69	2.63	20.30	3.31	11.76	2.98
37	15.68	18.77	2.55	20.22	3.21	11.68	2.89
38	15.80	18.84	2.47	20.15	3.11	11.61	2.80
39	15.91	18.92	2.39	20.07	3.02	11.53	2.72
40	16.02	18.99	2.31	20.00	2.93	11.46	2.63
41	16.13	19.07	2.24	19.93	2.84	11.39	2.55
42	16.23	19.14	2.17	19.86	2.76	11.32	2.47
43	16.33	19.21	2.10	19.79	2.67	11.25	2.39
44	16.43	19.28	2.03	19.73	2.59	11.18	2.32
45	16.53	19.34	1.96	19.66	2.51	11.12	2.25

THRESHOLD/NOISE AND REQUIRED SIGNAL/NOISE
==

PROBABILITY OF DETECTION= .5000
CONSTANT FALSE ALARM RATE
LOG10(PFA/N)=- 9

N	N DB	THRESH DB	NON-FLUC DB	SWER 1 DB	SWER 2 DB	SWER 3 DB	SWER 4 DB
1	0.00	13.16	13.06	14.61	14.61	13.78	13.78
2	3.01	13.66	10.36	11.91	11.08	11.08	10.70
3	4.77	14.06	8.84	10.39	9.31	9.57	9.06
4	6.02	14.41	7.80	9.35	8.14	8.53	7.96
5	6.99	14.72	7.01	8.56	7.28	7.74	7.14
6	7.78	14.99	6.38	7.93	6.60	7.10	6.48
7	8.45	15.24	5.85	7.40	6.04	6.58	5.94
8	9.03	15.47	5.40	6.95	5.56	6.13	5.48
9	9.54	15.69	5.01	6.56	5.15	5.74	5.08
10	10.00	15.89	4.67	6.22	4.79	5.39	4.72
11	10.41	16.08	4.36	5.91	4.47	5.08	4.41
12	10.79	16.25	4.07	5.63	4.18	4.80	4.12
13	11.14	16.42	3.82	5.37	3.91	4.54	3.86
14	11.46	16.58	3.58	5.13	3.67	4.31	3.62
15	11.76	16.73	3.37	4.92	3.45	4.09	3.40
16	12.04	16.87	3.16	4.71	3.24	3.89	3.20
17	12.30	17.01	2.98	4.53	3.04	3.70	3.01
18	12.55	17.15	2.80	4.35	2.86	3.52	2.83
19	12.79	17.27	2.63	4.18	2.69	3.36	2.66
20	13.01	17.40	2.48	4.03	2.53	3.20	2.50
21	13.22	17.51	2.33	3.88	2.38	3.05	2.35
22	13.42	17.63	2.19	3.74	2.24	2.91	2.21
23	13.62	17.74	2.05	3.60	2.10	2.78	2.07
24	13.80	17.85	1.92	3.47	1.97	2.65	1.95
25	13.98	17.95	1.80	3.35	1.85	2.53	1.82
26	14.15	18.05	1.68	3.23	1.73	2.41	1.7
27	14.31	18.15	1.57	3.12	1.61	2.30	1.59
28	14.47	18.25	1.46	3.01	1.50	2.19	1.48
29	14.62	18.34	1.36	2.91	1.40	2.08	1.38
30	14.77	18.43	1.26	2.81	1.30	1.98	1.28
31	14.91	18.52	1.16	2.71	1.20	1.89	1.19
32	15.05	18.60	1.07	2.62	1.10	1.79	1.09
33	15.19	18.69	.98	2.53	1.01	1.70	1.00
34	15.31	18.77	.89	2.44	.92	1.62	.91
35	15.44	18.85	.81	2.36	.84	1.53	.82
36	15.56	18.93	.73	2.28	.76	1.45	.74
37	15.68	19.00	.65	2.20	.68	1.37	.66
38	15.80	19.08	.57	2.12	.60	1.29	.58
39	15.91	19.15	.50	2.04	.52	1.22	.51
40	16.02	19.22	.42	1.97	.45	1.15	.44
41	16.13	19.30	.35	1.90	.38	1.07	.36
42	16.23	19.36	.28	1.83	.31	1.01	.29
43	16.33	19.43	.22	1.76	.24	.94	.23
44	16.43	19.50	.15	1.70	.17	.87	.16
45	16.53	19.56	.09	1.63	.11	.81	.10

Appendix: Radar Detection Tables 205

```
       THRESHOLD/NOISE AND REQUIRED SIGNAL/NOISE
       ==============================================
PROBABILITY OF DETECTION= .6000
CONSTANT FALSE ALARM RATE
LOG10(PFA/N)=- 9
```

N	N DB	THRESH DB	NON-FLUC DB	SWER 1 DB	SWER 2 DB	SWER 3 DB	SWER 4 DB
1	0.00	13.16	13.40	15.97	15.97	14.71	14.71
2	3.01	13.66	10.69	13.27	12.01	12.01	11.37
3	4.77	14.06	9.17	11.75	10.07	10.49	9.63
4	6.02	14.41	8.12	10.71	8.81	9.44	8.48
5	6.99	14.72	7.33	9.92	7.88	8.65	7.61
6	7.78	14.99	6.69	9.29	7.16	8.02	6.93
7	8.45	15.24	6.16	8.76	6.57	7.49	6.37
8	9.03	15.47	5.71	8.31	6.07	7.04	5.89
9	9.54	15.69	5.32	7.92	5.64	6.65	5.48
10	10.00	15.89	4.97	7.57	5.26	6.30	5.12
11	10.41	16.08	4.66	7.26	4.92	5.99	4.79
12	10.79	16.25	4.38	6.98	4.62	5.71	4.50
13	11.14	16.42	4.12	6.72	4.34	5.45	4.23
14	11.46	16.58	3.88	6.49	4.09	5.22	3.99
15	11.76	16.73	3.66	6.27	3.86	5.00	3.76
16	12.04	16.87	3.46	6.07	3.65	4.80	3.55
17	12.30	17.01	3.27	5.88	3.45	4.61	3.36
18	12.55	17.15	3.09	5.70	3.26	4.43	3.18
19	12.79	17.27	2.92	5.54	3.08	4.27	3.00
20	13.01	17.40	2.77	5.38	2.92	4.11	2.84
21	13.22	17.51	2.62	5.23	2.76	3.96	2.69
22	13.42	17.63	2.47	5.09	2.61	3.82	2.54
23	13.62	17.74	2.34	4.95	2.47	3.68	2.41
24	13.80	17.85	2.21	4.83	2.34	3.55	2.27
25	13.98	17.95	2.09	4.70	2.21	3.43	2.15
26	14.15	18.05	1.97	4.59	2.09	3.31	2.03
27	14.31	18.15	1.86	4.47	1.97	3.20	1.91
28	14.47	18.25	1.75	4.37	1.86	3.09	1.80
29	14.62	18.34	1.64	4.26	1.75	2.99	1.70
30	14.77	18.43	1.54	4.16	1.64	2.89	1.59
31	14.91	18.52	1.45	4.06	1.54	2.79	1.49
32	15.05	18.60	1.35	3.97	1.45	2.70	1.40
33	15.19	18.69	1.26	3.88	1.35	2.61	1.31
34	15.31	18.77	1.17	3.79	1.26	2.52	1.22
35	15.44	18.85	1.09	3.71	1.18	2.44	1.13
36	15.56	18.93	1.01	3.63	1.09	2.35	1.05
37	15.68	19.00	.93	3.55	1.01	2.27	.97
38	15.80	19.08	.85	3.47	.93	2.20	.89
39	15.91	19.15	.77	3.40	.85	2.12	.81
40	16.02	19.22	.70	3.32	.78	2.05	.74
41	16.13	19.30	.63	3.25	.70	1.98	.67
42	16.23	19.36	.56	3.18	.63	1.91	.60
43	16.33	19.43	.49	3.12	.56	1.84	.53
44	16.43	19.50	.43	3.05	.50	1.78	.46
45	16.53	19.56	.36	2.99	.43	1.71	.40

THRESHOLD/NOISE AND REQUIRED SIGNAL/NOISE

PROBABILITY OF DETECTION= .7000
CONSTANT FALSE ALARM RATE
LOG10(PFA/N)=- 9

N	N DB	THRESH DB	NON-FLUC DB	SWER 1 DB	SWER 2 DB	SWER 3 DB	SWER 4 DB
1	0.00	13.16	13.75	17.57	17.57	15.75	15.75
2	3.01	13.66	11.03	14.86	13.04	13.04	12.11
3	4.77	14.06	9.51	13.34	10.91	11.53	10.25
4	6.02	14.41	8.45	12.30	9.54	10.48	9.03
5	6.99	14.72	7.66	11.51	8.55	9.69	8.13
6	7.78	14.99	7.02	10.88	7.77	9.05	7.42
7	8.45	15.24	6.48	10.35	7.14	8.52	6.83
8	9.03	15.47	6.03	9.90	6.62	8.07	6.34
9	9.54	15.69	5.63	9.51	6.16	7.68	5.91
10	10.00	15.89	5.28	9.16	5.76	7.33	5.53
11	10.41	16.08	4.97	8.85	5.41	7.02	5.21
12	10.79	16.25	4.68	8.57	5.09	6.74	4.90
13	11.14	16.42	4.43	8.31	4.81	6.48	4.62
14	11.46	16.58	4.19	8.07	4.54	6.25	4.37
15	11.76	16.73	3.97	7.86	4.30	6.03	4.14
16	12.04	16.87	3.76	7.65	4.08	5.83	3.93
17	12.30	17.01	3.57	7.47	3.87	5.64	3.73
18	12.55	17.15	3.39	7.29	3.68	5.46	3.54
19	12.79	17.27	3.22	7.12	3.49	5.29	3.36
20	13.01	17.40	3.06	6.96	3.32	5.13	3.20
21	13.22	17.51	2.91	6.81	3.16	4.98	3.04
22	13.42	17.63	2.77	6.67	3.01	4.84	2.89
23	13.62	17.74	2.63	6.54	2.86	4.71	2.75
24	13.80	17.85	2.50	6.41	2.72	4.58	2.62
25	13.98	17.95	2.38	6.29	2.59	4.46	2.49
26	14.15	18.05	2.26	6.17	2.47	4.34	2.37
27	14.31	18.15	2.15	6.06	2.34	4.23	2.25
28	14.47	18.25	2.04	5.95	2.23	4.12	2.14
29	14.62	18.34	1.93	5.84	2.12	4.01	2.03
30	14.77	18.43	1.83	5.74	2.01	3.91	1.92
31	14.91	18.52	1.73	5.65	1.91	3.82	1.82
32	15.05	18.60	1.64	5.55	1.81	3.72	1.73
33	15.19	18.69	1.55	5.46	1.71	3.63	1.63
34	15.31	18.77	1.46	5.38	1.62	3.54	1.54
35	15.44	18.85	1.38	5.29	1.53	3.46	1.45
36	15.56	18.93	1.29	5.21	1.44	3.38	1.37
37	15.68	19.00	1.21	5.13	1.36	3.30	1.29
38	15.80	19.08	1.13	5.05	1.28	3.22	1.21
39	15.91	19.15	1.06	4.98	1.20	3.15	1.13
40	16.02	19.22	.98	4.90	1.12	3.07	1.05
41	16.13	19.30	.91	4.83	1.05	3.00	.98
42	16.23	19.36	.84	4.76	.97	2.93	.91
43	16.33	19.43	.78	4.70	.90	2.86	.84
44	16.43	19.50	.71	4.63	.83	2.80	.77
45	16.53	19.56	.64	4.57	.77	2.73	.71

Appendix: Radar Detection Tables 207

```
THRESHOLD/NOISE AND REQUIRED SIGNAL/NOISE
=============================================
PROBABILITY OF DETECTION= .8000
CONSTANT FALSE ALARM RATE
LOG10(PFA/N)=- 9
```

N	N DB	THRESH DB	NON-FLUC DB	SWER 1 DB	SWER 2 DB	SWER 3 DB	SWER 4 DB
1	0.00	13.16	14.14	19.63	19.63	17.05	17.05
2	3.01	13.66	11.42	16.93	14.34	14.34	13.00
3	4.77	14.06	9.88	15.41	11.93	12.82	10.99
4	6.02	14.41	8.82	14.36	10.42	11.77	9.69
5	6.99	14.72	8.02	13.57	9.34	10.98	8.73
6	7.78	14.99	7.38	12.94	8.51	10.34	7.98
7	8.45	15.24	6.84	12.41	7.83	9.81	7.37
8	9.03	15.47	6.39	11.96	7.26	9.36	6.85
9	9.54	15.69	5.99	11.57	6.78	8.97	6.41
10	10.00	15.89	5.63	11.22	6.36	8.62	6.02
11	10.41	16.08	5.32	10.91	5.98	8.31	5.67
12	10.79	16.25	5.03	10.63	5.65	8.03	5.36
13	11.14	16.42	4.77	10.37	5.35	7.77	5.07
14	11.46	16.58	4.53	10.13	5.07	7.53	4.81
15	11.76	16.73	4.31	9.92	4.82	7.31	4.57
16	12.04	16.87	4.10	9.71	4.58	7.11	4.35
17	12.30	17.01	3.91	9.52	4.36	6.92	4.15
18	12.55	17.15	3.73	9.35	4.16	6.74	3.96
19	12.79	17.27	3.56	9.18	3.97	6.58	3.77
20	13.01	17.40	3.40	9.02	3.79	6.42	3.6
21	13.22	17.51	3.25	8.87	3.62	6.27	3.44
22	13.42	17.63	3.10	8.73	3.47	6.13	3.29
23	13.62	17.74	2.97	8.60	3.31	5.99	3.15
24	13.80	17.85	2.84	8.47	3.17	5.86	3.01
25	13.98	17.95	2.71	8.34	3.03	5.74	2.88
26	14.15	18.05	2.59	8.23	2.90	5.62	2.75
27	14.31	18.15	2.48	8.11	2.78	5.51	2.63
28	14.47	18.25	2.37	8.01	2.66	5.40	2.52
29	14.62	18.34	2.26	7.90	2.54	5.29	2.41
30	14.77	18.43	2.16	7.80	2.43	5.19	2.30
31	14.91	18.52	2.06	7.70	2.33	5.10	2.20
32	15.05	18.60	1.96	7.61	2.22	5.00	2.10
33	15.19	18.69	1.87	7.52	2.12	4.91	2.00
34	15.31	18.77	1.78	7.43	2.03	4.83	1.91
35	15.44	18.85	1.70	7.35	1.94	4.74	1.82
36	15.56	18.93	1.61	7.27	1.85	4.66	1.73
37	15.68	19.00	1.53	7.19	1.76	4.58	1.65
38	15.80	19.08	1.45	7.11	1.68	4.50	1.57
39	15.91	19.15	1.38	7.03	1.59	4.43	1.49
40	16.02	19.22	1.30	6.96	1.52	4.35	1.41
41	16.13	19.30	1.23	6.89	1.44	4.28	1.34
42	16.23	19.36	1.16	6.82	1.36	4.21	1.26
43	16.33	19.43	1.09	6.75	1.29	4.14	1.19
44	16.43	19.50	1.03	6.69	1.22	4.08	1.12
45	16.53	19.56	.96	6.62	1.15	4.01	1.06

THRESHOLD/NOISE AND REQUIRED SIGNAL/NOISE
===

PROBABILITY OF DETECTION= .9000
CONSTANT FALSE ALARM RATE
LOG10(PFA/N) = - 9

N	N DB	THRESH DB	NON-FLUC DB	SWER 1 DB	SWER 2 DB	SWER 3 DB	SWER 4 DB
1	0.00	13.16	14.66	22.91	22.91	19.01	19.01
2	3.01	13.66	11.92	20.21	16.30	16.30	14.29
3	4.77	14.06	10.38	18.69	13.45	14.78	12.04
4	6.02	14.41	9.31	17.64	11.71	13.73	10.61
5	6.99	14.72	8.51	16.85	10.48	12.93	9.58
6	7.78	14.99	7.86	16.22	9.55	12.30	8.77
7	8.45	15.24	7.32	15.69	8.80	11.77	8.11
8	9.03	15.47	6.86	15.24	8.18	11.32	7.56
9	9.54	15.69	6.45	14.85	7.65	10.92	7.09
10	10.00	15.89	6.10	14.50	7.19	10.57	6.68
11	10.41	16.08	5.78	14.19	6.78	10.26	6.31
12	10.79	16.25	5.49	13.91	6.42	9.98	5.99
13	11.14	16.42	5.23	13.65	6.10	9.72	5.69
14	11.46	16.58	4.98	13.42	5.80	9.48	5.42
15	11.76	16.73	4.76	13.20	5.53	9.26	5.17
16	12.04	16.87	4.55	12.99	5.28	9.06	4.93
17	12.30	17.01	4.36	12.80	5.05	8.87	4.72
18	12.55	17.15	4.17	12.62	4.83	8.69	4.52
19	12.79	17.27	4.00	12.46	4.63	8.52	4.33
20	13.01	17.40	3.84	12.30	4.44	8.37	4.15
21	13.22	17.51	3.69	12.15	4.26	8.22	3.99
22	13.42	17.63	3.54	12.01	4.09	8.07	3.83
23	13.62	17.74	3.40	11.87	3.93	7.94	3.68
24	13.80	17.85	3.27	11.74	3.78	7.81	3.54
25	13.98	17.95	3.14	11.62	3.64	7.68	3.4
26	14.15	18.05	3.02	11.50	3.50	7.57	3.27
27	14.31	18.15	2.91	11.39	3.37	7.45	3.15
28	14.47	18.25	2.80	11.28	3.24	7.34	3.03
29	14.62	18.34	2.69	11.18	3.12	7.24	2.91
30	14.77	18.43	2.59	11.08	3.01	7.14	2.80
31	14.91	18.52	2.49	10.98	2.90	7.04	2.70
32	15.05	18.60	2.39	10.89	2.79	6.95	2.60
33	15.19	18.69	2.30	10.80	2.68	6.86	2.50
34	15.31	18.77	2.21	10.71	2.58	6.77	2.40
35	15.44	18.85	2.12	10.62	2.49	6.68	2.31
36	15.56	18.93	2.04	10.54	2.39	6.60	2.22
37	15.68	19.00	1.96	10.46	2.30	6.52	2.14
38	15.80	19.08	1.88	10.38	2.22	6.44	2.05
39	15.91	19.15	1.80	10.31	2.13	6.37	1.97
40	16.02	19.22	1.72	10.24	2.05	6.30	1.89
41	16.13	19.30	1.65	10.17	1.97	6.22	1.82
42	16.23	19.36	1.58	10.10	1.89	6.15	1.74
43	16.33	19.43	1.51	10.03	1.82	6.09	1.67
44	16.43	19.50	1.44	9.96	1.74	6.02	1.60
45	16.53	19.56	1.38	9.90	1.67	5.96	1.53

Appendix: Radar Detection Tables

THRESHOLD/NOISE AND REQUIRED SIGNAL/NOISE
==

PROBABILITY OF DETECTION= .9500
CONSTANT FALSE ALARM RATE
LOG10(PFA/N)=- 9

N	N DB	THRESH DB	NON-FLUC DB	SWER 1 DB	SWER 2 DB	SWER 3 DB	SWER 4 DB
1	0.00	13.16	15.07	26.05	26.05	20.80	20.80
2	3.01	13.66	12.32	23.35	18.09	18.08	15.42
3	4.77	14.06	10.77	21.83	14.79	16.56	12.95
4	6.02	14.41	9.70	20.78	12.84	15.51	11.47
5	6.99	14.72	8.88	20.00	11.47	14.72	10.28
6	7.78	14.99	8.23	19.36	10.45	14.08	9.42
7	8.45	15.24	7.69	18.83	9.63	13.55	8.73
8	9.03	15.47	7.22	18.38	8.95	13.10	8.15
9	9.54	15.69	6.82	17.99	8.38	12.70	7.66
10	10.00	15.89	6.46	17.64	7.89	12.35	7.22
11	10.41	16.08	6.14	17.33	7.46	12.04	6.84
12	10.79	16.25	5.85	17.05	7.07	11.76	6.5
13	11.14	16.42	5.58	16.79	6.72	11.50	6.19
14	11.46	16.58	5.34	16.55	6.41	11.26	5.9
15	11.76	16.73	5.11	16.33	6.12	11.04	5.65
16	12.04	16.87	4.90	16.13	5.86	10.84	5.41
17	12.30	17.01	4.71	15.94	5.61	10.65	5.18
18	12.55	17.15	4.52	15.76	5.38	10.47	4.98
19	12.79	17.27	4.35	15.60	5.17	10.30	4.78
20	13.01	17.40	4.19	15.44	4.97	10.14	4.6
21	13.22	17.51	4.03	15.29	4.79	9.99	4.43
22	13.42	17.63	3.89	15.15	4.61	9.85	4.26
23	13.62	17.74	3.75	15.01	4.44	9.71	4.11
24	13.80	17.85	3.61	14.89	4.28	9.58	3.96
25	13.98	17.95	3.49	14.76	4.13	9.46	3.82
26	14.15	18.05	3.36	14.64	3.99	9.34	3.69
27	14.31	18.15	3.25	14.53	3.85	9.23	3.56
28	14.47	18.25	3.13	14.42	3.72	9.12	3.44
29	14.62	18.34	3.03	14.32	3.59	9.02	3.32
30	14.77	18.43	2.92	14.22	3.47	8.91	3.21
31	14.91	18.52	2.82	14.12	3.36	8.82	3.10
32	15.05	18.60	2.73	14.03	3.25	8.72	3.00
33	15.19	18.69	2.63	13.94	3.14	8.63	2.90
34	15.31	18.77	2.54	13.84	3.04	8.54	2.80
35	15.44	18.85	2.45	13.76	2.94	8.46	2.70
36	15.56	18.93	2.37	13.68	2.84	8.38	2.61
37	15.68	19.00	2.29	13.60	2.74	8.30	2.52
38	15.80	19.08	2.21	13.52	2.65	8.22	2.44
39	15.91	19.15	2.13	13.44	2.57	8.14	2.36
40	16.02	19.22	2.05	13.37	2.48	8.07	2.27
41	16.13	19.30	1.98	13.30	2.40	8.00	2.20
42	16.23	19.36	1.91	13.23	2.32	7.93	2.12
43	16.33	19.43	1.84	13.16	2.24	7.86	2.05
44	16.43	19.50	1.77	13.10	2.16	7.79	1.97
45	16.53	19.56	1.70	13.03	2.09	7.73	1.90

THRESHOLD/NOISE AND REQUIRED SIGNAL/NOISE

PROBABILITY OF DETECTION= .9900
CONSTANT FALSE ALARM RATE
LOG10(PFA/N)=- 9

N	N DB	THRESH DB	NON-FLUC DB	SWER 1 DB	SWER 2 DB	SWER 3 DB	SWER 4 DB
1	0.00	13.16	15.78	33.15	33.15	24.62	24.62
2	3.01	13.66	13.02	30.43	21.91	21.91	17.72
3	4.77	14.06	11.45	28.90	17.59	20.39	14.73
4	6.02	14.41	10.37	27.86	15.12	19.34	12.92
5	6.99	14.72	9.55	27.06	13.46	18.54	11.64
6	7.78	14.99	8.89	26.43	12.23	17.90	10.68
7	8.45	15.24	8.34	25.90	11.26	17.37	9.90
8	9.03	15.47	7.87	25.45	10.47	16.92	9.26
9	9.54	15.69	7.46	25.05	9.80	16.52	8.72
10	10.00	15.89	7.10	24.70	9.24	16.17	8.24
11	10.41	16.08	6.78	24.39	8.75	15.88	7.83
12	10.79	16.25	6.48	24.11	8.30	15.60	7.45
13	11.14	16.42	6.21	23.85	7.92	15.34	7.11
14	11.46	16.58	5.96	23.62	7.56	15.10	6.81
15	11.76	16.73	5.73	23.40	7.24	14.88	6.53
16	12.04	16.87	5.52	23.20	6.95	14.68	6.27
17	12.30	17.01	5.32	23.01	6.68	14.49	6.04
18	12.55	17.15	5.14	22.83	6.43	14.29	5.82
19	12.79	17.27	4.96	22.66	6.19	14.12	5.61
20	13.01	17.40	4.80	22.50	5.97	13.96	5.42
21	13.22	17.51	4.64	22.35	5.77	13.81	5.23
22	13.42	17.63	4.49	22.21	5.57	13.67	5.06
23	13.62	17.74	4.35	22.08	5.39	13.53	4.89
24	13.80	17.85	4.21	21.95	5.22	13.4	4.74
25	13.98	17.95	4.08	21.83	5.05	13.28	4.59
26	14.15	18.05	3.96	21.71	4.9	13.16	4.45
27	14.31	18.15	3.84	21.60	4.75	13.04	4.32
28	14.47	18.25	3.73	21.49	4.60	12.94	4.19
29	14.62	18.34	3.62	21.38	4.47	12.83	4.06
30	14.77	18.43	3.51	21.28	4.34	12.73	3.95
31	14.91	18.52	3.41	21.19	4.21	12.63	3.83
32	15.05	18.60	3.32	21.09	4.09	12.54	3.72
33	15.19	18.69	3.22	21.00	3.98	12.45	3.62
34	15.31	18.77	3.13	20.91	3.87	12.36	3.51
35	15.44	18.85	3.04	20.83	3.76	12.27	3.41
36	15.56	18.93	2.95	20.75	3.65	12.19	3.32
37	15.68	19.00	2.87	20.67	3.55	12.11	3.23
38	15.80	19.08	2.79	20.59	3.46	12.03	3.14
39	15.91	19.15	2.71	20.52	3.36	11.96	3.05
40	16.02	19.22	2.63	20.44	3.27	11.88	2.97
41	16.13	19.30	2.56	20.37	3.18	11.81	2.88
42	16.23	19.36	2.49	20.30	3.10	11.74	2.80
43	16.33	19.43	2.42	20.24	3.01	11.67	2.73
44	16.43	19.50	2.35	20.17	2.93	11.61	2.65
45	16.53	19.56	2.28	20.10	2.85	11.55	2.58

Appendix: Radar Detection Tables

```
THRESHOLD/NOISE AND REQUIRED SIGNAL/NOISE
=========================================

PROBABILITY OF DETECTION= .5000
CONSTANT FALSE ALARM RATE
LOG10(PFA/N)=-10
```

N	N DB	THRESH DB	NON-FLUC DB	SWER 1 DB	SWER 2 DB	SWER 3 DB	SWER 4 DB
1	0.00	13.62	13.53	15.08	15.08	14.26	14.26
2	3.01	14.08	10.81	12.37	11.54	11.54	11.16
3	4.77	14.47	9.29	10.84	9.76	10.02	9.51
4	6.02	14.80	8.24	9.79	8.58	8.97	8.40
5	6.99	15.09	7.44	9.00	7.71	8.17	7.57
6	7.78	15.35	6.81	8.36	7.03	7.54	6.91
7	8.45	15.59	6.28	7.83	6.46	7.01	6.36
8	9.03	15.81	5.82	7.38	5.99	6.55	5.9
9	9.54	16.01	5.43	6.99	5.57	6.16	5.5
10	10.00	16.21	5.08	6.64	5.21	5.81	5.14
11	10.41	16.39	4.77	6.32	4.88	5.50	4.82
12	10.79	16.56	4.49	6.04	4.59	5.22	4.53
13	11.14	16.72	4.23	5.78	4.32	4.96	4.27
14	11.46	16.87	3.99	5.55	4.08	4.72	4.03
15	11.76	17.02	3.77	5.33	3.85	4.5	3.81
16	12.04	17.15	3.57	5.13	3.64	4.30	3.6
17	12.30	17.29	3.38	4.94	3.45	4.11	3.41
18	12.55	17.42	3.20	4.76	3.27	3.93	3.23
19	12.79	17.54	3.04	4.59	3.1	3.76	3.06
20	13.01	17.66	2.88	4.43	2.94	3.61	2.9
21	13.22	17.77	2.73	4.28	2.78	3.46	2.75
22	13.42	17.88	2.59	4.14	2.64	3.31	2.61
23	13.62	17.99	2.45	4.01	2.50	3.18	2.47
24	13.80	18.10	2.32	3.88	2.37	3.05	2.34
25	13.98	18.20	2.20	3.75	2.24	2.93	2.22
26	14.15	18.29	2.08	3.64	2.12	2.81	2.1
27	14.31	18.39	1.97	3.52	2.01	2.70	1.99
28	14.47	18.48	1.86	3.41	1.90	2.59	1.88
29	14.62	18.57	1.76	3.31	1.79	2.48	1.77
30	14.77	18.66	1.66	3.21	1.69	2.38	1.67
31	14.91	18.75	1.56	3.11	1.59	2.29	1.57
32	15.05	18.83	1.47	3.02	1.50	2.19	1.48
33	15.19	18.91	1.37	2.93	1.41	2.10	1.39
34	15.31	18.99	1.29	2.84	1.32	2.01	1.30
35	15.44	19.07	1.20	2.76	1.23	1.93	1.22
36	15.56	19.15	1.12	2.67	1.15	1.85	1.13
37	15.68	19.22	1.04	2.59	1.07	1.77	1.05
38	15.80	19.29	.96	2.52	.99	1.69	.98
39	15.91	19.37	.89	2.44	.91	1.62	.91
40	16.02	19.44	.82	2.37	.84	1.54	.83
41	16.13	19.50	.74	2.30	.77	1.47	.76
42	16.23	19.57	.67	2.23	.70	1.40	.69
43	16.33	19.64	.61	2.16	.63	1.33	.62
44	16.43	19.70	.54	2.09	.56	1.27	.55
45	16.53	19.77	.48	2.03	.50	1.20	.49

THRESHOLD/NOISE AND REQUIRED SIGNAL/NOISE

PROBABILITY OF DETECTION= .6000
CONSTANT FALSE ALARM RATE
LOG10(PFA/N)=-10

N	N DB	THRESH DB	NON-FLUC DB	SWER 1 DB	SWER 2 DB	SWER 3 DB	SWER 4 DB
1	0.00	13.62	13.85	16.44	16.44	15.17	15.17
2	3.01	14.08	11.13	13.73	12.46	12.46	11.82
3	4.77	14.47	9.60	12.20	10.51	10.93	10.07
4	6.02	14.80	8.54	11.15	9.24	9.88	8.91
5	6.99	15.09	7.74	10.35	8.31	9.08	8.04
6	7.78	15.35	7.10	9.72	7.58	8.45	7.35
7	8.45	15.59	6.57	9.19	6.99	7.91	6.79
8	9.03	15.81	6.12	8.73	6.48	7.46	6.31
9	9.54	16.01	5.72	8.34	6.05	7.07	5.89
10	10.00	16.21	5.37	7.99	5.67	6.72	5.52
11	10.41	16.39	5.06	7.68	5.33	6.40	5.20
12	10.79	16.56	4.77	7.39	5.02	6.12	4.90
13	11.14	16.72	4.51	7.14	4.75	5.86	4.63
14	11.46	16.87	4.27	6.90	4.49	5.63	4.38
15	11.76	17.02	4.05	6.68	4.26	5.41	4.16
16	12.04	17.15	3.85	6.48	4.04	5.20	3.95
17	12.30	17.29	3.66	6.29	3.84	5.01	3.75
18	12.55	17.42	3.48	6.11	3.65	4.83	3.57
19	12.79	17.54	3.31	5.94	3.47	4.67	3.39
20	13.01	17.66	3.15	5.78	3.31	4.51	3.23
21	13.22	17.77	3.00	5.63	3.15	4.36	3.08
22	13.42	17.88	2.86	5.49	3.00	4.22	2.93
23	13.62	17.99	2.72	5.36	2.86	4.08	2.79
24	13.80	18.10	2.59	5.23	2.72	3.95	2.66
25	13.98	18.20	2.47	5.10	2.59	3.83	2.53
26	14.15	18.29	2.35	4.99	2.47	3.71	2.41
27	14.31	18.39	2.24	4.87	2.35	3.60	2.30
28	14.47	18.48	2.13	4.76	2.24	3.49	2.19
29	14.62	18.57	2.02	4.66	2.13	3.38	2.08
30	14.77	18.66	1.92	4.56	2.03	3.28	1.98
31	14.91	18.75	1.82	4.46	1.93	3.19	1.88
32	15.05	18.83	1.73	4.37	1.83	3.09	1.78
33	15.19	18.91	1.64	4.28	1.74	3.00	1.69
34	15.31	18.99	1.55	4.19	1.64	2.91	1.6
35	15.44	19.07	1.47	4.11	1.56	2.83	1.51
36	15.56	19.15	1.38	4.02	1.47	2.75	1.43
37	15.68	19.22	1.30	3.94	1.39	2.67	1.35
38	15.80	19.29	1.23	3.87	1.31	2.59	1.27
39	15.91	19.37	1.15	3.79	1.23	2.51	1.19
40	16.02	19.44	1.08	3.72	1.16	2.44	1.12
41	16.13	19.50	1.00	3.65	1.08	2.37	1.04
42	16.23	19.57	.94	3.58	1.01	2.30	.97
43	16.33	19.64	.87	3.51	.94	2.23	.90
44	16.43	19.70	.80	3.44	.87	2.17	.84
45	16.53	19.77	.74	3.38	.81	2.10	.77

Appendix: Radar Detection Tables 213

```
THRESHOLD/NOISE AND REQUIRED SIGNAL/NOISE
===================================================

PROBABILITY OF DETECTION= .7000
CONSTANT FALSE ALARM RATE
LOG10(PFA/N)=-10
```

N	N DB	THRESH DB	NON-FLUC DB	SWER 1 DB	SWER 2 DB	SWER 3 DB	SWER 4 DB
1	0.00	13.62	14.18	18.03	18.03	16.21	16.21
2	3.01	14.08	11.45	15.31	13.49	13.49	12.55
3	4.77	14.47	9.92	13.79	11.34	11.96	10.68
4	6.02	14.80	8.86	12.74	9.96	10.91	9.45
5	6.99	15.09	8.06	11.94	8.96	10.11	8.54
6	7.78	15.35	7.41	11.30	8.19	9.47	7.82
7	8.45	15.59	6.88	10.77	7.55	8.94	7.24
8	9.03	15.81	6.42	10.32	7.02	8.49	6.74
9	9.54	16.01	6.02	9.92	6.56	8.09	6.31
10	10.00	16.21	5.67	9.57	6.16	7.74	5.93
11	10.41	16.39	5.35	9.26	5.81	7.43	5.59
12	10.79	16.56	5.07	8.98	5.49	7.15	5.29
13	11.14	16.72	4.80	8.72	5.20	6.89	5.01
14	11.46	16.87	4.56	8.48	4.93	6.65	4.76
15	11.76	17.02	4.34	8.26	4.69	6.43	4.52
16	12.04	17.15	4.14	8.06	4.46	6.23	4.31
17	12.30	17.29	3.95	7.87	4.25	6.04	4.11
18	12.55	17.42	3.76	7.69	4.06	5.86	3.92
19	12.79	17.54	3.60	7.52	3.87	5.69	3.74
20	13.01	17.66	3.44	7.36	3.70	5.53	3.57
21	13.22	17.77	3.28	7.21	3.54	5.38	3.42
22	13.42	17.88	3.14	7.07	3.38	5.24	3.27
23	13.62	17.99	3.00	6.94	3.24	5.10	3.12
24	13.80	18.10	2.87	6.81	3.10	4.97	2.99
25	13.98	18.20	2.75	6.68	2.97	4.85	2.86
26	14.15	18.29	2.63	6.57	2.84	4.73	2.74
27	14.31	18.39	2.51	6.45	2.72	4.62	2.62
28	14.47	18.48	2.40	6.34	2.60	4.51	2.51
29	14.62	18.57	2.30	6.24	2.49	4.40	2.40
30	14.77	18.66	2.20	6.14	2.38	4.30	2.29
31	14.91	18.75	2.10	6.04	2.28	4.21	2.19
32	15.05	18.83	2.00	5.95	2.18	4.11	2.09
33	15.19	18.91	1.91	5.86	2.08	4.02	2.00
34	15.31	18.99	1.82	5.77	1.99	3.93	1.91
35	15.44	19.07	1.74	5.69	1.90	3.85	1.82
36	15.56	19.15	1.66	5.60	1.81	3.77	1.74
37	15.68	19.22	1.57	5.52	1.73	3.69	1.65
38	15.80	19.29	1.50	5.45	1.64	3.61	1.57
39	15.91	19.37	1.42	5.37	1.56	3.53	1.49
40	16.02	19.44	1.35	5.30	1.49	3.46	1.42
41	16.13	19.50	1.27	5.23	1.41	3.39	1.34
42	16.23	19.57	1.20	5.16	1.34	3.32	1.27
43	16.33	19.64	1.14	5.09	1.27	3.25	1.20
44	16.43	19.70	1.07	5.02	1.20	3.18	1.13
45	16.53	19.77	1.00	4.96	1.13	3.12	1.07

THRESHOLD/NOISE AND REQUIRED SIGNAL/NOISE

PROBABILITY OF DETECTION= .8000
CONSTANT FALSE ALARM RATE
LOG10(PFA/N)=-10

N	N DB	THRESH DB	NON-FLUC DB	SWER 1 DB	SWER 2 DB	SWER 3 DB	SWER 4 DB
1	0.00	13.62	14.56	20.10	20.10	17.50	17.50
2	3.01	14.08	11.82	17.37	14.78	14.78	13.43
3	4.77	14.47	10.28	15.85	12.36	13.25	11.41
4	6.02	14.80	9.21	14.80	10.84	12.20	10.10
5	6.99	15.09	8.41	14.00	9.75	11.40	9.13
6	7.78	15.35	7.76	13.36	8.91	10.76	8.38
7	8.45	15.59	7.22	12.83	8.23	10.23	7.76
8	9.03	15.81	6.76	12.38	7.66	9.77	7.24
9	9.54	16.01	6.36	11.98	7.17	9.38	6.79
10	10.00	16.21	6.00	11.63	6.75	9.03	6.43
11	10.41	16.39	5.68	11.32	6.37	8.71	6.05
12	10.79	16.56	5.40	11.03	6.03	8.43	5.73
13	11.14	16.72	5.13	10.78	5.73	8.17	5.45
14	11.46	16.87	4.89	10.54	5.45	7.93	5.18
15	11.76	17.02	4.67	10.32	5.19	7.71	4.94
16	12.04	17.15	4.46	10.12	4.96	7.51	4.72
17	12.30	17.29	4.27	9.93	4.74	7.32	4.51
18	12.55	17.42	4.09	9.75	4.53	7.14	4.32
19	12.79	17.54	3.92	9.58	4.34	6.97	4.14
20	13.01	17.66	3.75	9.42	4.16	6.81	3.97
21	13.22	17.77	3.60	9.27	3.99	6.66	3.8-
22	13.42	17.88	3.46	9.13	3.83	6.52	3.65
23	13.62	17.99	3.32	8.99	3.68	6.38	3.51
24	13.80	18.10	3.19	8.86	3.53	6.25	3.37
25	13.98	18.20	3.06	8.74	3.40	6.13	3.24
26	14.15	18.29	2.94	8.62	3.26	6.01	3.11
27	14.31	18.39	2.83	8.51	3.14	5.90	2.99
28	14.47	18.48	2.72	8.40	3.02	5.79	2.87
29	14.62	18.57	2.61	8.29	2.90	5.68	2.76
30	14.77	18.66	2.51	8.19	2.79	5.58	2.65
31	14.91	18.75	2.41	8.10	2.68	5.48	2.55
32	15.05	18.83	2.31	8.00	2.58	5.39	2.45
33	15.19	18.91	2.22	7.91	2.48	5.30	2.35
34	15.31	18.99	2.13	7.82	2.38	5.21	2.26
35	15.44	19.07	2.04	7.74	2.29	5.13	2.17
36	15.56	19.15	1.96	7.66	2.20	5.04	2.08
37	15.68	19.22	1.88	7.58	2.11	4.96	2.00
38	15.80	19.29	1.80	7.50	2.03	4.88	1.92
39	15.91	19.37	1.72	7.42	1.95	4.81	1.84
40	16.02	19.44	1.65	7.35	1.87	4.74	1.76
41	16.13	19.50	1.58	7.28	1.79	4.66	1.69
42	16.23	19.57	1.51	7.21	1.72	4.59	1.61
43	16.33	19.64	1.44	7.14	1.64	4.53	1.54
44	16.43	19.70	1.37	7.08	1.57	4.46	1.47
45	16.53	19.77	1.30	7.01	1.50	4.40	1.41

Appendix: Radar Detection Tables

THRESHOLD/NOISE AND REQUIRED SIGNAL/NOISE
==

PROBABILITY OF DETECTION= .9000
CONSTANT FALSE ALARM RATE
LOG10(PFA/N)=-10

N	N DB	THRESH DB	NON-FLUC DB	SWER 1 DB	SWER 2 DB	SWER 3 DB	SWER 4 DB
1	0.00	13.62	15.95	23.37	23.37	19.46	19.46
2	3.01	14.08	12.30	20.66	16.74	16.74	14.71
3	4.77	14.47	10.75	19.13	13.87	15.20	12.45
4	6.02	14.80	9.68	18.08	12.12	14.15	11.01
5	6.99	15.09	8.87	17.28	10.89	13.35	9.96
6	7.78	15.35	8.21	16.64	9.94	12.71	9.15
7	8.45	15.59	7.67	16.11	9.19	12.18	8.49
8	9.03	15.81	7.21	15.65	8.56	11.72	7.94
9	9.54	16.01	6.80	15.26	8.03	11.33	7.46
10	10.00	16.21	6.44	14.91	7.57	10.97	7.04
11	10.41	16.39	6.12	14.60	7.16	10.66	6.68
12	10.79	16.56	5.83	14.32	6.79	10.38	6.34
13	11.14	16.72	5.57	14.06	6.46	10.12	6.04
14	11.46	16.87	5.32	13.82	6.17	9.88	5.77
15	11.76	17.02	5.10	13.60	5.89	9.66	5.52
16	12.04	17.15	4.89	13.40	5.64	9.45	5.29
17	12.30	17.29	4.69	13.20	5.41	9.26	5.07
18	12.55	17.42	4.51	13.02	5.19	9.08	4.87
19	12.79	17.54	4.34	12.85	4.98	8.91	4.68
20	13.01	17.66	4.17	12.70	4.79	8.75	4.5
21	13.22	17.77	4.02	12.55	4.61	8.60	4.33
22	13.42	17.88	3.87	12.40	4.44	8.46	4.17
23	13.62	17.99	3.74	12.27	4.28	8.33	4.02
24	13.80	18.10	3.60	12.14	4.13	8.20	3.88
25	13.98	18.20	3.48	12.01	3.99	8.07	3.74
26	14.15	18.29	3.35	11.90	3.85	7.95	3.61
27	14.31	18.39	3.24	11.78	3.71	7.84	3.49
28	14.47	18.48	3.13	11.67	3.59	7.73	3.37
29	14.62	18.57	3.02	11.57	3.47	7.62	3.25
30	14.77	18.66	2.92	11.47	3.35	7.52	3.14
31	14.91	18.75	2.82	11.37	3.24	7.42	3.03
32	15.05	18.83	2.72	11.28	3.13	7.33	2.93
33	15.19	18.91	2.63	11.19	3.03	7.24	2.83
34	15.31	18.99	2.54	11.10	2.92	7.15	2.74
35	15.44	19.07	2.45	11.01	2.83	7.06	2.64
36	15.56	19.15	2.36	10.93	2.73	6.98	2.56
37	15.68	19.22	2.28	10.85	2.64	6.90	2.47
38	15.80	19.29	2.20	10.77	2.55	6.82	2.38
39	15.91	19.37	2.12	10.70	2.47	6.75	2.30
40	16.02	19.44	2.05	10.62	2.39	6.67	2.22
41	16.13	19.50	1.98	10.55	2.31	6.60	2.15
42	16.23	19.57	1.90	10.48	2.23	6.53	2.07
43	16.33	19.64	1.83	10.42	2.15	6.47	2.00
44	16.43	19.70	1.77	10.35	2.08	6.40	1.93
45	16.53	19.77	1.70	10.28	2.00	6.33	1.86

THRESHOLD/NOISE AND REQUIRED SIGNAL/NOISE

PROBABILITY OF DETECTION= .9500
CONSTANT FALSE ALARM RATE
LOG10(PFA/N)=-10

N	N DB	THRESH DB	NON-FLUC DB	SWER 1 DB	SWER 2 DB	SWER 3 DB	SWER 4 DB
1	0.00	13.62	15.44	26.51	26.51	21.24	21.24
2	3.01	14.08	12.68	23.79	18.52	18.52	15.84
3	4.77	14.47	11.12	22.26	15.21	16.99	13.35
4	6.02	14.80	10.05	21.21	13.24	15.93	11.78
5	6.99	15.09	9.23	20.41	11.87	15.13	10.66
6	7.78	15.35	8.57	19.78	10.83	14.49	9.80
7	8.45	15.59	8.03	19.25	10.01	13.95	9.10
8	9.03	15.81	7.56	18.80	9.33	13.50	8.51
9	9.54	16.01	7.15	18.40	8.75	13.10	8.01
10	10.00	16.21	6.79	18.05	8.26	12.75	7.58
11	10.41	16.39	6.47	17.74	7.82	12.44	7.19
12	10.79	16.56	6.18	17.45	7.43	12.15	6.85
13	11.14	16.72	5.91	17.19	7.08	11.89	6.53
14	11.46	16.87	5.66	16.96	6.76	11.65	6.25
15	11.76	17.02	5.44	16.74	6.47	11.43	5.99
16	12.04	17.15	5.23	16.53	6.21	11.23	5.75
17	12.30	17.29	5.03	16.34	5.96	11.04	5.52
18	12.55	17.42	4.84	16.16	5.73	10.86	5.31
19	12.79	17.54	4.67	15.99	5.52	10.69	5.12
20	13.01	17.66	4.51	15.84	5.32	10.53	4.93
21	13.22	17.77	4.35	15.69	5.13	10.38	4.76
22	13.42	17.88	4.20	15.54	4.95	10.23	4.60
23	13.62	17.99	4.06	15.41	4.78	10.10	4.44
24	13.80	18.10	3.93	15.28	4.62	9.97	4.29
25	13.98	18.20	3.80	15.16	4.47	9.84	4.15
26	14.15	18.29	3.68	15.04	4.32	9.72	4.02
27	14.31	18.39	3.56	14.92	4.19	9.61	3.89
28	14.47	18.48	3.45	14.81	4.05	9.50	3.77
29	14.62	18.57	3.34	14.71	3.93	9.40	3.65
30	14.77	18.66	3.24	14.61	3.81	9.29	3.53
31	14.91	18.75	3.14	14.51	3.69	9.20	3.42
32	15.05	18.83	3.04	14.42	3.58	9.10	3.32
33	15.19	18.91	2.95	14.33	3.47	9.01	3.22
34	15.31	18.99	2.85	14.24	3.36	8.92	3.12
35	15.44	19.07	2.77	14.15	3.26	8.84	3.02
36	15.56	19.15	2.68	14.06	3.17	8.76	2.93
37	15.68	19.22	2.60	13.98	3.07	8.67	2.84
38	15.80	19.29	2.52	13.91	2.98	8.60	2.75
39	15.91	19.37	2.44	13.83	2.89	8.52	2.67
40	16.02	19.44	2.36	13.76	2.81	8.45	2.59
41	16.13	19.50	2.29	13.69	2.72	8.38	2.51
42	16.23	19.57	2.22	13.62	2.64	8.31	2.44
43	16.33	19.64	2.15	13.55	2.56	8.24	2.36
44	16.43	19.70	2.08	13.48	2.49	8.17	2.29
45	16.53	19.77	2.01	13.42	2.41	8.11	2.22

Appendix: Radar Detection Tables

THRESHOLD/NOISE AND REQUIRED SIGNAL/NOISE
=====================================

PROBABILITY OF DETECTION= .9900
CONSTANT FALSE ALARM RATE
LOG10(PFA/N)=-10

N	N DB	THRESH DB	NON-FLUC DB	SWER 1 DB	SWER 2 DB	SWER 3 DB	SWER 4 DB
1	0.00	13.62	16.12	33.60	33.60	25.06	25.06
2	3.01	14.08	13.35	30.87	22.34	22.34	18.13
3	4.77	14.47	11.78	29.34	18.00	20.81	15.12
4	6.02	14.80	10.70	28.29	15.52	19.75	13.29
5	6.99	15.09	9.87	27.49	13.85	18.95	12.01
6	7.78	15.35	9.21	26.85	12.61	18.31	11.04
7	8.45	15.59	8.66	26.31	11.63	17.77	10.25
8	9.03	15.81	8.18	25.86	10.83	17.32	9.61
9	9.54	16.01	7.77	25.46	10.16	16.92	9.05
10	10.00	16.21	7.41	25.11	9.60	16.59	8.58
11	10.41	16.39	7.08	24.80	9.10	16.27	8.16
12	10.79	16.56	6.78	24.52	8.65	15.99	7.78
13	11.14	16.72	6.51	24.26	8.26	15.73	7.44
14	11.46	16.87	6.26	24.02	7.90	15.49	7.14
15	11.76	17.02	6.03	23.80	7.58	15.27	6.86
16	12.04	17.15	5.82	23.60	7.28	15.06	6.60
17	12.30	17.29	5.62	23.40	7.01	14.87	6.36
18	12.55	17.42	5.43	23.23	6.76	14.69	6.13
19	12.79	17.54	5.25	23.06	6.52	14.52	5.92
20	13.01	17.66	5.09	22.90	6.30	14.34	5.73
21	13.22	17.77	4.93	22.75	6.09	14.19	5.55
22	13.42	17.88	4.78	22.61	5.90	14.05	5.37
23	13.62	17.99	4.64	22.47	5.71	13.91	5.20
24	13.80	18.10	4.51	22.34	5.54	13.78	5.05
25	13.98	18.20	4.38	22.22	5.37	13.66	4.90
26	14.15	18.29	4.25	22.10	5.21	13.54	4.76
27	14.31	18.39	4.13	21.99	5.06	13.42	4.62
28	14.47	18.48	4.02	21.88	4.92	13.31	4.49
29	14.62	18.57	3.91	21.77	4.78	13.21	4.37
30	14.77	18.66	3.80	21.67	4.65	13.11	4.25
31	14.91	18.75	3.70	21.58	4.53	13.01	4.13
32	15.05	18.83	3.60	21.48	4.41	12.91	4.02
33	15.19	18.91	3.51	21.39	4.29	12.82	3.91
34	15.31	18.99	3.42	21.30	4.18	12.73	3.81
35	15.44	19.07	3.33	21.22	4.07	12.65	3.71
36	15.56	19.15	3.24	21.14	3.97	12.57	3.62
37	15.68	19.22	3.16	21.06	3.86	12.49	3.52
38	15.80	19.29	3.08	20.98	3.77	12.41	3.43
39	15.91	19.37	3.00	20.90	3.67	12.33	3.35
40	16.02	19.44	2.92	20.83	3.58	12.26	3.26
41	16.13	19.50	2.85	20.76	3.49	12.19	3.18
42	16.23	19.57	2.77	20.69	3.41	12.12	3.10
43	16.33	19.64	2.70	20.62	3.32	12.05	3.02
44	16.43	19.70	2.63	20.55	3.24	11.98	2.95
45	16.53	19.77	2.56	20.49	3.16	11.92	2.87

PROBABILITY OF DETECTION= .5000
CONSTANT FALSE ALARM PROBABILITY
LOG10(PFA)=- 2

N	N DB	THRESH DB	NON-FLUC DB	SWER 1 DB	SWER 2 DB	SWER 3 DB	SWER 4 DB
1	0.00	6.63	6.12	7.52	7.52	6.76	6.76
2	3.01	8.22	4.08	5.47	4.71	4.71	4.37
3	4.77	9.25	2.72	4.30	3.31	3.54	3.10
4	6.02	10.02	2.11	3.49	2.39	2.73	2.24
5	6.99	10.65	1.50	2.88	1.72	2.12	1.60
6	7.78	11.18	1.01	2.38	1.18	1.62	1.09
7	8.45	11.63	.59	1.96	.74	1.20	.66
8	9.03	12.04	.24	1.61	.36	.84	.29
9	9.54	12.41	-.07	1.29	.03	.53	-.02
10	10.00	12.74	-.35	1.02	-.26	.25	-.31
11	10.41	13.04	-.60	.77	-.51	.00	-.56
12	10.79	13.32	-.82	.54	-.75	-.22	-.79
13	11.14	13.58	-1.03	.33	-.96	-.43	-1.00
14	11.46	13.83	-1.22	.14	-1.16	-.62	-1.19
15	11.76	14.06	-1.40	-.04	-1.34	-.80	-1.37
16	12.04	14.27	-1.56	-.20	-1.51	-.96	-1.53
17	12.30	14.48	-1.71	-.35	-1.66	-1.12	-1.69
18	12.55	14.67	-1.86	-.50	-1.81	-1.26	-1.84
19	12.79	14.85	-1.99	-.64	-1.95	-1.40	-1.97
20	13.01	15.03	-2.12	-.77	-2.08	-1.53	-2.10
21	13.22	15.20	-2.24	-.89	-2.21	-1.65	-2.23
22	13.42	15.36	-2.36	-1.01	-2.32	-1.77	-2.34
23	13.62	15.51	-2.47	-1.12	-2.44	-1.88	-2.45
24	13.80	15.66	-2.58	-1.22	-2.54	-1.99	-2.56
25	13.98	15.81	-2.68	-1.32	-2.65	-2.09	-2.66
26	14.15	15.94	-2.77	-1.42	-2.75	-2.18	-2.76
27	14.31	16.08	-2.87	-1.52	-2.84	-2.28	-2.85
28	14.47	16.21	-2.96	-1.61	-2.93	-2.37	-2.94
29	14.62	16.33	-3.04	-1.69	-3.02	-2.46	-3.03
30	14.77	16.45	-3.13	-1.78	-3.10	-2.54	-3.11
31	14.91	16.57	-3.21	-1.86	-3.18	-2.62	-3.20
32	15.05	16.68	-3.28	-1.94	-3.26	-2.70	-3.27
33	15.19	16.80	-3.36	-2.01	-3.34	-2.77	-3.35
34	15.31	16.90	-3.43	-2.08	-3.41	-2.85	-3.42
35	15.44	17.01	-3.50	-2.16	-3.48	-2.92	-3.49
36	15.56	17.11	-3.57	-2.22	-3.55	-2.99	-3.56
37	15.68	17.21	-3.64	-2.29	-3.62	-3.05	-3.63
38	15.80	17.31	-3.70	-2.36	-3.68	-3.12	-3.69
39	15.91	17.40	-3.77	-2.42	-3.75	-3.18	-3.76
40	16.02	17.49	-3.83	-2.48	-3.81	-3.24	-3.82
41	16.13	17.59	-3.89	-2.54	-3.87	-3.30	-3.88
42	16.23	17.67	-3.94	-2.60	-3.93	-3.36	-3.94
43	16.33	17.76	-4.00	-2.66	-3.99	-3.42	-3.99
44	16.43	17.85	-4.06	-2.71	-4.04	-3.48	-4.05
45	16.53	17.93	-4.11	-2.77	-4.10	-3.53	-4.10

Appendix: Radar Detection Tables

```
PROBABILITY OF DETECTION= .6000
CONSTANT FALSE ALARM PROBABILITY
LOG10(PFA)=- 2
```

N	N DB	THRESH DB	NON-FLUC DB	SWER 1 DB	SWER 2 DB	SWER 3 DB	SWER 4 DB
1	0.00	6.63	6.88	9.04	9.04	7.91	7.91
2	3.01	8.22	4.79	6.97	5.82	5.82	5.29
3	4.77	9.25	3.60	5.79	4.28	4.64	3.93
4	6.02	10.02	2.77	4.98	3.28	3.82	3.02
5	6.99	10.65	2.15	4.35	2.55	3.20	2.34
6	7.78	11.18	1.64	3.85	1.97	2.69	1.80
7	8.45	11.63	1.22	3.43	1.50	2.27	1.36
8	9.03	12.04	.85	3.07	1.10	1.91	.97
9	9.54	12.41	.54	2.76	.76	1.59	.64
10	10.00	12.74	.25	2.48	.45	1.31	.35
11	10.41	13.04	.00	2.23	.18	1.06	.09
12	10.79	13.32	-.23	2.00	-.07	.83	-.15
13	11.14	13.58	-.44	1.79	-.29	.62	-.36
14	11.46	13.83	-.63	1.60	-.49	.43	-.56
15	11.76	14.06	-.81	1.42	-.68	.25	-.75
16	12.04	14.27	-.98	1.25	-.86	.08	-.92
17	12.30	14.48	-1.13	1.10	-1.02	-.07	-1.08
18	12.55	14.67	-1.28	.95	-1.18	-.22	-1.23
19	12.79	14.85	-1.42	.81	-1.32	-.36	-1.37
20	13.01	15.03	-1.55	.68	-1.46	-.49	-1.50
21	13.22	15.20	-1.68	.56	-1.59	-.61	-1.63
22	13.42	15.36	-1.79	.44	-1.71	-.73	-1.75
23	13.62	15.51	-1.91	.33	-1.82	-.84	-1.87
24	13.80	15.66	-2.01	.22	-1.94	-.95	-1.97
25	13.98	15.81	-2.12	.12	-2.04	-1.05	-2.08
26	14.15	15.94	-2.21	.02	-2.14	-1.15	-2.18
27	14.31	16.08	-2.31	-.07	-2.24	-1.24	-2.28
28	14.47	16.21	-2.40	-.16	-2.33	-1.33	-2.37
29	14.62	16.33	-2.49	-.25	-2.42	-1.42	-2.46
30	14.77	16.45	-2.57	-.33	-2.51	-1.51	-2.54
31	14.91	16.57	-2.65	-.42	-2.59	-1.59	-2.63
32	15.05	16.68	-2.73	-.49	-2.68	-1.67	-2.71
33	15.19	16.80	-2.81	-.57	-2.75	-1.74	-2.78
34	15.31	16.90	-2.88	-.64	-2.83	-1.82	-2.86
35	15.44	17.01	-2.96	-.72	-2.90	-1.89	-2.93
36	15.56	17.11	-3.03	-.79	-2.97	-1.96	-3.00
37	15.68	17.21	-3.09	-.85	-3.04	-2.02	-3.07
38	15.8	17.31	-3.16	-.92	-3.11	-2.09	-3.14
39	15.91	17.40	-3.22	-.98	-3.18	-2.15	-3.20
40	16.02	17.49	-3.28	-1.04	-3.24	-2.22	-3.26
41	16.13	17.59	-3.35	-1.10	-3.30	-2.28	-3.32
42	16.23	17.67	-3.40	-1.16	-3.36	-2.34	-3.38
43	16.33	17.76	-3.46	-1.22	-3.42	-2.39	-3.44
44	16.43	17.85	-3.52	-1.28	-3.48	-2.45	-3.50
45	16.53	17.93	-3.57	-1.33	-3.53	-2.50	-3.55

PROBABILITY OF DETECTION= .7000
CONSTANT FALSE ALARM PROBABILITY
LOG10(PFA)=- 2

N	N DB	THRESH DB	NON-FLUC DB	SWER 1 DB	SWER 2 DB	SWER 3 DB	SWER 4 DB
1	0.00	6.63	7.62	10.76	10.76	9.14	9.14
2	3.01	8.22	5.48	8.67	7.03	7.03	6.25
3	4.77	9.25	4.26	7.48	5.30	5.84	4.78
4	6.02	10.02	3.42	6.66	4.21	5.01	3.81
5	6.99	10.65	2.78	6.03	3.41	4.37	3.09
6	7.78	11.18	2.26	5.53	2.79	3.87	2.53
7	8.45	11.63	1.83	5.11	2.29	3.44	2.06
8	9.03	12.04	1.46	4.74	1.86	3.07	1.66
9	9.54	12.41	1.14	4.43	1.49	2.76	1.31
10	10.00	12.74	.85	4.15	1.17	2.47	1.01
11	10.41	13.04	.59	3.89	.88	2.22	.74
12	10.79	13.32	.36	3.66	.63	1.99	.49
13	11.14	13.58	.14	3.45	.39	1.77	.27
14	11.46	13.83	-.05	3.26	.18	1.58	.06
15	11.76	14.06	-.24	3.08	-.02	1.40	-.13
16	12.04	14.27	-.41	2.91	-.20	1.23	-.31
17	12.30	14.48	-.57	2.76	-.38	1.07	-.47
18	12.55	14.67	-.72	2.61	-.54	.93	-.63
19	12.79	14.85	-.86	2.47	-.69	.79	-.77
20	13.01	15.03	-.99	2.34	-.83	.66	-.91
21	13.22	15.20	-1.12	2.22	-.96	.53	-1.04
22	13.42	15.36	-1.24	2.10	-1.09	.41	-1.16
23	13.62	15.51	-1.35	1.99	-1.21	.30	-1.28
24	13.80	15.66	-1.46	1.88	-1.33	.19	-1.39
25	13.98	15.81	-1.57	1.78	-1.44	.09	-1.50
26	14.15	15.94	-1.67	1.68	-1.54	-.01	-1.60
27	14.31	16.08	-1.76	1.58	-1.64	-.11	-1.70
28	14.47	16.21	-1.86	1.49	-1.74	-.20	-1.80
29	14.62	16.33	-1.95	1.40	-1.83	-.29	-1.89
30	14.77	16.45	-2.03	1.32	-1.92	-.37	-1.98
31	14.91	16.57	-2.11	1.24	-2.01	-.45	-2.06
32	15.05	16.68	-2.19	1.16	-2.09	-.53	-2.14
33	15.19	16.80	-2.27	1.08	-2.17	-.61	-2.22
34	15.31	16.90	-2.35	1.01	-2.25	-.68	-2.30
35	15.44	17.01	-2.42	.94	-2.33	-.76	-2.37
36	15.56	17.11	-2.49	.87	-2.40	-.83	-2.45
37	15.68	17.21	-2.56	.80	-2.47	-.89	-2.52
38	15.80	17.31	-2.63	.73	-2.54	-.96	-2.58
39	15.91	17.40	-2.69	.67	-2.61	-1.03	-2.65
40	16.02	17.49	-2.75	.61	-2.67	-1.09	-2.71
41	16.13	17.59	-2.82	.55	-2.74	-1.15	-2.78
42	16.23	17.67	-2.88	.49	-2.80	-1.21	-2.84
43	16.33	17.76	-2.93	.43	-2.86	-1.27	-2.90
44	16.43	17.85	-2.99	.38	-2.92	-1.32	-2.95
45	16.53	17.93	-3.05	.32	-2.97	-1.38	-3.01

Appendix: Radar Detection Tables

PROBABILITY OF DETECTION= .8000
CONSTANT FALSE ALARM PROBABILITY
LOG10(PFA)=- 2

N	N DB	THRESH DB	NON-FLUC DB	SWER 1 DB	SWER 2 DB	SWER 3 DB	SWER 4 DB
1	0.00	6.63	8.41	12.93	12.93	10.62	10.62
2	3.01	8.22	6.22	10.83	8.48	8.48	7.35
3	4.77	9.25	4.98	9.63	6.51	7.27	5.75
4	6.02	10.02	4.12	8.81	5.28	6.44	4.70
5	6.99	10.65	3.46	8.17	4.40	5.80	3.93
6	7.78	11.18	2.93	7.67	3.73	5.28	3.33
7	8.45	11.63	2.49	7.24	3.18	4.85	2.84
8	9.03	12.04	2.11	6.88	2.72	4.48	2.42
9	9.54	12.41	1.78	6.56	2.32	4.16	2.05
10	10.00	12.74	1.49	6.27	1.98	3.88	1.73
11	10.41	13.04	1.22	6.02	1.67	3.62	1.45
12	10.79	13.32	.98	5.79	1.40	3.39	1.19
13	11.14	13.58	.76	5.58	1.15	3.17	.96
14	11.46	13.83	.56	5.38	.92	2.98	.74
15	11.76	14.06	.38	5.20	.71	2.79	.55
16	12.04	14.27	.20	5.04	.52	2.62	.36
17	12.30	14.48	.04	4.88	.34	2.47	.19
18	12.55	14.67	-.11	4.73	.17	2.32	.03
19	12.79	14.85	-.26	4.59	.01	2.18	-.12
20	13.01	15.03	-.39	4.46	-.14	2.04	-.26
21	13.22	15.20	-.52	4.34	-.28	1.92	-.40
22	13.42	15.36	-.64	4.22	-.41	1.80	-.53
23	13.62	15.51	-.76	4.11	-.54	1.68	-.65
24	13.80	15.66	-.87	4.00	-.66	1.58	-.76
25	13.98	15.81	-.98	3.89	-.77	1.47	-.87
26	14.15	15.94	-1.08	3.80	-.88	1.37	-.98
27	14.31	16.08	-1.18	3.70	-.99	1.28	-1.08
28	14.47	16.21	-1.27	3.61	-1.09	1.18	-1.18
29	14.62	16.33	-1.36	3.52	-1.18	1.09	-1.27
30	14.77	16.45	-1.45	3.44	-1.28	1.01	-1.36
31	14.91	16.57	-1.54	3.35	-1.37	.93	-1.45
32	15.05	16.68	-1.62	3.27	-1.45	.85	-1.54
33	15.19	16.80	-1.70	3.20	-1.54	.77	-1.62
34	15.31	16.90	-1.77	3.12	-1.62	.69	-1.70
35	15.44	17.01	-1.85	3.05	-1.70	.62	-1.77
36	15.56	17.11	-1.92	2.98	-1.77	.55	-1.85
37	15.68	17.21	-1.99	2.91	-1.85	.48	-1.92
38	15.80	17.31	-2.06	2.85	-1.92	.42	-1.99
39	15.91	17.41	-2.12	2.78	-1.99	.35	-2.05
40	16.02	17.49	-2.19	2.72	-2.05	.29	-2.12
41	16.13	17.59	-2.25	2.66	-2.12	.23	-2.18
42	16.23	17.67	-2.31	2.60	-2.18	.17	-2.25
43	16.33	17.76	-2.37	2.54	-2.25	.11	-2.31
44	16.43	17.85	-2.43	2.49	-2.31	.05	-2.37
45	16.53	17.93	-2.48	2.43	-2.37	-.00	-2.43

PROBABILITY OF DETECTION= .9000
CONSTANT FALSE ALARM PROBABILITY
LOG10(PFA)=- 2

N	N DB	THRESH DB	NON-FLUC DB	SWER 1 DB	SWER 2 DB	SWER 3 DB	SWER 4 DB
1	0.00	6.63	9.40	16.30	16.30	12.76	12.76
2	3.01	8.22	7.15	14.18	10.60	10.60	8.88
3	4.77	9.25	5.87	12.98	8.21	9.38	7.05
4	6.02	10.02	4.99	12.15	6.78	8.53	5.89
5	6.99	10.65	4.31	11.52	5.76	7.88	5.05
6	7.78	11.18	3.77	11.01	4.99	7.36	4.39
7	8.45	11.63	3.31	10.58	4.38	6.93	3.86
8	9.03	12.04	2.93	10.22	3.87	6.56	3.40
9	9.54	12.41	2.59	9.90	3.43	6.23	3.02
10	10.00	12.74	2.29	9.62	3.05	5.94	2.67
11	10.41	13.04	2.02	9.36	2.72	5.68	2.37
12	10.79	13.32	1.77	9.13	2.42	5.45	2.10
13	11.14	13.58	1.55	8.91	2.15	5.23	1.85
14	11.46	13.83	1.34	8.72	1.90	5.04	1.62
15	11.76	14.06	1.15	8.54	1.67	4.85	1.41
16	12.04	14.27	.97	8.37	1.47	4.68	1.22
17	12.30	14.48	.80	8.21	1.27	4.52	1.04
18	12.55	14.67	.65	8.06	1.09	4.37	.87
19	12.79	14.85	.50	7.93	.92	4.23	.71
20	13.01	15.03	.36	7.79	.76	4.10	.56
21	13.22	15.20	.23	7.67	.61	3.97	.42
22	13.42	15.36	.10	7.55	.47	3.85	.29
23	13.62	15.51	-.02	7.44	.34	3.73	.16
24	13.80	15.66	-.13	7.33	.21	3.62	.04
25	13.98	15.81	-.24	7.23	.09	3.52	-.07
26	14.15	15.94	-.34	7.13	-.03	3.42	-.18
27	14.31	16.08	-.44	7.03	-.14	3.32	-.29
28	14.47	16.21	-.54	6.94	-.25	3.23	-.39
29	14.62	16.33	-.63	6.85	-.35	3.14	-.49
30	14.77	16.45	-.72	6.77	-.45	3.05	-.58
31	14.91	16.57	-.81	6.68	-.54	2.97	-.67
32	15.05	16.68	-.89	6.60	-.63	2.89	-.76
33	15.19	16.80	-.97	6.52	-.72	2.81	-.85
34	15.31	16.90	-1.05	6.45	-.81	2.74	-.93
35	15.44	17.01	-1.13	6.38	-.89	2.66	-1.01
36	15.56	17.11	-1.20	6.31	-.97	2.59	-1.08
37	15.68	17.21	-1.27	6.24	-1.05	2.52	-1.16
38	15.80	17.31	-1.34	6.17	-1.12	2.46	-1.23
39	15.91	17.40	-1.41	6.11	-1.19	2.39	-1.30
40	16.02	17.49	-1.47	6.05	-1.26	2.33	-1.37
41	16.13	17.59	-1.54	5.99	-1.33	2.27	-1.44
42	16.23	17.67	-1.60	5.93	-1.40	2.21	-1.50
43	16.33	17.76	-1.66	5.87	-1.46	2.15	-1.56
44	16.43	17.85	-1.72	5.81	-1.53	2.09	-1.62
45	16.53	17.93	-1.78	5.76	-1.59	2.04	-1.68

```
PROBABILITY OF DETECTION= .9500
CONSTANT FALSE ALARM PROBABILITY
LOG10(PFA)=- 2
```

N	N DB	THRESH DB	NON-FLUC DB	SWER 1 DB	SWER 2 DB	SWER 3 DB	SWER 4 DB
1	0.00	6.63	10.14	19.48	19.48	14.65	14.65
2	3.01	8.22	7.84	17.36	12.48	12.48	10.15
3	4.77	9.25	6.54	16.16	9.67	11.24	8.12
4	6.02	10.02	5.64	15.32	8.03	10.39	6.85
5	6.99	10.65	4.95	14.69	6.89	9.74	5.93
6	7.78	11.18	4.40	14.18	6.04	9.22	5.23
7	8.45	11.63	3.94	13.75	5.36	8.78	4.66
8	9.03	12.04	3.54	13.38	4.80	8.41	4.18
9	9.54	12.41	3.19	13.06	4.32	8.08	3.77
10	10.00	12.74	2.89	12.77	3.91	7.79	3.41
11	10.41	13.04	2.61	12.52	3.55	7.53	3.09
12	10.79	13.32	2.36	12.29	3.23	7.29	2.80
13	11.14	13.58	2.13	12.08	2.94	7.08	2.54
14	11.46	13.83	1.92	11.88	2.67	6.88	2.30
15	11.76	14.06	1.73	11.70	2.43	6.69	2.09
16	12.04	14.27	1.55	11.53	2.21	6.52	1.88
17	12.30	14.48	1.38	11.37	2.01	6.36	1.70
18	12.55	14.67	1.22	11.23	1.82	6.21	1.52
19	12.79	14.85	1.07	11.09	1.64	6.07	1.35
20	13.01	15.03	.92	10.95	1.47	5.93	1.20
21	13.22	15.20	.79	10.83	1.31	5.81	1.05
22	13.42	15.36	.66	10.71	1.16	5.69	.92
23	13.62	15.51	.54	10.60	1.02	5.57	.78
24	13.80	15.66	.43	10.49	.89	5.46	.66
25	13.98	15.81	.31	10.39	.76	5.35	.54
26	14.15	15.94	.21	10.29	.64	5.25	.42
27	14.31	16.08	.11	10.19	.52	5.16	.31
28	14.47	16.21	.01	10.10	.41	5.06	.21
29	14.62	16.33	-.09	10.01	.30	4.97	.11
30	14.77	16.45	-.18	9.93	.20	4.89	.01
31	14.91	16.57	-.27	9.84	.10	4.80	-.08
32	15.05	16.68	-.35	9.76	.00	4.72	-.17
33	15.19	16.80	-.43	9.69	-.09	4.64	-.26
34	15.31	16.90	-.51	9.61	-.18	4.57	-.34
35	15.44	17.01	-.59	9.54	-.26	4.50	-.43
36	15.56	17.11	-.66	9.47	-.35	4.42	-.50
37	15.68	17.21	-.74	9.40	-.43	4.36	-.58
38	15.80	17.31	-.81	9.33	-.50	4.28	-.65
39	15.91	17.40	-.87	9.26	-.58	4.22	-.73
40	16.02	17.49	-.94	9.20	-.66	4.16	-.80
41	16.13	17.59	-1.01	9.14	-.73	4.09	-.87
42	16.23	17.67	-1.07	9.08	-.80	4.03	-.93
43	16.33	17.76	-1.13	9.02	-.86	3.97	-1.00
44	16.43	17.85	-1.19	8.97	-.93	3.92	-1.06
45	16.53	17.93	-1.25	8.91	-.99	3.86	-1.12

PROBABILITY OF DETECTION= .9900
CONSTANT FALSE ALARM PROBABILITY
LOG10(PFA)=- 2

N	N DB	THRESH DB	NON-FLUC DB	SWER 1 DB	SWER 2 DB	SWER 3 DB	SWER 4 DB
1	0.00	6.63	11.37	26.60	26.60	18.61	18.61
2	3.01	8.22	9.00	24.51	16.42	16.41	12.65
3	4.77	9.25	7.66	23.27	12.62	15.17	10.13
4	6.02	10.02	6.74	22.43	10.49	14.31	8.63
5	6.99	10.65	6.03	21.80	9.07	13.66	7.56
6	7.78	11.18	5.46	21.28	8.02	13.14	6.75
7	8.45	11.63	4.98	20.86	7.21	12.70	6.10
8	9.03	12.04	4.57	20.49	6.54	12.32	5.56
9	9.54	12.41	4.21	20.17	5.98	12.00	5.10
10	10.00	12.74	3.89	19.88	5.51	11.70	4.71
11	10.41	13.04	3.61	19.63	5.08	11.44	4.35
12	10.79	13.32	3.35	19.40	4.71	11.21	4.04
13	11.14	13.58	3.12	19.18	4.38	10.99	3.76
14	11.46	13.83	2.90	18.99	4.08	10.77	3.50
15	11.76	14.06	2.70	18.81	3.80	10.58	3.26
16	12.04	14.27	2.51	18.64	3.56	10.41	3.04
17	12.30	14.48	2.34	18.48	3.33	10.25	2.84
18	12.55	14.67	2.17	18.33	3.11	10.10	2.65
19	12.79	14.85	2.02	18.19	2.91	9.95	2.47
20	13.01	15.03	1.88	18.06	2.73	9.82	2.31
21	13.22	15.20	1.74	17.94	2.55	9.69	2.15
22	13.42	15.36	1.61	17.82	2.39	9.57	2.00
23	13.62	15.51	1.48	17.70	2.23	9.46	1.86
24	13.80	15.66	1.36	17.60	2.08	9.35	1.73
25	13.98	15.81	1.25	17.49	1.94	9.24	1.60
26	14.15	15.94	1.14	17.37	1.81	9.14	1.48
27	14.31	16.08	1.03	17.27	1.68	9.04	1.36
28	14.47	16.21	.93	17.18	1.56	8.95	1.25
29	14.62	16.33	.83	17.09	1.44	8.86	1.14
30	14.77	16.45	.74	17.01	1.33	8.77	1.04
31	14.91	16.57	.65	16.92	1.22	8.69	.94
32	15.05	16.68	.56	16.85	1.12	8.60	.84
33	15.19	16.80	.48	16.77	1.02	8.53	.75
34	15.31	16.90	.40	16.69	.92	8.45	.66
35	15.44	17.01	.32	16.62	.83	8.38	.58
36	15.56	17.11	.24	16.55	.74	8.30	.49
37	15.68	17.21	.16	16.48	.66	8.24	.41
38	15.80	17.31	.09	16.42	.57	8.17	.33
39	15.91	17.40	.02	16.35	.49	8.10	.26
40	16.02	17.49	-.05	16.29	.41	8.04	.18
41	16.13	17.59	-.11	16.23	.33	7.98	.11
42	16.23	17.67	-.18	16.17	.25	7.92	.04
43	16.33	17.76	-.24	16.11	.18	7.86	-.03
44	16.43	17.85	-.30	16.06	.11	7.80	-.09
45	16.53	17.93	-.36	16.00	.04	7.74	-.16

PROBABILITY OF DETECTION= .5000
CONSTANT FALSE ALARM PROBABILITY
LOG10(PFA)=- 3

N	N DB	THRESH DB	NON-FLUC DB	SWER 1 DB	SWER 2 DB	SWER 3 DB	SWER 4 DB
1	0.00	8.39	8.06	9.53	9.53	8.73	8.73
2	3.01	9.65	5.86	7.33	6.53	6.53	6.17
3	4.77	10.50	4.63	6.09	5.05	5.30	4.82
4	6.02	11.16	3.77	5.24	4.08	4.44	3.91
5	6.99	11.70	3.12	4.59	3.36	3.79	3.23
6	7.78	12.16	2.60	4.06	2.79	3.26	2.69
7	8.45	12.57	2.17	3.63	2.32	2.83	2.24
8	9.03	12.93	1.79	3.25	1.93	2.45	1.86
9	9.54	13.25	1.47	2.93	1.58	2.13	1.52
10	10.00	13.55	1.18	2.64	1.28	1.84	1.23
11	10.41	13.83	.92	2.38	1.01	1.58	.96
12	10.79	14.08	.68	2.14	.77	1.34	.72
13	11.14	14.32	.47	1.93	.54	1.13	.50
14	11.46	14.54	.27	1.73	.34	.93	.30
15	11.76	14.75	.09	1.54	.15	.74	.12
16	12.04	14.95	-.09	1.37	-.03	.57	-.06
17	12.30	15.14	-.25	1.21	-.19	.41	-.22
18	12.55	15.31	-.40	1.06	-.34	.26	-.37
19	12.79	15.48	-.54	.92	-.49	.12	-.51
20	13.01	15.65	-.67	.79	-.62	-.01	-.65
21	13.22	15.80	-.80	.66	-.75	-.14	-.78
22	13.42	15.95	-.92	.54	-.88	-.26	-.90
23	13.62	16.10	-1.03	.43	-.99	-.38	-1.01
24	13.80	16.23	-1.14	.32	-1.10	-.49	-1.12
25	13.98	16.37	-1.25	.21	-1.21	-.59	-1.23
26	14.15	16.50	-1.35	.11	-1.31	-.69	-1.33
27	14.31	16.62	-1.44	.01	-1.41	-.79	-1.43
28	14.47	16.74	-1.54	-.08	-1.51	-.88	-1.52
29	14.62	16.86	-1.62	-.17	-1.60	-.97	-1.61
30	14.77	16.97	-1.71	-.26	-1.68	-1.06	-1.70
31	14.91	17.08	-1.79	-.34	-1.77	-1.14	-1.78
32	15.05	17.19	-1.87	-.42	-1.85	-1.22	-1.86
33	15.19	17.29	-1.95	-.50	-1.93	-1.30	-1.94
34	15.31	17.40	-2.03	-.57	-2.00	-1.38	-2.02
35	15.44	17.49	-2.10	-.65	-2.08	-1.45	-2.09
36	15.56	17.59	-2.17	-.72	-2.15	-1.52	-2.16
37	15.68	17.68	-2.24	-.79	-2.22	-1.59	-2.23
38	15.80	17.78	-2.31	-.85	-2.28	-1.66	-2.30
39	15.91	17.87	-2.37	-.92	-2.35	-1.72	-2.36
40	16.02	17.95	-2.43	-.98	-2.41	-1.78	-2.42
41	16.13	18.04	-2.49	-1.04	-2.48	-1.85	-2.49
42	16.23	18.12	-2.55	-1.10	-2.54	-1.91	-2.55
43	16.33	18.20	-2.61	-1.16	-2.60	-1.96	-2.61
44	16.43	18.28	-2.67	-1.22	-2.65	-2.02	-2.66
45	16.53	18.36	-2.73	-1.27	-2.71	-2.08	-2.72

PROBABILITY OF DETECTION= .6000
CONSTANT FALSE ALARM PROBABILITY
LOG10(PFA)=- 3

N	N DB	THRESH DB	NON-FLUC DB	SWER 1 DB	SWER 2 DB	SWER 3 DB	SWER 4 DB
1	0.00	8.39	8.67	10.98	10.98	9.79	9.79
2	3.01	9.65	6.43	8.77	7.57	7.57	6.99
3	4.77	10.50	5.17	7.52	5.93	6.32	5.54
4	6.02	11.16	4.30	6.66	4.87	5.45	4.58
5	6.99	11.70	3.64	6.01	4.09	4.80	3.87
6	7.78	12.16	3.11	5.48	3.49	4.27	3.30
7	8.45	12.57	2.67	5.04	2.99	3.83	2.83
8	9.03	12.93	2.29	4.67	2.57	3.45	2.43
9	9.54	13.25	1.96	4.34	2.21	3.12	2.08
10	10.00	13.55	1.66	4.05	1.89	2.83	1.77
11	10.41	13.83	1.40	3.79	1.60	2.57	1.50
12	10.79	14.08	1.16	3.55	1.35	2.33	1.25
13	11.14	14.32	.94	3.33	1.11	2.11	1.03
14	11.46	14.54	.74	3.13	.90	1.91	.82
15	11.76	14.75	.55	2.95	.70	1.73	.63
16	12.04	14.95	.38	2.78	.52	1.55	.45
17	12.30	15.14	.22	2.62	.35	1.39	.28
18	12.55	15.31	.07	2.46	.19	1.24	.13
19	12.79	15.48	-.08	2.32	.04	1.10	-.02
20	13.01	15.65	-.21	2.19	-.10	.96	-.16
21	13.22	15.80	-.34	2.06	-.24	.83	-.29
22	13.42	15.95	-.47	1.94	-.36	.71	-.41
23	13.62	16.10	-.58	1.82	-.48	.60	-.53
24	13.80	16.23	-.69	1.71	-.60	.49	-.65
25	13.98	16.37	-.80	1.61	-.71	.38	-.75
26	14.15	16.50	-.90	1.51	-.82	.28	-.86
27	14.31	16.62	-1.00	1.41	-.92	.18	-.96
28	14.47	16.74	-1.09	1.32	-1.01	.09	-1.05
29	14.62	16.86	-1.18	1.23	-1.11	-.00	-1.15
30	14.77	16.97	-1.27	1.14	-1.20	-.09	-1.23
31	14.91	17.08	-1.35	1.06	-1.28	-.17	-1.32
32	15.05	17.19	-1.44	.98	-1.37	-.25	-1.40
33	15.19	17.29	-1.51	.90	-1.45	-.33	-1.48
34	15.31	17.40	-1.59	.82	-1.53	-.41	-1.56
35	15.44	17.49	-1.66	.75	-1.60	-.48	-1.63
36	15.56	17.59	-1.74	.68	-1.67	-.55	-1.71
37	15.68	17.68	-1.81	.61	-1.75	-.62	-1.78
38	15.80	17.78	-1.87	.54	-1.82	-.69	-1.84
39	15.91	17.87	-1.94	.48	-1.88	-.75	-1.91
40	16.02	17.95	-2.00	.41	-1.95	-.82	-1.98
41	16.13	18.04	-2.07	.35	-2.01	-.88	-2.04
42	16.23	18.12	-2.13	.29	-2.07	-.94	-2.10
43	16.33	18.20	-2.19	.23	-2.13	-1.00	-2.16
44	16.43	18.28	-2.24	.17	-2.19	-1.06	-2.22
45	16.53	18.36	-2.30	.12	-2.25	-1.11	-2.28

Appendix: Radar Detection Tables

PROBABILITY OF DETECTION= .7000
CONSTANT FALSE ALARM PROBABILITY
LOG10(PFA)=- 3

N	N DB	THRESH DB	NON-FLUC DB	SWER 1 DB	SWER 2 DB	SWER 3 DB	SWER 4 DB
1	0.00	8.39	9.27	12.64	12.64	10.94	10.94
2	3.01	9.65	7.00	10.42	8.70	8.70	7.86
3	4.77	10.50	5.72	9.17	6.87	7.44	6.30
4	6.02	11.16	4.83	8.30	5.71	6.57	5.28
5	6.99	11.70	4.16	7.65	4.87	5.91	4.52
6	7.78	12.16	3.62	7.12	4.22	5.38	3.92
7	8.45	12.57	3.17	6.68	3.69	4.93	3.43
8	9.03	12.93	2.78	6.30	3.24	4.55	3.02
9	9.54	13.25	2.45	5.97	2.86	4.22	2.65
10	10.00	13.55	2.15	5.68	2.52	3.93	2.34
11	10.41	13.83	1.88	5.41	2.22	3.66	2.05
12	10.79	14.08	1.64	5.18	1.95	3.42	1.79
13	11.14	14.32	1.42	4.96	1.70	3.21	1.56
14	11.46	14.54	1.21	4.76	1.48	3.00	1.35
15	11.76	14.75	1.02	4.57	1.27	2.82	1.15
16	12.04	14.95	.85	4.40	1.08	2.64	.96
17	12.30	15.14	.68	4.24	.90	2.48	.79
18	12.55	15.31	.53	4.09	.74	2.33	.63
19	12.79	15.48	.38	3.94	.58	2.19	.48
20	13.01	15.65	.24	3.81	.43	2.05	.34
21	13.22	15.80	.11	3.68	.29	1.92	.20
22	13.42	15.95	-.01	3.56	.16	1.80	.07
23	13.62	16.10	-.13	3.44	.04	1.68	-.05
24	13.80	16.23	-.24	3.33	-.08	1.57	-.16
25	13.98	16.37	-.35	3.23	-.20	1.46	-.27
26	14.15	16.50	-.45	3.13	-.31	1.36	-.38
27	14.31	16.62	-.55	3.03	-.41	1.26	-.48
28	14.47	16.74	-.65	2.94	-.51	1.17	-.58
29	14.62	16.86	-.74	2.85	-.61	1.08	-.67
30	14.77	16.97	-.83	2.76	-.70	.99	-.76
31	14.91	17.08	-.92	2.67	-.79	.91	-.85
32	15.05	17.19	-1.00	2.59	-.88	.83	-.94
33	15.19	17.29	-1.08	2.52	-.96	.75	-1.02
34	15.31	17.40	-1.15	2.44	-1.04	.67	-1.10
35	15.44	17.49	-1.23	2.37	-1.12	.60	-1.17
36	15.56	17.59	-1.30	2.29	-1.19	.53	-1.25
37	15.68	17.68	-1.37	2.23	-1.27	.46	-1.32
38	15.80	17.78	-1.44	2.16	-1.34	.39	-1.39
39	15.91	17.87	-1.51	2.09	-1.41	.32	-1.46
40	16.02	17.95	-1.57	2.03	-1.47	.26	-1.52
41	16.13	18.04	-1.64	1.97	-1.54	.20	-1.59
42	16.23	18.12	-1.70	1.91	-1.60	.14	-1.65
43	16.33	18.20	-1.76	1.85	-1.67	.08	-1.71
44	16.43	18.28	-1.82	1.79	-1.73	.02	-1.77
45	16.53	18.36	-1.87	1.73	-1.79	-.04	-1.83

PROBABILITY OF DETECTION= .8000
CONSTANT FALSE ALARM PROBABILITY
LOG10(PFA)=- 3

K	N DB	THRESH DB	NON-FLUC DB	SWER 1 DB	SWER 2 DB	SWER 3 DB	SWER 4 DB
1	0.00	8.39	9.93	14.76	14.76	12.34	12.34
2	3.01	9.65	7.62	12.53	10.09	10.09	8.87
3	4.77	10.50	6.31	11.27	8.00	8.82	7.18
4	6.02	11.16	5.41	10.41	6.71	7.94	6.08
5	6.99	11.70	4.73	9.75	5.78	7.27	5.27
6	7.78	12.16	4.18	9.22	5.07	6.74	4.64
7	8.45	12.57	3.72	8.77	4.50	6.29	4.12
8	9.03	12.93	3.33	8.40	4.01	5.91	3.68
9	9.54	13.25	2.98	8.06	3.60	5.58	3.30
10	10.00	13.55	2.68	7.77	3.24	5.28	2.96
11	10.41	13.83	2.41	7.51	2.92	5.01	2.67
12	10.79	14.08	2.16	7.27	2.63	4.77	2.40
13	11.14	14.32	1.93	7.05	2.37	4.55	2.16
14	11.46	14.54	1.73	6.85	2.14	4.35	1.93
15	11.76	14.75	1.53	6.66	1.92	4.16	1.73
16	12.04	14.95	1.35	6.49	1.72	3.99	1.54
17	12.30	15.14	1.19	6.33	1.53	3.82	1.36
18	12.55	15.31	1.03	6.18	1.35	3.67	1.19
19	12.79	15.48	.88	6.03	1.19	3.53	1.04
20	13.01	15.65	.74	5.90	1.04	3.39	.89
21	13.22	15.80	.61	5.77	.89	3.26	.75
22	13.42	15.95	.48	5.65	.75	3.14	.62
23	13.62	16.10	.36	5.53	.62	3.02	.49
24	13.80	16.23	.25	5.42	.50	2.91	.37
25	13.98	16.37	.14	5.32	.38	2.80	.26
26	14.15	16.50	.03	5.21	.26	2.70	.15
27	14.31	16.62	-.07	5.12	.15	2.60	.04
28	14.47	16.74	-.17	5.02	.05	2.51	-.06
29	14.62	16.86	-.26	4.93	-.05	2.41	-.15
30	14.77	16.97	-.35	4.84	-.15	2.33	-.25
31	14.91	17.08	-.44	4.76	-.24	2.24	-.34
32	15.05	17.19	-.52	4.68	-.33	2.16	-.42
33	15.19	17.29	-.60	4.60	-.41	2.08	-.51
34	15.31	17.40	-.68	4.53	-.50	2.00	-.59
35	15.44	17.49	-.75	4.45	-.58	1.93	-.67
36	15.56	17.59	-.83	4.38	-.66	1.86	-.74
37	15.68	17.68	-.90	4.31	-.73	1.79	-.82
38	15.80	17.78	-.97	4.24	-.81	1.72	-.89
39	15.91	17.87	-1.04	4.18	-.88	1.65	-.96
40	16.02	17.95	-1.10	4.11	-.95	1.59	-1.02
41	16.13	18.04	-1.17	4.05	-1.02	1.53	-1.09
42	16.23	18.12	-1.23	3.99	-1.08	1.47	-1.16
43	16.33	18.20	-1.29	3.93	-1.15	1.41	-1.22
44	16.43	18.28	-1.35	3.87	-1.21	1.35	-1.28
45	16.53	18.36	-1.41	3.82	-1.27	1.29	-1.34

Appendix: Radar Detection Tables

PROBABILITY OF DETECTION= .9000
CONSTANT FALSE ALARM PROBABILITY
LOG10(PFA)=- 3

N	N DB	THRESH DB	NON-FLUC DB	SWER 1 DB	SWER 2 DB	SWER 3 DB	SWER 4 DB
1	0.00	8.39	10.76	18.10	18.10	14.41	14.41
2	3.01	9.65	8.40	15.86	12.14	12.14	10.31
3	4.77	10.50	7.07	14.60	9.63	10.86	8.39
4	6.02	11.16	6.15	13.73	8.12	9.98	7.17
5	6.99	11.70	5.45	13.07	7.06	9.30	6.28
6	7.78	12.16	4.89	12.53	6.25	8.76	5.59
7	8.45	12.57	4.42	12.09	5.61	8.31	5.03
8	9.03	12.93	4.02	11.71	5.07	7.93	4.56
9	9.54	13.25	3.67	11.38	4.62	7.59	4.16
10	10.00	13.55	3.36	11.08	4.22	7.30	3.80
11	10.41	13.83	3.08	10.82	3.87	7.03	3.48
12	10.79	14.08	2.83	10.58	3.56	6.79	3.20
13	11.14	14.32	2.60	10.36	3.28	6.56	2.94
14	11.46	14.54	2.38	10.16	3.02	6.36	2.71
15	11.76	14.75	2.19	9.97	2.78	6.17	2.49
16	12.04	14.95	2.00	9.80	2.57	6.00	2.29
17	12.30	15.14	1.83	9.64	2.37	5.83	2.10
18	12.55	15.31	1.67	9.48	2.18	5.68	1.93
19	12.79	15.48	1.52	9.34	2.00	5.53	1.77
20	13.01	15.65	1.38	9.20	1.84	5.39	1.61
21	13.22	15.80	1.24	9.07	1.68	5.26	1.47
22	13.42	15.95	1.11	8.95	1.54	5.14	1.33
23	13.62	16.10	.99	8.84	1.40	5.02	1.20
24	13.80	16.23	.87	8.72	1.26	4.91	1.07
25	13.98	16.37	.76	8.62	1.14	4.80	.95
26	14.15	16.50	.66	8.52	1.02	4.70	.84
27	14.31	16.62	.55	8.42	.90	4.60	.73
28	14.47	16.74	.45	8.32	.79	4.51	.62
29	14.62	16.86	.36	8.23	.69	4.41	.52
30	14.77	16.97	.27	8.14	.58	4.33	.43
31	14.91	17.08	.18	8.06	.49	4.24	.33
32	15.05	17.19	.09	7.98	.39	4.16	.24
33	15.19	17.29	.01	7.90	.30	4.08	.16
34	15.31	17.40	-.07	7.82	.21	4.00	.07
35	15.44	17.49	-.15	7.75	.13	3.93	-.01
36	15.56	17.59	-.22	7.68	.05	3.86	-.09
37	15.68	17.68	-.30	7.61	-.03	3.78	-.16
38	15.80	17.78	-.37	7.54	-.11	3.72	-.24
39	15.91	17.87	-.44	7.47	-.19	3.65	-.31
40	16.02	17.95	-.50	7.41	-.26	3.59	-.38
41	16.13	18.04	-.57	7.35	-.33	3.52	-.45
42	16.23	18.12	-.63	7.29	-.40	3.46	-.51
43	16.33	18.20	-.69	7.23	-.47	3.40	-.58
44	16.43	18.28	-.75	7.17	-.53	3.34	-.64
45	16.53	18.36	-.81	7.11	-.59	3.29	-.70

PROBABILITY OF DETECTION= .9500
CONSTANT FALSE ALARM PROBABILITY
LOG10(PFA)=- 3

N	N DB	THRESH DB	NON-FLUC DB	SWER 1 DB	SWER 2 DB	SWER 3 DB	SWER 4 DB
1	0.00	8.39	11.39	21.25	21.25	16.26	16.26
2	3.01	9.05	9.00	19.02	13.98	13.98	11.53
3	4.77	10.50	7.65	17.75	11.05	12.69	9.39
4	6.02	11.16	6.72	16.88	9.33	11.80	8.06
5	6.99	11.70	6.01	16.22	8.14	11.13	7.10
6	7.78	12.16	5.43	15.68	7.24	10.59	6.37
7	8.45	12.57	4.96	15.24	6.53	10.14	5.77
8	9.03	12.93	4.55	14.86	5.94	9.75	5.27
9	9.54	13.25	4.19	14.53	5.45	9.41	4.84
10	10.00	13.55	3.88	14.23	5.02	9.11	4.46
11	10.41	13.83	3.59	13.96	4.64	8.85	4.13
12	10.79	14.08	3.34	13.72	4.31	8.60	3.84
13	11.14	14.32	3.10	13.50	4.00	8.38	3.57
14	11.46	14.54	2.89	13.30	3.73	8.18	3.32
15	11.76	14.75	2.69	13.11	3.48	7.99	3.09
16	12.04	14.95	2.50	12.94	3.25	7.81	2.88
17	12.30	15.14	2.33	12.78	3.04	7.65	2.69
18	12.55	15.31	2.16	12.62	2.84	7.49	2.51
19	12.79	15.48	2.01	12.48	2.65	7.35	2.34
20	13.01	15.65	1.86	12.34	2.48	7.21	2.18
21	13.22	15.80	1.73	12.22	2.31	7.07	2.03
22	13.42	15.95	1.60	12.09	2.16	6.95	1.88
23	13.62	16.10	1.47	11.98	2.01	6.83	1.75
24	13.80	16.23	1.35	11.87	1.87	6.72	1.62
25	13.98	16.37	1.24	11.76	1.74	6.61	1.49
26	14.15	16.50	1.13	11.66	1.61	6.51	1.38
27	14.31	16.62	1.02	11.56	1.49	6.41	1.26
28	14.47	16.74	.92	11.47	1.38	6.31	1.16
29	14.62	16.86	.83	11.38	1.27	6.22	1.05
30	14.77	16.97	.74	11.29	1.16	6.13	.95
31	14.91	17.08	.65	11.20	1.06	6.05	.86
32	15.05	17.19	.56	11.12	.96	5.96	.76
33	15.19	17.29	.48	11.04	.86	5.88	.67
34	15.31	17.40	.39	10.97	.77	5.81	.59
35	15.44	17.49	.32	10.89	.68	5.73	.50
36	15.56	17.59	.24	10.82	.60	5.66	.42
37	15.68	17.68	.17	10.75	.52	5.59	.34
38	15.80	17.78	.09	10.69	.44	5.52	.27
39	15.91	17.87	.02	10.62	.36	5.45	.19
40	16.02	17.95	-.04	10.56	.28	5.39	.12
41	16.13	18.04	-.11	10.49	.21	5.33	.05
42	16.23	18.12	-.18	10.43	.14	5.26	-.02
43	16.33	18.20	-.24	10.37	.07	5.20	-.08
44	16.43	18.28	-.30	10.32	-.00	5.15	-.15
45	16.53	18.36	-.36	10.26	-.07	5.09	-.21

Appendix: Radar Detection Tables

```
PROBABILITY OF DETECTION= .9900
CONSTANT FALSE ALARM PROBABILITY
LOG10(PFA)=- 3
```

N	N DB	THRESH DB	NON-FLUC DB	SWER 1 DB	SWER 2 DB	SWER 3 DB	SWER 4 DB
1	0.00	8.39	12.46	28.36	28.36	20.15	20.15
2	3.01	9.65	10.02	26.11	17.85	17.84	13.94
3	4.77	10.50	8.83	24.85	13.94	16.57	11.33
4	6.02	11.16	7.67	23.97	11.71	15.68	9.75
5	6.99	11.70	6.95	23.31	10.24	15.01	8.64
6	7.78	12.16	6.36	22.78	9.15	14.46	7.79
7	8.45	12.57	5.88	22.33	8.29	14.01	7.12
8	9.03	12.93	5.46	21.95	7.60	13.62	6.56
9	9.54	13.25	5.09	21.62	7.01	13.26	6.08
10	10.00	13.55	4.76	21.32	6.52	12.96	5.67
11	10.41	13.83	4.47	21.06	6.08	12.69	5.30
12	10.79	14.08	4.21	20.82	5.70	12.44	4.98
13	11.14	14.32	3.97	20.60	5.36	12.22	4.68
14	11.46	14.54	3.75	20.40	5.05	12.02	4.42
15	11.76	14.75	3.54	20.21	4.77	11.83	4.17
16	12.04	14.95	3.35	20.04	4.50	11.65	3.94
17	12.30	15.14	3.18	19.88	4.26	11.49	3.73
18	12.55	15.31	3.01	19.72	4.04	11.33	3.53
19	12.79	15.48	2.85	19.58	3.83	11.19	3.35
20	13.01	15.65	2.70	19.44	3.64	11.05	3.18
21	13.22	15.80	2.56	19.36	3.45	10.92	3.02
22	13.42	15.95	2.42	19.23	3.28	10.80	2.87
23	13.62	16.10	2.30	19.11	3.12	10.68	2.72
24	13.80	16.23	2.17	19.00	2.97	10.57	2.58
25	13.98	16.37	2.06	18.89	2.83	10.46	2.45
26	14.15	16.50	1.95	18.78	2.69	10.35	2.33
27	14.31	16.62	1.84	18.68	2.56	10.26	2.21
28	14.47	16.74	1.74	18.58	2.43	10.16	2.09
29	14.62	16.86	1.64	18.49	2.31	10.07	1.98
30	14.77	16.97	1.54	18.40	2.19	9.98	1.87
31	14.91	17.08	1.45	18.31	2.08	9.90	1.77
32	15.05	17.19	1.36	18.23	1.98	9.81	1.67
33	15.19	17.29	1.27	18.15	1.87	9.73	1.58
34	15.31	17.40	1.19	18.07	1.77	9.66	1.49
35	15.44	17.49	1.11	17.99	1.68	9.58	1.40
36	15.56	17.59	1.03	17.92	1.59	9.51	1.31
37	15.68	17.68	.96	17.85	1.50	9.44	1.23
38	15.8	17.78	.88	17.79	1.41	9.37	1.15
39	15.91	17.87	.81	17.73	1.33	9.30	1.07
40	16.02	17.95	.74	17.67	1.25	9.24	1.00
41	16.13	18.04	.67	17.61	1.17	9.17	.92
42	16.23	18.12	.61	17.55	1.09	9.11	.85
43	16.33	18.20	.54	17.49	1.01	9.05	.78
44	16.43	18.28	.48	17.44	.94	8.99	.71
45	16.53	18.36	.42	17.38	.87	8.96	.65

PROBABILITY OF DETECTION= .5000
CONSTANT FALSE ALARM PROBABILITY
LOG10(PFA)=- 4

N	N DB	THRESH DB	NON-FLUC DB	SWER 1 DB	SWER 2 DB	SWER 3 DB	SWER 4 DB
1	0.00	9.64	9.40	10.89	10.89	10.09	10.09
2	3.01	10.70	7.09	8.59	7.79	7.78	7.41
3	4.77	11.44	5.80	7.30	6.24	6.49	6.01
4	6.02	12.02	4.91	6.41	5.23	5.60	5.06
5	6.99	12.50	4.23	5.74	4.48	4.92	4.35
6	7.78	12.92	3.69	5.19	3.89	4.38	3.79
7	8.45	13.28	3.24	4.74	3.41	3.93	3.32
8	9.03	13.61	2.85	4.35	3.00	3.54	2.92
9	9.54	13.91	2.51	4.02	2.64	3.20	2.57
10	10.00	14.18	2.22	3.72	2.33	2.90	2.27
11	10.41	14.43	1.95	3.45	2.05	2.63	1.99
12	10.79	14.67	1.70	3.20	1.79	2.39	1.75
13	11.14	14.89	1.48	2.98	1.56	2.17	1.52
14	11.46	15.10	1.28	2.78	1.35	1.96	1.31
15	11.76	15.29	1.09	2.59	1.16	1.77	1.12
16	12.04	15.48	.91	2.41	.98	1.60	.94
17	12.30	15.65	.75	2.25	.81	1.43	.78
18	12.55	15.82	.59	2.09	.65	1.28	.62
19	12.79	15.98	.45	1.95	.50	1.13	.47
20	13.01	16.13	.31	1.81	.36	.99	.33
21	13.22	16.28	.18	1.68	.23	.86	.20
22	13.42	16.42	.06	1.55	.10	.74	.08
23	13.62	16.55	-.06	1.44	-.02	.62	-.04
24	13.80	16.68	-.17	1.32	-.14	.51	-.16
25	13.98	16.81	-.28	1.22	-.24	.40	-.26
26	14.15	16.93	-.39	1.11	-.35	.30	-.37
27	14.31	17.05	-.48	1.01	-.45	.20	-.47
28	14.47	17.17	-.58	.92	-.55	.10	-.56
29	14.62	17.28	-.67	.83	-.64	.01	-.66
30	14.77	17.38	-.76	.74	-.73	-.08	-.74
31	14.91	17.49	-.84	.65	-.82	-.16	-.83
32	15.05	17.59	-.93	.57	-.90	-.24	-.91
33	15.19	17.69	-1.01	.49	-.98	-.32	-.99
34	15.31	17.79	-1.08	.42	-1.06	-.40	-1.07
35	15.44	17.88	-1.16	.34	-1.13	-.47	-1.15
36	15.56	17.97	-1.23	.27	-1.21	-.55	-1.22
37	15.68	18.06	-1.30	.20	-1.28	-.62	-1.29
38	15.80	18.15	-1.37	.13	-1.35	-.69	-1.36
39	15.91	18.23	-1.43	.06	-1.41	-.75	-1.42
40	16.02	18.32	-1.50	-.00	-1.48	-.82	-1.49
41	16.13	18.40	-1.56	-.06	-1.54	-.88	-1.55
42	16.23	18.48	-1.62	-.13	-1.60	-.94	-1.61
43	16.33	18.56	-1.68	-.19	-1.66	-1.00	-1.67
44	16.43	18.64	-1.74	-.24	-1.72	-1.06	-1.73
45	16.53	18.71	-1.80	-.30	-1.78	-1.12	-1.79

PROBABILITY OF DETECTION= .6000
CONSTANT FALSE ALARM PROBABILITY
LOG10(PFA)=- 4

N	N DB	THRESH DB	NON-FLUC DB	SWER 1 DB	SWER 2 DB	SWER 3 DB	SWER 4 DB
1	0.00	9.64	9.92	12.31	12.31	11.09	11.09
2	3.01	10.70	7.58	10.00	8.77	8.77	8.18
3	4.77	11.44	6.27	8.70	7.07	7.47	6.67
4	6.02	12.02	5.37	7.81	5.97	6.58	5.67
5	6.99	12.50	4.68	7.13	5.17	5.90	4.93
6	7.78	12.92	4.13	6.59	4.54	5.35	4.34
7	8.45	13.28	3.67	6.13	4.02	4.89	3.85
8	9.03	13.61	3.28	5.74	3.59	4.50	3.43
9	9.54	13.91	2.94	5.40	3.21	4.16	3.08
10	10.00	14.18	2.64	5.10	2.88	3.86	2.76
11	10.41	14.43	2.36	4.83	2.59	3.59	2.47
12	10.79	14.67	2.12	4.59	2.32	3.35	2.22
13	11.14	14.89	1.89	4.37	2.08	3.12	1.99
14	11.46	15.10	1.68	4.16	1.86	2.92	1.77
15	11.76	15.29	1.49	3.97	1.66	2.73	1.57
16	12.04	15.48	1.31	3.79	1.47	2.55	1.39
17	12.30	15.65	1.15	3.63	1.29	2.38	1.22
18	12.55	15.82	.99	3.47	1.13	2.23	1.06
19	12.79	15.98	.84	3.33	.97	2.08	.91
20	13.01	16.13	.70	3.19	.83	1.94	.77
21	13.22	16.28	.57	3.06	.69	1.81	.63
22	13.42	16.42	.45	2.93	.56	1.69	.50
23	13.62	16.55	.33	2.82	.43	1.57	.38
24	13.80	16.68	.21	2.70	.32	1.45	.26
25	13.98	16.81	.10	2.59	.20	1.35	.15
26	14.15	16.93	-.00	2.49	.09	1.24	.05
27	14.31	17.05	-.10	2.39	-.01	1.14	-.06
28	14.47	17.17	-.20	2.30	-.11	1.05	-.15
29	14.62	17.28	-.29	2.20	-.20	.96	-.25
30	14.77	17.38	-.38	2.12	-.30	.87	-.34
31	14.91	17.49	-.47	2.03	-.39	.78	-.43
32	15.05	17.59	-.55	1.95	-.47	.70	-.51
33	15.19	17.69	-.63	1.87	-.55	.62	-.59
34	15.31	17.79	-.71	1.79	-.63	.54	-.67
35	15.44	17.88	-.78	1.72	-.71	.47	-.75
36	15.56	17.97	-.86	1.64	-.79	.39	-.82
37	15.68	18.06	-.93	1.57	-.86	.32	-.89
38	15.80	18.15	-1.00	1.50	-.93	.25	-.96
39	15.91	18.23	-1.06	1.44	-1.00	.19	-1.03
40	16.02	18.32	-1.13	1.37	-1.07	.12	-1.10
41	16.13	18.40	-1.19	1.31	-1.13	.06	-1.16
42	16.23	18.48	-1.25	1.25	-1.20	-.00	-1.23
43	16.33	18.56	-1.31	1.19	-1.26	-.06	-1.29
44	16.43	18.64	-1.37	1.13	-1.32	-.12	-1.35
45	16.53	18.71	-1.43	1.07	-1.38	-.18	-1.40

PROBABILITY OF DETECTION= .7000
CONSTANT FALSE ALARM PROBABILITY
LOG10(PFA)=- 4

N	N DB	THRESH DB	NON-FLUC DB	SWER 1 DB	SWER 2 DB	SWER 3 DB	SWER 4 DB
1	0.00	9.64	10.44	13.95	13.95	12.20	12.20
2	3.01	10.70	8.08	11.63	9.87	9.87	9.00
3	4.77	11.44	6.75	10.33	7.98	8.57	7.38
4	6.02	12.02	5.83	9.43	6.77	7.66	6.32
5	6.99	12.50	5.14	8.75	5.90	6.98	5.53
6	7.78	12.92	4.58	8.20	5.23	6.43	4.91
7	8.45	13.28	4.12	7.75	4.68	5.97	4.40
8	9.03	13.61	3.72	7.36	4.21	5.58	3.97
9	9.54	13.91	3.37	7.02	3.81	5.24	3.60
10	10.00	14.18	3.06	6.72	3.46	4.93	3.27
11	10.41	14.43	2.79	6.45	3.15	4.66	2.97
12	10.79	14.67	2.54	6.20	2.88	4.42	2.71
13	11.14	14.89	2.31	5.98	2.62	4.19	2.47
14	11.46	15.10	2.10	5.77	2.39	3.98	2.25
15	11.76	15.29	1.90	5.58	2.18	3.79	2.04
16	12.04	15.48	1.72	5.40	1.98	3.61	1.85
17	12.30	15.65	1.55	5.24	1.80	3.45	1.68
18	12.55	15.82	1.39	5.08	1.63	3.29	1.51
19	12.79	15.98	1.24	4.93	1.47	3.14	1.36
20	13.01	16.13	1.10	4.79	1.31	3.00	1.21
21	13.22	16.28	.97	4.66	1.17	2.87	1.07
22	13.42	16.42	.84	4.54	1.03	2.75	.94
23	13.62	16.55	.72	4.42	.91	2.63	.81
24	13.80	16.68	.61	4.31	.78	2.51	.69
25	13.98	16.81	.49	4.20	.67	2.41	.58
26	14.15	16.93	.39	4.09	.55	2.30	.47
27	14.31	17.05	.29	4.00	.45	2.20	.37
28	14.47	17.17	.19	3.90	.34	2.10	.27
29	14.62	17.28	.10	3.81	.24	2.01	.17
30	14.77	17.38	.01	3.72	.15	1.92	.08
31	14.91	17.49	-.08	3.63	.06	1.84	-.01
32	15.05	17.59	-.17	3.55	-.03	1.75	-.10
33	15.19	17.69	-.25	3.47	-.12	1.67	-.18
34	15.31	17.79	-.33	3.39	-.20	1.60	-.26
35	15.44	17.88	-.40	3.32	-.28	1.52	-.34
36	15.56	17.97	-.48	3.24	-.36	1.45	-.42
37	15.68	18.06	-.55	3.17	-.43	1.38	-.49
38	15.80	18.15	-.62	3.10	-.50	1.31	-.56
39	15.91	18.23	-.69	3.04	-.58	1.24	-.63
40	16.02	18.32	-.75	2.97	-.64	1.17	-.70
41	16.13	18.40	-.82	2.91	-.71	1.11	-.76
42	16.23	18.48	-.88	2.85	-.78	1.05	-.83
43	16.33	18.56	-.94	2.79	-.84	.99	-.89
44	16.43	18.64	-1.00	2.73	-.90	.93	-.95
45	16.53	18.71	-1.06	2.67	-.96	.87	-1.01

Appendix: Radar Detection Tables

PROBABILITY OF DETECTION= .8000
CONSTANT FALSE ALARM PROBABILITY
LOG10(PFA)=- 4

N	N DB	THRESH DB	NON-FLUC DB	SWER 1 DB	SWER 2 DB	SWER 3 DB	SWER 4 DB
1	0.00	9.64	11.01	16.05	16.05	13.57	13.57
2	3.01	10.79	8.62	13.73	11.23	11.23	9.97
3	4.77	11.44	7.27	12.42	9.07	9.91	8.21
4	6.02	12.02	6.35	11.52	7.73	9.00	7.06
5	6.99	12.50	5.64	10.84	6.77	8.31	6.23
6	7.78	12.92	5.07	10.29	6.03	7.76	5.57
7	8.45	13.28	4.60	9.83	5.44	7.30	5.03
8	9.03	13.61	4.20	9.44	4.94	6.91	4.58
9	9.54	13.91	3.84	9.10	4.51	6.56	4.19
10	10.00	14.18	3.53	8.80	4.14	6.26	3.84
11	10.41	14.43	3.25	8.52	3.81	5.99	3.54
12	10.79	14.67	3.00	8.28	3.51	5.74	3.26
13	11.14	14.89	2.76	8.05	3.24	5.51	3.01
14	11.46	15.10	2.55	7.85	3.00	5.30	2.78
15	11.76	15.29	2.35	7.66	2.78	5.11	2.57
16	12.04	15.48	2.17	7.48	2.57	4.93	2.37
17	12.30	15.65	2.00	7.31	2.37	4.77	2.19
18	12.55	15.82	1.84	7.16	2.19	4.61	2.02
19	12.79	15.98	1.69	7.01	2.03	4.46	1.86
20	13.01	16.13	1.54	6.87	1.87	4.32	1.71
21	13.22	16.28	1.41	6.74	1.72	4.19	1.56
22	13.42	16.42	1.28	6.61	1.58	4.06	1.43
23	13.62	16.55	1.15	6.49	1.44	3.94	1.30
24	13.81	16.68	1.04	6.38	1.31	3.83	1.18
25	13.98	16.81	.93	6.27	1.19	3.72	1.06
26	14.15	16.93	.82	6.17	1.07	3.61	.95
27	14.31	17.05	.72	6.07	.96	3.51	.84
28	14.47	17.17	.62	5.97	.86	3.42	.74
29	14.62	17.28	.52	5.88	.75	3.32	.64
30	14.77	17.38	.43	5.79	.65	3.23	.54
31	14.91	17.49	.34	5.71	.56	3.15	.45
32	15.05	17.59	.26	5.62	.47	3.06	.36
33	15.19	17.69	.17	5.54	.38	2.98	.28
34	15.31	17.79	.09	5.46	.29	2.91	.19
35	15.44	17.88	.01	5.39	.21	2.83	.11
36	15.56	17.97	-.06	5.32	.13	2.76	.04
37	15.68	18.06	-.13	5.25	.05	2.69	-.04
38	15.8	18.15	-.20	5.18	-.02	2.62	-.11
39	15.91	18.23	-.27	5.11	-.10	2.55	-.18
40	16.02	18.32	-.34	5.04	-.17	2.48	-.25
41	16.13	18.40	-.41	4.98	-.24	2.42	-.32
42	16.23	18.48	-.47	4.92	-.30	2.36	-.39
43	16.33	18.56	-.53	4.86	-.37	2.30	-.45
44	16.43	18.64	-.59	4.80	-.43	2.24	-.51
45	16.53	18.71	-.65	4.74	-.50	2.18	-.57

PROBABILITY OF DETECTION= .9000
CONSTANT FALSE ALARM PROBABILITY
LOG10(PFA)=- 4

N	N DB	THRESH DB	NON-FLUC DB	SWER 1 DB	SWER 2 DB	SWER 3 DB	SWER 4 DB
1	0.00	9.64	11.75	19.37	19.37	15.60	15.60
2	3.01	10.70	9.32	17.03	13.24	13.24	11.35
3	4.77	11.44	7.95	15.72	10.66	11.92	9.36
4	6.02	12.02	7.00	14.82	9.10	11.01	8.10
5	6.99	12.50	6.29	14.14	8.00	10.32	7.18
6	7.78	12.92	5.71	13.59	7.17	9.76	6.47
7	8.45	13.28	5.23	13.13	6.50	9.30	5.89
8	9.03	13.61	4.82	12.74	5.95	8.90	5.40
9	9.54	13.91	4.46	12.39	5.47	8.56	4.98
10	10.00	14.18	4.14	12.09	5.07	8.25	4.62
11	10.41	14.43	3.85	11.82	4.71	7.98	4.29
12	10.79	14.67	3.59	11.57	4.38	7.73	4.00
13	11.14	14.89	3.36	11.35	4.09	7.50	3.74
14	11.46	15.10	3.14	11.14	3.83	7.29	3.49
15	11.76	15.29	2.94	10.95	3.59	7.10	3.27
16	12.04	15.48	2.75	10.77	3.36	6.92	3.07
17	12.30	15.65	2.58	10.61	3.16	6.75	2.87
18	12.55	15.82	2.41	10.45	2.96	6.59	2.69
19	12.79	15.98	2.26	10.30	2.78	6.44	2.53
20	13.01	16.13	2.11	10.16	2.61	6.30	2.37
21	13.22	16.28	1.97	10.03	2.45	6.17	2.22
22	13.42	16.42	1.84	9.91	2.30	6.04	2.08
23	13.62	16.55	1.72	9.79	2.16	5.92	1.94
24	13.80	16.68	1.60	9.67	2.02	5.81	1.81
25	13.98	16.81	1.48	9.56	1.89	5.70	1.69
26	14.15	16.93	1.37	9.46	1.77	5.59	1.58
27	14.31	17.05	1.27	9.36	1.65	5.49	1.46
28	14.47	17.17	1.17	9.26	1.54	5.39	1.36
29	14.62	17.28	1.07	9.17	1.43	5.30	1.25
30	14.77	17.38	.98	9.08	1.33	5.21	1.16
31	14.91	17.49	.89	9.00	1.23	5.12	1.06
32	15.05	17.59	.80	8.91	1.13	5.04	.97
33	15.19	17.69	.72	8.83	1.04	4.96	.88
34	15.31	17.79	.63	8.75	.95	4.88	.79
35	15.44	17.88	.56	8.68	.86	4.80	.71
36	15.56	17.97	.48	8.61	.77	4.73	.63
37	15.68	18.06	.40	8.54	.69	4.66	.55
38	15.80	18.15	.33	8.47	.61	4.59	.48
39	15.91	18.23	.26	8.40	.54	4.52	.40
40	16.02	18.32	.19	8.33	.46	4.46	.33
41	16.13	18.40	.13	8.27	.39	4.39	.26
42	16.23	18.48	.06	8.21	.32	4.33	.19
43	16.33	18.56	-.00	8.15	.25	4.27	.13
44	16.43	18.64	-.06	8.09	.18	4.21	.06
45	16.53	18.71	-.12	8.03	.12	4.15	.00

Appendix: Radar Detection Tables

```
PROBABILITY OF DETECTION= .9500
CONSTANT FALSE ALARM PROBABILITY
LOG10(PFA)=- 4
```

N	N DB	THRESH DB	NON-FLUC DB	SWER 1 DB	SWER 2 DB	SWER 3 DB	SWER 4 DB
1	0.00	9.64	12.31	22.52	22.52	17.43	17.43
2	3.01	10.70	9.06	20.18	15.06	15.06	12.54
3	4.77	11.44	8.47	18.87	12.05	13.74	10.33
4	6.02	12.02	7.51	17.97	10.27	12.82	8.95
5	6.99	12.50	6.78	17.28	9.04	12.13	7.96
6	7.78	12.92	6.20	16.73	8.12	11.57	7.20
7	8.45	13.28	5.71	16.27	7.39	11.10	6.58
8	9.03	13.61	5.29	15.88	6.79	10.71	6.07
9	9.54	13.91	4.93	15.54	6.27	10.36	5.63
10	10.00	14.18	4.61	15.24	5.83	10.05	5.24
11	10.41	14.43	4.32	14.96	5.44	9.78	4.90
12	10.79	14.67	4.05	14.72	5.10	9.53	4.59
13	11.14	14.89	3.81	14.49	4.79	9.30	4.32
14	11.46	15.10	3.59	14.29	4.50	9.09	4.06
15	11.76	15.29	3.39	14.09	4.24	8.90	3.83
16	12.04	15.48	3.20	13.91	4.01	8.71	3.62
17	12.30	15.65	3.02	13.76	3.79	8.55	3.42
18	12.55	15.82	2.85	13.60	3.58	8.39	3.23
19	12.79	15.98	2.70	13.45	3.39	8.24	3.06
20	13.01	16.13	2.55	13.31	3.21	8.10	2.89
21	13.22	16.28	2.41	13.18	3.04	7.96	2.74
22	13.42	16.42	2.28	13.05	2.89	7.84	2.59
23	13.62	16.55	2.15	12.93	2.74	7.72	2.45
24	13.80	16.68	2.03	12.81	2.59	7.60	2.32
25	13.98	16.81	1.91	12.71	2.46	7.49	2.19
26	14.15	16.93	1.80	12.60	2.33	7.39	2.07
27	14.31	17.05	1.70	12.50	2.20	7.29	1.96
28	14.47	17.17	1.59	12.40	2.08	7.19	1.85
29	14.62	17.28	1.50	12.31	1.97	7.10	1.74
30	14.77	17.38	1.40	12.22	1.86	7.01	1.64
31	14.91	17.49	1.31	12.14	1.76	6.92	1.54
32	15.05	17.59	1.22	12.05	1.66	6.83	1.44
33	15.19	17.69	1.14	11.97	1.56	6.75	1.35
34	15.31	17.79	1.05	11.90	1.47	6.67	1.26
35	15.44	17.88	.97	11.82	1.38	6.60	1.18
36	15.56	17.97	.90	11.75	1.29	6.52	1.10
37	15.68	18.06	.82	11.68	1.20	6.45	1.02
38	15.80	18.15	.75	11.61	1.12	6.38	.94
39	15.91	18.23	.68	11.54	1.04	6.32	.86
40	16.02	18.32	.61	11.48	.96	6.25	.79
41	16.13	18.40	.54	11.41	.89	6.18	.72
42	16.23	18.48	.47	11.35	.82	6.12	.65
43	16.33	18.56	.41	11.29	.74	6.06	.58
44	16.43	18.64	.35	11.23	.67	6.00	.51
45	16.53	18.71	.29	11.17	.61	5.94	.45

PROBABILITY OF DETECTION= .9900
CONSTANT FALSE ALARM PROBABILITY
LOG10(PFA)=- 4

N	N DB	THRESH DB	NON-FLUC DB	SWER 1 DB	SWER 2 DB	SWER 3 DB	SWER 4 DB
1	0.00	9.64	13.28	29.62	29.62	21.31	21.31
2	3.01	10.70	10.78	27.32	18.93	18.93	14.92
3	4.77	11.44	9.36	25.99	14.91	17.61	12.22
4	6.02	12.02	8.39	25.08	12.63	16.68	10.58
5	6.99	12.50	7.64	24.39	11.11	15.98	9.44
6	7.78	12.92	7.04	23.84	9.99	15.42	8.57
7	8.45	13.28	6.55	23.39	9.10	14.95	7.87
8	9.03	13.61	6.12	23.00	8.39	14.56	7.30
9	9.54	13.91	5.74	22.66	7.79	14.21	6.81
10	10.00	14.18	5.41	22.36	7.28	13.90	6.38
11	10.41	14.43	5.11	22.09	6.83	13.64	6.00
12	10.79	14.67	4.84	21.85	6.43	13.39	5.67
13	11.14	14.89	4.60	21.62	6.08	13.15	5.36
14	11.46	15.10	4.37	21.42	5.76	12.94	5.09
15	11.76	15.29	4.16	21.19	5.47	12.74	4.84
16	12.04	15.48	3.97	21.01	5.20	12.57	4.61
17	12.30	15.65	3.79	20.84	4.96	12.40	4.39
18	12.55	15.82	3.62	20.68	4.73	12.25	4.19
19	12.79	15.98	3.46	20.54	4.52	12.08	4.01
20	13.01	16.13	3.31	20.40	4.32	11.94	3.83
21	13.22	16.28	3.16	20.27	4.14	11.81	3.66
22	13.42	16.42	3.03	20.14	3.96	11.68	3.51
23	13.62	16.55	2.90	20.02	3.79	11.56	3.36
24	13.80	16.68	2.77	19.91	3.63	11.44	3.22
25	13.98	16.81	2.66	19.80	3.48	11.33	3.08
26	14.15	16.93	2.54	19.69	3.34	11.23	2.96
27	14.31	17.05	2.43	19.59	3.21	11.13	2.83
28	14.47	17.17	2.33	19.50	3.08	11.03	2.72
29	14.62	17.28	2.23	19.41	2.95	10.94	2.60
30	14.77	17.38	2.13	19.32	2.83	10.85	2.49
31	14.91	17.49	2.04	19.23	2.72	10.76	2.39
32	15.05	17.59	1.95	19.15	2.61	10.68	2.29
33	15.19	17.69	1.86	19.07	2.51	10.59	2.19
34	15.31	17.79	1.78	18.99	2.40	10.52	2.10
35	15.44	17.88	1.69	18.91	2.31	10.44	2.01
36	15.56	17.97	1.61	18.84	2.21	10.37	1.92
37	15.68	18.06	1.54	18.77	2.12	10.29	1.84
38	15.80	18.15	1.46	18.70	2.03	10.22	1.76
39	15.91	18.23	1.39	18.63	1.95	10.16	1.68
40	16.02	18.32	1.32	18.57	1.86	10.09	1.60
41	16.13	18.40	1.25	18.51	1.78	10.03	1.52
42	16.23	18.48	1.18	18.44	1.70	9.97	1.45
43	16.33	18.56	1.12	18.38	1.63	9.90	1.38
44	16.43	18.64	1.05	18.32	1.55	9.85	1.31
45	16.53	18.71	.99	18.27	1.48	9.79	1.24

Appendix: Radar Detection Tables

```
PROBABILITY OF DETECTION= .5000
CONSTANT FALSE ALARM PROBABILITY
LOG10(PFA)=- 5
```

N	N DB	THRESH DB	NON-FLUC DB	SWER 1 DB	SWER 2 DB	SWER 3 DB	SWER 4 DB
1	0.00	10.01	10.42	11.93	11.93	11.12	11.12
2	3.01	11.53	8.04	9.56	8.74	8.74	8.30
3	4.77	12.19	6.70	8.22	7.15	7.41	6.91
4	6.02	12.71	5.78	7.30	6.11	6.49	5.93
5	6.99	13.15	5.09	6.61	5.34	5.79	5.20
6	7.78	13.53	4.53	6.05	4.73	5.23	4.62
7	8.45	13.87	4.06	5.58	4.24	4.76	4.14
8	9.03	14.17	3.66	5.18	3.81	4.37	3.73
9	9.54	14.45	3.32	4.84	3.45	4.02	3.38
10	10.00	14.70	3.01	4.53	3.12	3.71	3.06
11	10.41	14.94	2.73	4.25	2.84	3.44	2.78
12	10.79	15.16	2.48	4.00	2.58	3.19	2.53
13	11.14	15.36	2.25	3.78	2.34	2.96	2.29
14	11.46	15.56	2.04	3.57	2.12	2.75	2.08
15	11.76	15.74	1.85	3.37	1.92	2.55	1.88
16	12.04	15.92	1.67	3.19	1.74	2.37	1.70
17	12.30	16.08	1.50	3.02	1.56	2.20	1.53
18	12.55	16.24	1.34	2.87	1.40	2.05	1.37
19	12.79	16.39	1.19	2.72	1.25	1.90	1.22
20	13.01	16.54	1.05	2.58	1.11	1.76	1.08
21	13.22	16.68	.92	2.44	.97	1.62	.94
22	13.42	16.81	.79	2.32	.84	1.50	.82
23	13.62	16.94	.67	2.20	.72	1.37	.69
24	13.80	17.06	.56	2.08	.60	1.26	.58
25	13.98	17.18	.45	1.97	.49	1.15	.47
26	14.15	17.30	.34	1.86	.38	1.04	.36
27	14.31	17.41	.24	1.76	.28	.94	.26
28	14.47	17.52	.14	1.67	.18	.85	.16
29	14.62	17.63	.05	1.57	.08	.75	.07
30	14.77	17.73	-.04	1.48	-.01	.66	-.02
31	14.91	17.83	-.13	1.40	-.09	.58	-.11
32	15.05	17.93	-.21	1.31	-.18	.49	-.20
33	15.19	18.02	-.29	1.23	-.26	.41	-.28
34	15.31	18.12	-.37	1.15	-.34	.33	-.36
35	15.44	18.21	-.45	1.08	-.42	.26	-.43
36	15.56	18.29	-.52	1.00	-.49	.18	-.51
37	15.68	18.38	-.59	.93	-.57	.11	-.58
38	15.80	18.46	-.66	.86	-.64	.04	-.65
39	15.91	18.55	-.73	.79	-.70	-.03	-.72
40	16.02	18.63	-.79	.73	-.77	-.09	-.78
41	16.13	18.70	-.86	.66	-.84	-.16	-.85
42	16.23	18.78	-.92	.60	-.90	-.22	-.91
43	16.33	18.86	-.98	.54	-.96	-.28	-.97
44	16.43	18.93	-1.04	.48	-1.02	-.34	-1.03
45	16.53	19.00	-1.10	.42	-1.08	-.40	-1.09

PROBABILITY OF DETECTION= .6000
CONSTANT FALSE ALARM PROBABILITY
LOG10(PFA)=- 5

N	N DB	THRESH DB	NON-FLUC DB	SWER 1 DB	SWER 2 DB	SWER 3 DB	SWER 4 DB
1	0.00	10.61	10.88	13.33	13.33	12.10	12.10
2	3.01	11.53	8.47	10.95	9.71	9.70	9.09
3	4.77	12.19	7.13	9.61	7.95	8.36	7.54
4	6.02	12.71	6.19	8.69	6.82	7.44	6.51
5	6.99	13.15	5.49	7.99	6.00	6.74	5.74
6	7.78	13.53	4.92	7.43	5.35	6.18	5.14
7	8.45	13.87	4.45	6.96	4.82	5.71	4.64
8	9.03	14.17	4.05	6.56	4.37	5.31	4.21
9	9.54	14.45	3.70	6.21	3.99	4.96	3.84
10	10.00	14.70	3.39	5.91	3.65	4.65	3.52
11	10.41	14.94	3.11	5.63	3.34	4.37	3.23
12	10.79	15.16	2.85	5.38	3.07	4.12	2.96
13	11.14	15.36	2.62	5.15	2.83	3.89	2.72
14	11.46	15.56	2.41	4.94	2.60	3.68	2.50
15	11.76	15.74	2.21	4.74	2.39	3.49	2.30
16	12.04	15.92	2.03	4.56	2.20	3.31	2.11
17	12.30	16.08	1.86	4.39	2.02	3.14	1.94
18	12.55	16.24	1.70	4.24	1.85	2.98	1.77
19	12.79	16.39	1.55	4.09	1.69	2.83	1.62
20	13.01	16.54	1.41	3.94	1.54	2.69	1.47
21	13.22	16.68	1.27	3.81	1.40	2.55	1.34
22	13.42	16.81	1.14	3.68	1.27	2.43	1.20
23	13.62	16.94	1.02	3.56	1.14	2.30	1.08
24	13.80	17.06	.91	3.45	1.02	2.19	.96
25	13.98	17.18	.79	3.34	.90	2.08	.85
26	14.15	17.30	.69	3.23	.79	1.97	.74
27	14.31	17.41	.59	3.13	.69	1.87	.64
28	14.47	17.52	.49	3.03	.58	1.77	.54
29	14.62	17.63	.39	2.94	.49	1.68	.44
30	14.77	17.73	.30	2.85	.39	1.59	.35
31	14.91	17.83	.21	2.76	.30	1.50	.26
32	15.05	17.93	.13	2.68	.21	1.42	.17
33	15.19	18.02	.05	2.60	.13	1.33	.09
34	15.31	18.12	-.03	2.52	.05	1.26	.01
35	15.44	18.21	-.11	2.44	-.03	1.18	-.07
36	15.56	18.29	-.18	2.37	-.11	1.11	-.15
37	15.68	18.38	-.26	2.29	-.18	1.03	-.22
38	15.80	18.46	-.33	2.23	-.26	.96	-.29
39	15.91	18.55	-.39	2.16	-.33	.90	-.36
40	16.02	18.63	-.46	2.09	-.39	.83	-.43
41	16.13	18.70	-.53	2.03	-.46	.76	-.49
42	16.23	18.78	-.59	1.96	-.53	.70	-.56
43	16.33	18.86	-.65	1.90	-.59	.64	-.62
44	16.43	18.93	-.71	1.84	-.65	.58	-.68
45	16.53	19.00	-.77	1.78	-.71	.52	-.74

Appendix: Radar Detection Tables

```
PROBABILITY OF DETECTION= .7000
CONSTANT FALSE ALARM PROBABILITY
LOG10(PFA)=- 5
```

K	N DB	THRESH DB	NON-FLUC DB	SWER 1 DB	SWER 2 DB	SWER 3 DB	SWER 4 DB
1	0.00	10.61	11.35	14.95	14.95	13.18	13.18
2	3.01	11.53	8.92	12.56	10.78	10.78	9.88
3	4.77	12.19	7.55	11.22	8.84	9.44	8.22
4	6.02	12.71	6.61	10.30	7.60	8.51	7.12
5	6.99	13.15	5.90	9.60	6.70	7.80	6.32
6	7.78	13.53	5.33	9.03	6.01	7.24	5.68
7	8.45	13.87	4.85	8.57	5.44	6.77	5.16
8	9.03	14.17	4.44	8.16	4.97	6.37	4.71
9	9.54	14.45	4.09	7.82	4.56	6.02	4.33
10	10.00	14.70	3.77	7.51	4.20	5.71	3.99
11	10.41	14.94	3.49	7.23	3.88	5.43	3.69
12	10.79	15.16	3.23	6.98	3.60	5.18	3.42
13	11.14	15.36	3.00	6.75	3.34	4.95	3.17
14	11.46	15.56	2.78	6.54	3.10	4.73	2.94
15	11.76	15.74	2.58	6.34	2.88	4.54	2.74
16	12.04	15.92	2.40	6.16	2.68	4.36	2.54
17	12.30	16.08	2.23	5.99	2.49	4.19	2.36
18	12.55	16.24	2.06	5.83	2.31	4.03	2.19
19	12.79	16.39	1.91	5.68	2.15	3.88	2.03
20	13.01	16.54	1.77	5.54	1.99	3.73	1.88
21	13.22	16.68	1.63	5.41	1.85	3.60	1.74
22	13.42	16.81	1.50	5.28	1.71	3.47	1.61
23	13.62	16.94	1.38	5.16	1.58	3.35	1.48
24	13.80	17.06	1.26	5.04	1.45	3.23	1.36
25	13.98	17.18	1.15	4.93	1.33	3.12	1.24
26	14.15	17.30	1.04	4.83	1.22	3.02	1.13
27	14.31	17.41	.94	4.72	1.11	2.91	1.02
28	14.47	17.52	.84	4.63	1.00	2.82	.92
29	14.62	17.63	.74	4.53	.90	2.72	.82
30	14.77	17.73	.65	4.44	.81	2.63	.73
31	14.91	17.83	.56	4.36	.71	2.54	.64
32	15.05	17.93	.47	4.27	.62	2.46	.55
33	15.19	18.02	.39	4.19	.53	2.38	.46
34	15.31	18.12	.31	4.11	.45	2.30	.38
35	15.44	18.21	.23	4.03	.37	2.22	.30
36	15.56	18.29	.16	3.96	.29	2.15	.22
37	15.68	18.38	.08	3.89	.21	2.07	.15
38	15.81	18.46	.01	3.82	.14	2.00	.08
39	15.91	18.55	-.06	3.75	.07	1.94	.01
40	16.02	18.63	-.12	3.68	-.00	1.87	-.06
41	16.13	18.70	-.19	3.62	-.07	1.80	-.13
42	16.23	18.78	-.25	3.56	-.14	1.74	-.20
43	16.33	18.86	-.31	3.49	-.20	1.68	-.26
44	16.43	18.93	-.38	3.44	-.27	1.62	-.32
45	16.53	19.00	-.43	3.38	-.33	1.56	-.38

PROBABILITY OF DETECTION= .8000
CONSTANT FALSE ALARM PROBABILITY
LOG10(PFA)=- 5

N	N DB	THRESH DB	NON-FLUC DB	SWER 1 DB	SWER 2 DB	SWER 3 DB	SWER 4 DB
1	0.00	10.61	11.86	17.04	17.04	14.52	14.52
2	3.01	11.53	9.41	14.65	12.11	12.11	10.83
3	4.77	12.19	8.03	13.30	9.90	10.76	9.01
4	6.02	12.71	7.08	12.37	8.53	9.83	7.84
5	6.99	13.15	6.36	11.67	7.55	9.12	6.98
6	7.78	13.53	5.78	11.11	6.79	8.56	6.30
7	8.45	13.87	5.29	10.64	6.18	8.08	5.75
8	9.03	14.17	4.88	10.24	5.66	7.68	5.29
9	9.54	14.45	4.52	9.89	5.23	7.33	4.89
10	10.00	14.70	4.20	9.58	4.84	7.02	4.53
11	10.41	14.94	3.91	9.30	4.50	6.74	4.22
12	10.79	15.16	3.65	9.05	4.20	6.48	3.94
13	11.14	15.36	3.42	8.82	3.93	6.25	3.68
14	11.46	15.56	3.20	8.61	3.68	6.04	3.44
15	11.76	15.74	3.00	8.41	3.45	5.84	3.23
16	12.04	15.92	2.81	8.23	3.23	5.66	3.03
17	12.30	16.08	2.63	8.06	3.04	5.49	2.84
18	12.55	16.24	2.47	7.90	2.85	5.33	2.67
19	12.79	16.39	2.31	7.75	2.68	5.18	2.50
20	13.01	16.54	2.17	7.61	2.52	5.04	2.35
21	13.22	16.68	2.03	7.47	2.36	4.90	2.20
22	13.42	16.81	1.90	7.35	2.22	4.77	2.06
23	13.62	16.94	1.77	7.23	2.08	4.65	1.93
24	13.80	17.06	1.65	7.11	1.95	4.53	1.81
25	13.98	17.18	1.54	7.00	1.82	4.42	1.69
26	14.15	17.30	1.43	6.89	1.71	4.32	1.57
27	14.31	17.41	1.33	6.79	1.59	4.21	1.46
28	14.47	17.52	1.23	6.69	1.48	4.12	1.36
29	14.62	17.63	1.13	6.60	1.38	4.02	1.26
30	14.77	17.73	1.04	6.51	1.28	3.93	1.16
31	14.91	17.83	.94	6.42	1.18	3.84	1.06
32	15.05	17.93	.86	6.34	1.09	3.76	.97
33	15.19	18.02	.77	6.25	1.00	3.67	.89
34	15.31	18.12	.69	6.18	.91	3.59	.80
35	15.44	18.21	.61	6.10	.82	3.52	.72
36	15.56	18.29	.54	6.02	.74	3.44	.64
37	15.68	18.38	.46	5.95	.66	3.37	.56
38	15.80	18.46	.39	5.88	.59	3.30	.49
39	15.91	18.55	.32	5.81	.51	3.23	.42
40	16.02	18.63	.25	5.75	.44	3.17	.35
41	16.13	18.70	.19	5.68	.37	3.10	.28
42	16.23	18.78	.12	5.62	.30	3.04	.21
43	16.33	18.86	.06	5.56	.23	2.98	.15
44	16.43	18.93	-.00	5.50	.17	2.92	.08
45	16.53	19.00	-.06	5.44	.10	2.86	.02

Appendix: Radar Detection Tables

```
PROBABILITY OF DETECTION= .9000
CONSTANT FALSE ALARM PROBABILITY
LOG10(PFA)=- 5
```

N	N DB	THRESH DB	NON-FLUC DB	SWER 1 DB	SWER 2 DB	SWER 3 DB	SWER 4 DB
1	0.00	10.61	12.53	20.35	20.35	16.53	16.53
2	3.01	11.53	10.05	17.95	14.11	14.11	12.18
3	4.77	12.19	8.65	16.60	11.47	12.75	10.13
4	6.02	12.71	7.68	15.67	9.87	11.82	8.83
5	6.99	13.15	6.95	14.97	8.74	11.11	7.89
6	7.78	13.53	6.36	14.40	7.89	10.54	7.16
7	8.45	13.87	5.87	13.93	7.20	10.06	6.57
8	9.03	14.17	5.45	13.53	6.64	9.66	6.07
9	9.54	14.45	5.08	13.18	6.15	9.30	5.64
10	10.00	14.70	4.75	12.87	5.73	8.99	5.27
11	10.41	14.94	4.46	12.59	5.37	8.71	4.93
12	10.79	15.16	4.20	12.34	5.04	8.46	4.63
13	11.14	15.36	3.96	12.11	4.74	8.22	4.36
14	11.46	15.56	3.74	11.90	4.47	8.01	4.12
15	11.76	15.74	3.53	11.70	4.22	7.81	3.89
16	12.04	15.92	3.34	11.52	3.99	7.63	3.68
17	12.30	16.08	3.16	11.35	3.78	7.46	3.48
18	12.55	16.24	3.00	11.19	3.58	7.30	3.30
19	12.79	16.39	2.84	11.04	3.40	7.15	3.13
20	13.01	16.54	2.69	10.90	3.22	7.00	2.97
21	13.22	16.68	2.55	10.76	3.06	6.87	2.81
22	13.42	16.81	2.42	10.63	2.91	6.74	2.67
23	13.62	16.94	2.29	10.51	2.76	6.62	2.53
24	13.80	17.06	2.17	10.40	2.62	6.50	2.40
25	13.98	17.18	2.05	10.29	2.49	6.39	2.28
26	14.15	17.30	1.94	10.18	2.36	6.28	2.16
27	14.31	17.41	1.83	10.08	2.24	6.18	2.04
28	14.47	17.52	1.73	9.98	2.13	6.08	1.93
29	14.62	17.63	1.63	9.89	2.02	5.98	1.83
30	14.77	17.73	1.54	9.79	1.91	5.89	1.73
31	14.91	17.83	1.45	9.71	1.81	5.80	1.63
32	15.05	17.93	1.36	9.62	1.71	5.72	1.54
33	15.19	18.02	1.27	9.54	1.61	5.64	1.45
34	15.31	18.12	1.19	9.46	1.52	5.56	1.36
35	15.44	18.21	1.11	9.39	1.43	5.48	1.28
36	15.56	18.29	1.03	9.31	1.35	5.40	1.19
37	15.68	18.38	.96	9.24	1.26	5.33	1.11
38	15.8	18.46	.88	9.17	1.18	5.26	1.04
39	15.91	18.55	.81	9.10	1.11	5.19	.96
40	16.02	18.63	.74	9.03	1.03	5.13	.89
41	16.13	18.70	.67	8.97	.96	5.06	.82
42	16.23	18.78	.61	8.91	.88	5.00	.75
43	16.33	18.86	.54	8.85	.81	4.94	.68
44	16.43	18.93	.48	8.79	.75	4.88	.62
45	16.53	19.00	.42	8.73	.68	4.82	.55

PROBABILITY OF DETECTION= .9500
CONSTANT FALSE ALARM PROBABILITY
LOG10(PFA)=- 5

N	N DB	THRESH DB	NON-FLUC DB	SWER 1 DB	SWER 2 DB	SWER 3 DB	SWER 4 DB
1	0.00	10.61	13.05	23.50	23.50	18.34	18.34
2	3.01	11.53	10.54	21.09	15.92	15.92	13.35
3	4.77	12.19	9.13	19.74	12.84	14.56	11.08
4	6.02	12.71	8.15	18.81	11.02	13.62	9.67
5	6.99	13.15	7.41	18.11	9.77	12.91	8.65
6	7.78	13.53	6.81	17.54	8.82	12.34	7.87
7	8.45	13.87	6.31	17.07	8.07	11.86	7.24
8	9.03	14.17	5.89	16.67	7.45	11.46	6.71
9	9.54	14.45	5.51	16.32	6.93	11.10	6.25
10	10.00	14.70	5.19	16.01	6.48	10.79	5.86
11	10.41	14.94	4.89	15.73	6.08	10.51	5.51
12	10.79	15.16	4.62	15.48	5.72	10.25	5.20
13	11.14	15.36	4.38	15.25	5.41	10.02	4.91
14	11.46	15.56	4.16	15.04	5.12	9.80	4.66
15	11.76	15.74	3.95	14.84	4.85	9.61	4.42
16	12.04	15.92	3.75	14.66	4.61	9.42	4.20
17	12.30	16.08	3.57	14.49	4.39	9.25	3.99
18	12.55	16.24	3.40	14.33	4.18	9.09	3.80
19	12.79	16.39	3.25	14.18	3.98	8.94	3.63
20	13.01	16.54	3.09	14.04	3.80	8.79	3.46
21	13.22	16.68	2.95	13.90	3.63	8.66	3.30
22	13.42	16.81	2.82	13.77	3.46	8.53	3.15
23	13.62	16.94	2.69	13.65	3.31	8.40	3.01
24	13.80	17.06	2.57	13.54	3.16	8.29	2.87
25	13.98	17.18	2.45	13.43	3.03	8.18	2.75
26	14.15	17.30	2.34	13.32	2.89	8.07	2.62
27	14.31	17.41	2.23	13.22	2.77	7.97	2.51
28	14.47	17.52	2.12	13.12	2.65	7.87	2.39
29	14.62	17.63	2.02	13.03	2.53	7.77	2.28
30	14.77	17.73	1.93	12.93	2.42	7.68	2.18
31	14.91	17.83	1.83	12.85	2.31	7.59	2.08
32	15.05	17.93	1.75	12.76	2.21	7.51	1.98
33	15.19	18.02	1.66	12.68	2.11	7.42	1.89
34	15.31	18.12	1.57	12.60	2.02	7.34	1.80
35	15.44	18.21	1.49	12.52	1.92	7.27	1.71
36	15.56	18.29	1.41	12.45	1.83	7.19	1.63
37	15.68	18.38	1.34	12.38	1.75	7.12	1.55
38	15.80	18.46	1.26	12.31	1.66	7.05	1.47
39	15.91	18.55	1.19	12.24	1.58	6.98	1.39
40	16.02	18.63	1.12	12.17	1.50	6.91	1.32
41	16.13	18.70	1.05	12.11	1.43	6.84	1.24
42	16.23	18.78	.99	12.05	1.35	6.78	1.17
43	16.33	18.86	.92	11.99	1.28	6.72	1.10
44	16.43	18.93	.86	11.93	1.21	6.66	1.04
45	16.53	19.00	.80	11.87	1.14	6.60	.97

Appendix: Radar Detection Tables

```
PROBABILITY OF DETECTION= .9900
CONSTANT FALSE ALARM PROBABILITY
LOG10(PFA)=- 5
```

N	N DB	THRESH DB	NON-FLUC DB	SWER 1 DB	SWER 2 DB	SWER 3 DB	SWER 4 DB
1	0.00	10.61	13.94	30.57	30.57	22.20	22.20
2	3.01	11.53	11.39	28.19	19.79	19.79	15.70
3	4.77	12.19	9.96	26.84	15.68	18.42	12.93
4	6.02	12.71	8.96	25.91	13.35	17.47	11.26
5	6.99	13.15	8.20	25.20	11.79	16.76	10.09
6	7.78	13.53	7.59	24.64	10.65	16.18	9.20
7	8.45	13.87	7.08	24.17	9.76	15.71	8.48
8	9.03	14.17	6.65	23.76	9.02	15.30	7.90
9	9.54	14.45	6.27	23.41	8.41	14.94	7.39
10	10.00	14.70	5.94	23.10	7.89	14.63	6.96
11	10.41	14.94	5.63	22.82	7.43	14.35	6.58
12	10.79	15.16	5.36	22.57	7.03	14.09	6.23
13	11.14	15.36	5.11	22.34	6.66	13.86	5.92
14	11.46	15.56	4.88	22.13	6.33	13.64	5.64
15	11.76	15.74	4.67	21.93	6.04	13.45	5.38
16	12.04	15.92	4.47	21.75	5.77	13.26	5.14
17	12.30	16.08	4.29	21.58	5.52	13.09	4.92
18	12.55	16.24	4.12	21.42	5.28	12.93	4.72
19	12.79	16.39	3.96	21.27	5.07	12.78	4.53
20	13.01	16.54	3.80	21.13	4.86	12.63	4.35
21	13.22	16.68	3.65	20.99	4.67	12.50	4.18
22	13.42	16.81	3.52	20.87	4.50	12.37	4.02
23	13.62	16.94	3.38	20.75	4.33	12.25	3.87
24	13.80	17.06	3.26	20.63	4.17	12.13	3.73
25	13.98	17.18	3.14	20.52	4.02	12.02	3.59
26	14.15	17.30	3.02	20.41	3.87	11.91	3.46
27	14.31	17.41	2.91	20.31	3.73	11.81	3.34
28	14.47	17.52	2.81	20.21	3.60	11.71	3.22
29	14.62	17.63	2.70	20.12	3.47	11.61	3.10
30	14.77	17.73	2.61	20.03	3.35	11.52	2.99
31	14.91	17.83	2.51	19.94	3.24	11.43	2.89
32	15.05	17.93	2.42	19.85	3.12	11.35	2.78
33	15.19	18.02	2.33	19.77	3.02	11.27	2.68
34	15.31	18.12	2.25	19.69	2.91	11.19	2.59
35	15.44	18.21	2.16	19.62	2.81	11.11	2.50
36	15.56	18.29	2.08	19.54	2.72	11.04	2.41
37	15.68	18.38	2.01	19.47	2.62	10.96	2.32
38	15.8	18.46	1.93	19.40	2.53	10.89	2.24
39	15.91	18.55	1.86	19.33	2.45	10.82	2.16
40	16.02	18.63	1.78	19.27	2.36	10.76	2.08
41	16.13	18.70	1.71	19.20	2.28	10.69	2.00
42	16.23	18.78	1.65	19.11	2.20	10.63	1.93
43	16.33	18.86	1.58	19.05	2.12	10.54	1.85
44	16.43	18.93	1.52	18.99	2.05	10.48	1.78
45	16.53	19.00	1.45	18.93	1.97	10.42	1.72

PROBABILITY OF DETECTION= .5000
CONSTANT FALSE ALARM PROBABILITY
LOG10(PFA)=- 6

N	N DB	THRESH DB	NON-FLUC DB	SWER 1 DB	SWER 2 DB	SWER 3 DB	SWER 4 DB
1	0.00	11.40	11.24	12.77	12.77	11.95	11.95
2	3.01	12.22	8.80	10.33	9.52	9.52	9.14
3	4.77	12.82	7.43	8.97	7.89	8.15	7.65
4	6.02	13.29	6.49	8.03	6.83	7.21	6.65
5	6.99	13.70	5.78	7.31	6.04	6.49	5.90
6	7.78	14.05	5.21	6.74	5.42	5.92	5.30
7	8.45	14.36	4.73	6.26	4.91	5.44	4.81
8	9.03	14.65	4.32	5.86	4.48	5.03	4.39
9	9.54	14.91	3.97	5.50	4.10	4.68	4.03
10	10.00	15.15	3.65	5.19	3.77	4.36	3.71
11	10.41	15.37	3.37	4.91	3.48	4.08	3.42
12	10.79	15.58	3.11	4.65	3.21	3.83	3.16
13	11.14	15.77	2.88	4.42	2.97	3.59	2.92
14	11.46	15.96	2.67	4.20	2.75	3.38	2.71
15	11.76	16.13	2.47	4.00	2.54	3.18	2.50
16	12.04	16.30	2.29	3.82	2.36	3.00	2.32
17	12.31	16.45	2.11	3.65	2.18	2.83	2.14
18	12.55	16.60	1.95	3.49	2.01	2.66	1.98
19	12.79	16.75	1.80	3.34	1.86	2.51	1.83
20	13.01	16.89	1.66	3.19	1.71	2.37	1.68
21	13.22	17.02	1.52	3.06	1.57	2.23	1.54
22	13.42	17.15	1.39	2.93	1.44	2.10	1.41
23	13.62	17.27	1.27	2.80	1.31	1.98	1.29
24	13.80	17.39	1.15	2.69	1.19	1.86	1.17
25	13.98	17.51	1.04	2.57	1.08	1.75	1.06
26	14.15	17.62	.93	2.47	.97	1.64	.95
27	14.31	17.73	.83	2.36	.87	1.54	.85
28	14.47	17.83	.73	2.27	.77	1.44	.75
29	14.62	17.93	.63	2.17	.67	1.35	.65
30	14.77	18.03	.54	2.08	.58	1.26	.56
31	14.91	18.13	.45	1.99	.49	1.17	.47
32	15.05	18.22	.37	1.90	.40	1.08	.38
33	15.19	18.31	.29	1.82	.32	1.00	.30
34	15.31	18.40	.21	1.74	.24	.92	.22
35	15.44	18.49	.13	1.66	.16	.84	.14
36	15.56	18.57	.05	1.59	.08	.77	.07
37	15.68	18.66	-.02	1.52	.01	.69	-.01
38	15.80	18.74	-.09	1.45	-.06	.62	-.08
39	15.91	18.82	-.16	1.38	-.13	.55	-.15
40	16.02	18.90	-.22	1.31	-.20	.49	-.21
41	16.13	18.97	-.29	1.25	-.27	.42	-.28
42	16.23	19.05	-.35	1.18	-.33	.36	-.34
43	16.33	19.12	-.42	1.12	-.39	.30	-.41
44	16.43	19.19	-.48	1.06	-.45	.24	-.47
45	16.53	19.26	-.53	1.00	-.51	.18	-.52

Appendix: Radar Detection Tables

```
PROBABILITY OF DETECTION= .6000
CONSTANT FALSE ALARM PROBABILITY
LOG10(PFA)=- 6
```

N	N DB	THRESH DB	NON-FLUC DB	SWER 1 DB	SWER 2 DB	SWER 3 DB	SWER 4 DB
1	0.00	11.40	11.66	14.16	14.16	12.91	12.91
2	3.01	12.22	9.20	11.71	10.46	10.46	9.84
3	4.77	12.82	7.82	10.34	8.68	9.09	8.26
4	6.02	13.29	6.87	9.40	7.52	8.14	7.20
5	6.99	13.70	6.15	8.68	6.67	7.43	6.41
6	7.78	14.05	5.57	8.11	6.01	6.85	5.79
7	8.45	14.36	5.09	7.63	5.47	6.37	5.28
8	9.03	14.65	4.68	7.22	5.01	5.96	4.85
9	9.54	14.91	4.32	6.87	4.62	5.61	4.47
10	10.00	15.15	4.00	6.55	4.27	5.29	4.14
11	10.41	15.37	3.71	6.27	3.96	5.01	3.84
12	10.79	15.58	3.46	6.02	3.68	4.75	3.57
13	11.14	15.77	3.22	5.78	3.43	4.52	3.33
14	11.46	15.96	3.00	5.57	3.20	4.30	3.10
15	11.76	16.13	2.80	5.37	2.99	4.11	2.90
16	12.04	16.30	2.62	5.18	2.79	3.92	2.70
17	12.30	16.45	2.44	5.01	2.61	3.75	2.53
18	12.55	16.60	2.28	4.85	2.43	3.58	2.36
19	12.79	16.75	2.13	4.70	2.27	3.43	2.20
20	13.01	16.89	1.98	4.55	2.12	3.29	2.05
21	13.22	17.02	1.84	4.42	1.98	3.15	1.91
22	13.42	17.15	1.71	4.29	1.84	3.02	1.78
23	13.62	17.27	1.59	4.16	1.71	2.90	1.65
24	13.80	17.39	1.47	4.05	1.59	2.78	1.53
25	13.98	17.51	1.36	3.93	1.47	2.67	1.41
26	14.15	17.62	1.25	3.83	1.36	2.56	1.30
27	14.31	17.73	1.14	3.72	1.25	2.46	1.20
28	14.47	17.83	1.04	3.62	1.15	2.36	1.10
29	14.62	17.93	.95	3.53	1.05	2.26	1.00
30	14.77	18.03	.86	3.44	.95	2.17	.90
31	14.91	18.13	.77	3.35	.86	2.08	.81
32	15.05	18.22	.68	3.26	.77	2.00	.72
33	15.19	18.31	.60	3.18	.68	1.91	.64
34	15.31	18.40	.52	3.10	.60	1.83	.56
35	15.44	18.49	.44	3.02	.52	1.76	.48
36	15.56	18.57	.36	2.95	.44	1.68	.40
37	15.68	18.66	.29	2.87	.37	1.61	.33
38	15.80	18.74	.22	2.80	.29	1.54	.25
39	15.91	18.82	.15	2.73	.22	1.47	.18
40	16.02	18.90	.08	2.67	.15	1.40	.12
41	16.13	18.97	.01	2.60	.08	1.33	.05
42	16.23	19.05	-.05	2.54	.02	1.27	-.02
43	16.33	19.12	-.11	2.48	-.05	1.21	-.08
44	16.43	19.19	-.17	2.42	-.11	1.15	-.14
45	16.53	19.26	-.23	2.36	-.17	1.09	-.20

PROBABILITY OF DETECTION= .7000
CONSTANT FALSE ALARM PROBABILITY
LOG10(PFA)=- 6

N	N DB	THRESH DB	NON-FLUC DB	SWER 1 DB	SWER 2 DB	SWER 3 DB	SWER 4 DB
1	0.00	11.40	12.09	15.77	15.77	13.98	13.98
2	3.01	12.22	9.01	13.32	11.53	11.53	10.61
3	4.77	12.82	8.22	11.95	9.54	10.15	8.91
4	6.02	13.29	7.26	11.00	8.28	9.20	7.79
5	6.99	13.70	6.53	10.28	7.36	8.48	6.96
6	7.78	14.05	5.94	9.71	6.65	7.90	6.31
7	8.45	14.36	5.45	9.23	6.07	7.42	5.78
8	9.03	14.65	5.04	8.82	5.59	7.01	5.32
9	9.54	14.91	4.68	8.46	5.17	6.65	4.93
10	10.00	15.15	4.35	8.15	4.80	6.34	4.59
11	10.41	15.37	4.07	7.86	4.48	6.05	4.28
12	10.79	15.58	3.80	7.61	4.19	5.80	4.00
13	11.14	15.77	3.57	7.37	3.92	5.56	3.75
14	11.46	15.96	3.35	7.16	3.68	5.34	3.52
15	11.76	16.13	3.14	6.96	3.46	5.15	3.30
16	12.04	16.30	2.96	6.78	3.25	4.96	3.11
17	12.30	16.45	2.78	6.60	3.06	4.79	2.92
18	12.55	16.60	2.62	6.44	2.88	4.62	2.75
19	12.79	16.75	2.46	6.29	2.71	4.47	2.59
20	13.01	16.89	2.31	6.14	2.55	4.33	2.44
21	13.22	17.02	2.17	6.01	2.40	4.19	2.29
22	13.42	17.15	2.04	5.88	2.26	4.06	2.15
23	13.62	17.27	1.92	5.75	2.13	3.93	2.02
24	13.80	17.39	1.80	5.64	2.00	3.82	1.90
25	13.98	17.51	1.68	5.52	1.88	3.70	1.78
26	14.15	17.62	1.57	5.42	1.76	3.60	1.67
27	14.31	17.73	1.47	5.31	1.65	3.49	1.56
28	14.47	17.83	1.37	5.21	1.54	3.39	1.46
29	14.62	17.93	1.27	5.12	1.44	3.30	1.36
30	14.77	18.03	1.18	5.03	1.34	3.20	1.26
31	14.91	18.13	1.09	4.94	1.25	3.12	1.17
32	15.05	18.22	1.00	4.85	1.15	3.03	1.08
33	15.19	18.31	.91	4.77	1.07	2.95	.99
34	15.31	18.40	.83	4.69	.98	2.87	.91
35	15.44	18.49	.75	4.61	.90	2.79	.83
36	15.56	18.57	.68	4.54	.82	2.71	.75
37	15.68	18.66	.60	4.46	.74	2.64	.67
38	15.80	18.74	.53	4.39	.66	2.57	.60
39	15.91	18.82	.46	4.32	.59	2.50	.53
40	16.02	18.90	.39	4.25	.52	2.43	.46
41	16.13	18.97	.33	4.19	.45	2.37	.39
42	16.23	19.05	.26	4.13	.38	2.30	.32
43	16.33	19.12	.20	4.06	.32	2.24	.26
44	16.43	19.19	.14	4.00	.25	2.18	.19
45	16.53	19.26	.08	3.94	.19	2.12	.13

Appendix: Radar Detection Tables 249

```
PROBABILITY OF DETECTION= .8000
CONSTANT FALSE ALARM PROBABILITY
LOG10(PFA)=- 6
```

N	N DB	THRESH DB	NON-FLUC DB	SWER 1 DB	SWER 2 DB	SWER 3 DB	SWER 4 DB
1	0.00	11.40	12.57	17.85	17.85	15.31	15.31
2	3.01	12.22	10.06	15.40	12.84	12.84	11.53
3	4.77	12.82	8.06	14.02	10.59	11.40	9.68
4	6.02	13.29	7.68	13.07	9.19	10.51	8.48
5	6.99	13.71	6.95	12.35	8.19	9.79	7.60
6	7.78	14.05	6.36	11.78	7.41	9.21	6.91
7	8.45	14.36	5.86	11.30	6.79	8.73	6.35
8	9.03	14.65	5.44	10.89	6.26	8.31	5.87
9	9.54	14.91	5.07	10.53	5.82	7.95	5.46
10	10.00	15.15	4.75	10.21	5.43	7.64	5.10
11	10.41	15.37	4.46	9.93	5.08	7.35	4.78
12	10.79	15.58	4.19	9.67	4.77	7.09	4.49
13	11.14	15.77	3.95	9.44	4.49	6.86	4.23
14	11.46	15.96	3.73	9.22	4.23	6.64	3.99
15	11.76	16.13	3.53	9.03	4.00	6.44	3.77
16	12.04	16.30	3.34	8.84	3.78	6.25	3.57
17	12.30	16.45	3.16	8.67	3.58	6.08	3.38
18	12.55	16.60	2.99	8.50	3.39	5.92	3.20
19	12.79	16.75	2.83	8.35	3.22	5.76	3.03
20	13.01	16.89	2.68	8.21	3.05	5.62	2.87
21	13.22	17.02	2.54	8.07	2.90	5.48	2.73
22	13.42	17.15	2.41	7.94	2.75	5.35	2.58
23	13.62	17.27	2.28	7.82	2.61	5.23	2.45
24	13.80	17.39	2.16	7.70	2.47	5.11	2.32
25	13.98	17.51	2.05	7.59	2.35	5.00	2.20
26	14.15	17.62	1.94	7.48	2.23	4.89	2.08
27	14.31	17.73	1.83	7.37	2.11	4.78	1.97
28	14.47	17.83	1.73	7.28	2.00	4.68	1.87
29	14.62	17.93	1.63	7.18	1.89	4.59	1.76
30	14.77	18.03	1.53	7.09	1.79	4.49	1.66
31	14.91	18.13	1.44	7.00	1.69	4.40	1.57
32	15.05	18.22	1.35	6.91	1.60	4.32	1.48
33	15.19	18.31	1.27	6.83	1.50	4.24	1.39
34	15.31	18.40	1.19	6.75	1.41	4.15	1.30
35	15.44	18.49	1.11	6.67	1.33	4.08	1.22
36	15.56	18.57	1.03	6.60	1.25	4.00	1.14
37	15.68	18.66	.95	6.52	1.18	3.93	1.06
38	15.8	18.74	.88	6.45	1.09	3.86	.99
39	15.91	18.82	.81	6.38	1.01	3.79	.91
40	16.02	18.90	.74	6.31	.94	3.72	.84
41	16.13	18.97	.67	6.25	.87	3.65	.77
42	16.23	19.05	.61	6.19	.80	3.59	.70
43	16.33	19.12	.54	6.12	.73	3.53	.64
44	16.43	19.19	.48	6.06	.66	3.46	.57
45	16.53	19.26	.42	6.00	.60	3.41	.51

PROBABILITY OF DETECTION= .9000
CONSTANT FALSE ALARM PROBABILITY
LOG10(PFA)=- 6

N	N DB	THRESH DB	NON-FLUC DB	SWER 1 DB	SWER 2 DB	SWER 3 DB	SWER 4 DB
1	0.00	11.40	13.18	21.14	21.14	17.30	17.30
2	3.01	12.22	10.65	18.69	14.83	14.83	12.86
3	4.77	12.82	9.23	17.31	12.14	13.44	10.78
4	6.02	13.29	8.25	16.36	10.51	12.48	9.45
5	6.99	13.70	7.50	15.64	9.36	11.76	8.49
6	7.78	14.05	6.90	15.07	8.49	11.18	7.74
7	8.45	14.36	6.40	14.59	7.79	10.69	7.14
8	9.03	14.65	5.97	14.18	7.21	10.28	6.63
9	9.54	14.91	5.60	13.82	6.72	9.92	6.19
10	10.00	15.15	5.27	13.50	6.29	9.60	5.81
11	10.41	15.37	4.97	13.22	5.92	9.32	5.47
12	10.79	15.58	4.70	12.96	5.58	9.06	5.16
13	11.14	15.77	4.46	12.73	5.27	8.82	4.89
14	11.46	15.96	4.23	12.51	5.00	8.60	4.63
15	11.76	16.13	4.03	12.31	4.75	8.40	4.40
16	12.04	16.30	3.83	12.12	4.51	8.21	4.19
17	12.30	16.45	3.65	11.95	4.30	8.04	3.99
18	12.55	16.60	3.48	11.79	4.10	7.88	3.80
19	12.79	16.75	3.32	11.64	3.91	7.72	3.63
20	13.01	16.89	3.17	11.49	3.73	7.58	3.46
21	13.22	17.02	3.03	11.35	3.57	7.44	3.31
22	13.42	17.15	2.89	11.22	3.41	7.31	3.16
23	13.62	17.27	2.76	11.10	3.26	7.18	3.02
24	13.80	17.39	2.64	10.98	3.12	7.07	2.89
25	13.98	17.51	2.52	10.87	2.98	6.95	2.76
26	14.15	17.62	2.41	10.76	2.86	6.84	2.64
27	14.31	17.73	2.30	10.66	2.73	6.74	2.52
28	14.47	17.83	2.20	10.56	2.62	6.64	2.41
29	14.62	17.93	2.10	10.46	2.50	6.54	2.31
30	14.77	18.03	2.00	10.37	2.39	6.45	2.20
31	14.91	18.13	1.91	10.28	2.29	6.36	2.11
32	15.05	18.22	1.82	10.19	2.19	6.27	2.01
33	15.19	18.31	1.73	10.11	2.09	6.19	1.92
34	15.31	18.40	1.65	10.03	2.00	6.11	1.83
35	15.44	18.49	1.57	9.95	1.91	6.03	1.74
36	15.56	18.57	1.49	9.88	1.82	5.95	1.66
37	15.68	18.66	1.41	9.80	1.74	5.88	1.58
38	15.80	18.74	1.34	9.73	1.66	5.81	1.50
39	15.91	18.82	1.27	9.66	1.58	5.74	1.43
40	16.02	18.90	1.20	9.59	1.50	5.67	1.35
41	16.13	18.97	1.13	9.53	1.43	5.60	1.28
42	16.23	19.05	1.06	9.46	1.35	5.54	1.21
43	16.33	19.12	1.00	9.40	1.28	5.48	1.14
44	16.43	19.19	.93	9.34	1.21	5.42	1.08
45	16.53	19.26	.87	9.28	1.15	5.36	1.01

Appendix: Radar Detection Tables

PROBABILITY OF DETECTION= .9500
CONSTANT FALSE ALARM PROBABILITY
LOG10(PFA)=- 6

N	N DB	THRESH DB	NON-FLUC DB	SWER 1 DB	SWER 2 DB	SWER 3 DB	SWER 4 DB
1	0.00	11.40	13.66	24.29	24.29	19.10	19.10
2	3.01	12.22	11.11	21.84	16.62	16.62	14.02
3	4.77	12.82	9.67	20.46	13.50	15.24	11.71
4	6.02	13.29	8.68	19.51	11.65	14.28	10.26
5	6.99	13.70	7.93	18.79	10.37	13.55	9.23
6	7.78	14.05	7.32	18.21	9.41	12.97	8.43
7	8.45	14.36	6.81	17.73	8.64	12.49	7.78
8	9.03	14.65	6.38	17.32	8.01	12.07	7.24
9	9.54	14.91	6.01	16.96	7.48	11.71	6.78
10	10.00	15.15	5.67	16.64	7.02	11.39	6.38
11	10.41	15.37	5.37	16.36	6.61	11.10	6.02
12	10.79	15.58	5.10	16.10	6.25	10.84	5.70
13	11.14	15.77	4.85	15.86	5.92	10.61	5.41
14	11.46	15.96	4.62	15.65	5.63	10.39	5.15
15	11.76	16.13	4.41	15.45	5.36	10.19	4.91
16	12.04	16.30	4.22	15.26	5.11	10.00	4.69
17	12.30	16.45	4.04	15.09	4.89	9.82	4.48
18	12.55	16.60	3.86	14.93	4.67	9.66	4.29
19	12.79	16.75	3.70	14.77	4.47	9.51	4.10
20	13.01	16.89	3.55	14.63	4.29	9.36	3.93
21	13.22	17.02	3.41	14.49	4.11	9.22	3.77
22	13.42	17.15	3.27	14.36	3.95	9.09	3.62
23	13.62	17.27	3.14	14.24	3.79	8.97	3.48
24	13.80	17.39	3.01	14.12	3.64	8.85	3.34
25	13.98	17.51	2.89	14.01	3.50	8.73	3.21
26	14.15	17.62	2.78	13.90	3.37	8.62	3.08
27	14.31	17.73	2.67	13.80	3.24	8.52	2.96
28	14.47	17.83	2.57	13.70	3.12	8.42	2.85
29	14.62	17.93	2.46	13.60	3.00	8.32	2.74
30	14.77	18.03	2.37	13.51	2.88	8.23	2.63
31	14.91	18.13	2.27	13.42	2.78	8.14	2.53
32	15.05	18.22	2.18	13.34	2.67	8.05	2.43
33	15.19	18.31	2.09	13.25	2.57	7.97	2.34
34	15.31	18.40	2.01	13.17	2.47	7.89	2.25
35	15.44	18.49	1.93	13.09	2.38	7.81	2.16
36	15.56	18.57	1.85	13.02	2.29	7.73	2.07
37	15.68	18.66	1.77	12.95	2.20	7.66	1.99
38	15.8	18.74	1.69	12.87	2.12	7.59	1.91
39	15.91	18.82	1.62	12.80	2.03	7.52	1.83
40	16.02	18.90	1.55	12.74	1.95	7.45	1.76
41	16.13	18.97	1.48	12.67	1.88	7.38	1.68
42	16.23	19.05	1.41	12.61	1.80	7.32	1.61
43	16.33	19.12	1.35	12.54	1.73	7.26	1.54
44	16.43	19.19	1.28	12.48	1.66	7.19	1.47
45	16.53	19.26	1.22	12.42	1.59	7.14	1.41

PROBABILITY OF DETECTION= .9900
CONSTANT FALSE ALARM PROBABILITY
LOG10(PFA)=- 6

N	N DB	THRESH DB	NON-FLUC DB	SWER 1 DB	SWER 2 DB	SWER 3 DB	SWER 4 DB
1	0.00	11.40	14.50	31.37	31.37	22.95	22.95
2	3.01	12.22	11.91	28.91	20.47	20.47	16.35
3	4.77	12.82	10.45	27.55	16.32	19.07	13.54
4	6.02	13.29	9.44	26.60	13.97	18.13	11.83
5	6.99	13.70	8.67	25.88	12.39	17.40	10.64
6	7.78	14.05	8.06	25.30	11.22	16.82	9.72
7	8.45	14.36	7.54	24.82	10.31	16.33	9.00
8	9.03	14.65	7.10	24.41	9.56	15.92	8.40
9	9.54	14.91	6.72	24.05	8.93	15.55	7.89
10	10.00	15.15	6.38	23.73	8.40	15.23	7.44
11	10.41	15.37	6.07	23.45	7.94	14.95	7.05
12	10.79	15.58	5.80	23.19	7.52	14.69	6.71
13	11.14	15.77	5.54	22.96	7.16	14.45	6.39
14	11.46	15.96	5.31	22.74	6.83	14.23	6.10
15	11.76	16.13	5.10	22.54	6.52	14.03	5.84
16	12.04	16.30	4.90	22.36	6.24	13.84	5.60
17	12.30	16.45	4.71	22.18	5.99	13.67	5.37
18	12.55	16.60	4.53	22.02	5.75	13.50	5.16
19	12.79	16.75	4.37	21.87	5.53	13.35	4.97
20	13.01	16.89	4.21	21.72	5.32	13.20	4.79
21	13.22	17.02	4.07	21.59	5.13	13.07	4.62
22	13.42	17.15	3.93	21.46	4.95	12.93	4.46
23	13.62	17.27	3.79	21.33	4.78	12.81	4.30
24	13.80	17.39	3.66	21.21	4.61	12.69	4.16
25	13.98	17.51	3.54	21.10	4.46	12.58	4.02
26	14.15	17.62	3.43	20.99	4.31	12.47	3.89
27	14.31	17.73	3.31	20.86	4.17	12.36	3.76
28	14.47	17.83	3.21	20.76	4.04	12.24	3.64
29	14.62	17.93	3.10	20.67	3.91	12.14	3.52
30	14.77	18.03	3.01	20.57	3.78	12.05	3.41
31	14.91	18.13	2.91	20.49	3.67	11.96	3.30
32	15.05	18.22	2.82	20.40	3.55	11.87	3.20
33	15.19	18.31	2.73	20.32	3.44	11.79	3.10
34	15.31	18.40	2.64	20.24	3.34	11.71	3.00
35	15.44	18.49	2.56	20.16	3.24	11.63	2.91
36	15.56	18.57	2.48	20.08	3.14	11.55	2.82
37	15.68	18.66	2.40	20.01	3.05	11.48	2.73
38	15.80	18.74	2.32	19.94	2.95	11.41	2.65
39	15.91	18.82	2.25	19.87	2.87	11.34	2.57
40	16.02	18.90	2.17	19.80	2.78	11.27	2.49
41	16.13	18.97	2.10	19.74	2.70	11.20	2.41
42	16.23	19.05	2.03	19.67	2.61	11.14	2.33
43	16.33	19.12	1.97	19.61	2.54	11.07	2.26
44	16.43	19.19	1.90	19.55	2.46	11.01	2.19
45	16.53	19.26	1.84	19.49	2.38	10.95	2.12

Appendix: Radar Detection Tables

```
PROBABILITY OF DETECTION= .5000
CONSTANT FALSE ALARM PROBABILITY
.OG10(PFA)=- 7
```

N	N DB	THRESH DB	NON-FLUC DB	SWER 1 DB	SWER 2 DB	SWER 3 DB	SWER 4 DB
1	0.00	12.07	11.94	13.47	13.47	12.65	12.65
2	3.01	12.81	9.45	10.99	10.17	10.17	9.79
3	4.77	13.36	8.05	9.59	8.51	8.77	8.27
4	6.02	13.80	7.09	8.63	7.43	7.81	7.25
5	6.99	14.18	6.36	7.91	6.63	7.08	6.49
6	7.78	14.51	5.78	7.32	6.00	6.50	5.88
7	8.45	14.80	5.29	6.84	5.47	6.01	5.38
8	9.03	15.07	4.88	6.42	5.03	5.60	4.95
9	9.54	15.31	4.51	6.06	4.65	5.24	4.58
10	10.00	15.54	4.19	5.74	4.32	4.91	4.25
11	10.41	15.75	3.91	5.45	4.02	4.63	3.96
12	10.79	15.95	3.65	5.19	3.75	4.37	3.69
13	11.14	16.13	3.41	4.95	3.50	4.13	3.45
14	11.46	16.31	3.19	4.74	3.28	3.91	3.23
15	11.76	16.47	2.99	4.53	3.07	3.71	3.03
16	12.04	16.63	2.80	4.35	2.87	3.52	2.84
17	12.30	16.78	2.63	4.17	2.69	3.35	2.66
18	12.55	16.93	2.46	4.01	2.53	3.18	2.49
19	12.79	17.06	2.31	3.85	2.37	3.03	2.34
20	13.01	17.20	2.16	3.71	2.22	2.88	2.19
21	13.22	17.33	2.02	3.57	2.08	2.75	2.05
22	13.42	17.45	1.89	3.44	1.94	2.61	1.92
23	13.62	17.57	1.77	3.31	1.82	2.49	1.79
24	13.80	17.68	1.65	3.19	1.69	2.37	1.67
25	13.98	17.79	1.54	3.08	1.58	2.26	1.56
26	14.15	17.90	1.43	2.97	1.47	2.15	1.45
27	14.31	18.00	1.32	2.87	1.36	2.04	1.34
28	14.47	18.11	1.22	2.77	1.26	1.94	1.24
29	14.62	18.20	1.13	2.67	1.16	1.85	1.14
30	14.77	18.30	1.03	2.58	1.07	1.75	1.05
31	14.91	18.39	.94	2.49	.96	1.66	.96
32	15.05	18.48	.86	2.40	.89	1.58	.87
33	15.19	18.57	.77	2.32	.80	1.49	.79
34	15.31	18.66	.69	2.24	.72	1.41	.70
35	15.44	18.74	.61	2.16	.64	1.33	.63
36	15.56	18.82	.54	2.08	.56	1.26	.55
37	15.68	18.90	.46	2.01	.49	1.18	.47
38	15.8	18.98	.39	1.94	.42	1.11	.40
39	15.91	19.06	.32	1.87	.35	1.04	.33
40	16.02	19.14	.25	1.80	.28	.97	.26
41	16.13	19.21	.19	1.73	.21	.91	.20
42	16.23	19.28	.12	1.67	.15	.84	.13
43	16.33	19.35	.06	1.60	.08	.78	.07
44	16.43	19.42	-.00	1.54	.02	.72	.01
45	16.53	19.49	-.06	1.48	-.04	.66	-.05

PROBABILITY OF DETECTION= .6000
CONSTANT FALSE ALARM PROBABILITY
LOG10(PFA)=- 7

N	N DB	THRESH DB	NON-FLUC DB	SWER 1 DB	SWER 2 DB	SWER 3 DB	SWER 4 DB
1	0.00	12.07	12.32	14.85	14.85	13.60	13.60
2	3.01	12.81	9.82	12.36	11.10	11.10	10.48
3	4.77	13.36	8.41	10.96	9.29	9.70	8.86
4	6.02	13.80	7.44	10.00	8.11	8.74	7.78
5	6.99	14.18	6.71	9.27	7.25	8.01	6.98
6	7.78	14.51	6.12	8.69	6.57	7.42	6.35
7	8.45	14.80	5.63	8.20	6.02	6.93	5.83
8	9.03	15.07	5.21	7.78	5.55	6.52	5.38
9	9.54	15.31	4.84	7.42	5.15	6.15	5.00
10	10.00	15.54	4.52	7.10	4.80	5.83	4.66
11	10.41	15.75	4.23	6.81	4.49	5.55	4.36
12	10.79	15.95	3.97	6.55	4.20	5.28	4.09
13	11.14	16.13	3.73	6.31	3.95	5.05	3.84
14	11.46	16.31	3.51	6.10	3.71	4.83	3.61
15	11.76	16.47	3.30	5.89	3.49	4.63	3.40
16	12.04	16.63	3.11	5.70	3.29	4.44	3.20
17	12.30	16.78	2.94	5.53	3.11	4.26	3.02
18	12.55	16.93	2.77	5.36	2.93	4.10	2.85
19	12.79	17.06	2.61	5.21	2.77	3.94	2.69
20	13.01	17.20	2.47	5.06	2.61	3.80	2.54
21	13.22	17.33	2.33	4.93	2.47	3.66	2.40
22	13.42	17.45	2.19	4.79	2.33	3.52	2.26
23	13.62	17.57	2.07	4.67	2.20	3.40	2.13
24	13.80	17.68	1.95	4.55	2.07	3.28	2.01
25	13.98	17.79	1.83	4.44	1.95	3.17	1.89
26	14.15	17.90	1.72	4.33	1.84	3.06	1.78
27	14.31	18.00	1.62	4.22	1.73	2.95	1.67
28	14.47	18.11	1.52	4.12	1.62	2.85	1.57
29	14.62	18.20	1.42	4.02	1.52	2.75	1.47
30	14.77	18.30	1.32	3.93	1.42	2.66	1.37
31	14.91	18.39	1.23	3.84	1.33	2.57	1.28
32	15.05	18.48	1.15	3.75	1.24	2.48	1.19
33	15.19	18.57	1.06	3.67	1.15	2.40	1.11
34	15.31	18.66	.98	3.59	1.07	2.32	1.02
35	15.44	18.74	.90	3.51	.99	2.24	.94
36	15.56	18.82	.82	3.43	.91	2.16	.86
37	15.68	18.90	.75	3.36	.83	2.09	.79
38	15.80	18.98	.68	3.29	.75	2.02	.72
39	15.91	19.06	.61	3.22	.68	1.95	.64
40	16.02	19.14	.54	3.15	.61	1.88	.57
41	16.13	19.21	.47	3.08	.54	1.81	.51
42	16.23	19.28	.41	3.02	.48	1.75	.44
43	16.33	19.35	.34	2.96	.41	1.68	.38
44	16.43	19.42	.28	2.90	.35	1.62	.31
45	16.53	19.49	.22	2.84	.29	1.56	.25

Appendix: Radar Detection Tables

PROBABILITY OF DETECTION= .7000
CONSTANT FALSE ALARM PROBABILITY
LOG10(PFA)=- 7

N	N DB	THRESH DB	NON-FLUC DB	SWER 1 DB	SWER 2 DB	SWER 3 DB	SWER 4 DB
1	0.00	12.07	12.72	16.45	16.45	14.65	14.65
2	3.01	12.81	10.20	13.96	12.15	12.15	11.23
3	4.77	13.36	8.78	12.56	10.14	10.75	9.50
4	6.02	13.80	7.80	11.60	8.85	9.78	8.36
5	6.99	14.18	7.06	10.87	7.92	9.05	7.51
6	7.78	14.51	6.46	10.28	7.20	8.46	6.85
7	8.45	14.80	5.97	9.79	6.61	7.97	6.30
8	9.03	15.07	5.55	9.37	6.11	7.56	5.84
9	9.54	15.31	5.18	9.01	5.69	7.19	5.44
10	10.00	15.54	4.85	8.69	5.32	6.87	5.09
11	10.41	15.75	4.56	8.40	4.99	6.58	4.78
12	10.79	15.95	4.29	8.14	4.69	6.32	4.50
13	11.14	16.13	4.05	7.90	4.42	6.08	4.24
14	11.46	16.31	3.83	7.68	4.17	5.86	4.01
15	11.76	16.47	3.62	7.48	3.95	5.66	3.79
16	12.04	16.63	3.43	7.29	3.74	5.47	3.59
17	12.31	16.78	3.25	7.12	3.54	5.29	3.40
18	12.55	16.93	3.08	6.95	3.36	5.13	3.23
19	12.79	17.06	2.93	6.80	3.19	4.97	3.06
20	13.01	17.20	2.78	6.65	3.03	4.83	2.91
21	13.22	17.33	2.64	6.51	2.88	4.69	2.76
22	13.42	17.45	2.50	6.38	2.73	4.55	2.62
23	13.62	17.57	2.37	6.26	2.60	4.43	2.49
24	13.80	17.68	2.25	6.14	2.47	4.31	2.36
25	13.98	17.79	2.14	6.02	2.34	4.19	2.24
26	14.15	17.90	2.03	5.91	2.22	4.08	2.13
27	14.31	18.00	1.92	5.81	2.11	3.98	2.02
28	14.47	18.11	1.82	5.71	2.00	3.88	1.91
29	14.62	18.20	1.72	5.61	1.90	3.78	1.81
30	14.77	18.30	1.62	5.52	1.80	3.69	1.71
31	14.91	18.39	1.53	5.43	1.70	3.60	1.62
32	15.05	18.48	1.44	5.34	1.61	3.51	1.53
33	15.19	18.57	1.36	5.25	1.52	3.43	1.44
34	15.31	18.66	1.27	5.17	1.43	3.34	1.35
35	15.44	18.74	1.19	5.09	1.35	3.26	1.27
36	15.56	18.82	1.12	5.02	1.26	3.19	1.19
37	15.68	18.90	1.04	4.94	1.19	3.11	1.11
38	15.81	18.98	.97	4.87	1.11	3.04	1.04
39	15.91	19.06	.90	4.80	1.03	2.97	.97
40	16.02	19.14	.83	4.73	.96	2.90	.90
41	16.13	19.21	.76	4.67	.89	2.84	.83
42	16.23	19.28	.70	4.60	.82	2.77	.76
43	16.33	19.35	.63	4.54	.76	2.71	.69
44	16.43	19.42	.57	4.48	.69	2.65	.63
45	16.53	19.49	.51	4.42	.63	2.59	.57

PROBABILITY OF DETECTION= .8000
CONSTANT FALSE ALARM PROBABILITY
LOG10(PFA)=- 7

N	N DB	THRESH DB	NON-FLUC DB	SWER 1 DB	SWER 2 DB	SWER 3 DB	SWER 4 DB
1	0.00	12.07	13.16	18.53	18.53	15.97	15.97
2	3.01	12.81	10.02	16.03	13.46	13.46	12.14
3	4.77	13.36	9.19	14.63	11.18	12.06	10.25
4	6.02	13.80	8.20	13.66	9.75	11.09	9.03
5	6.99	14.18	7.45	12.93	8.73	10.35	8.13
6	7.78	14.51	6.85	12.34	7.94	9.76	7.43
7	8.45	14.80	6.35	11.86	7.31	9.27	6.86
8	9.03	15.07	5.92	11.44	6.78	8.85	6.37
9	9.54	15.31	5.55	11.07	6.32	8.49	5.96
10	10.00	15.54	5.22	10.75	5.92	8.16	5.59
11	10.41	15.75	4.92	10.46	5.57	7.87	5.26
12	10.79	15.95	4.66	10.20	5.26	7.61	4.97
13	11.14	16.13	4.41	9.96	4.97	7.37	4.70
14	11.46	16.31	4.19	9.74	4.71	7.15	4.46
15	11.76	16.47	3.98	9.54	4.47	6.95	4.24
16	12.04	16.63	3.79	9.35	4.25	6.76	4.03
17	12.31	16.78	3.60	9.18	4.05	6.58	3.83
18	12.55	16.93	3.44	9.01	3.86	6.42	3.65
19	12.79	17.06	3.28	8.86	3.68	6.26	3.48
20	13.01	17.20	3.13	8.71	3.51	6.11	3.32
21	13.22	17.33	2.98	8.57	3.35	5.97	3.17
22	13.42	17.45	2.85	8.44	3.20	5.84	3.03
23	13.62	17.57	2.72	8.31	3.06	5.71	2.89
24	13.80	17.68	2.60	8.19	2.92	5.59	2.76
25	13.98	17.79	2.48	8.08	2.79	5.48	2.64
26	14.15	17.90	2.37	7.97	2.67	5.37	2.52
27	14.31	18.00	2.26	7.87	2.55	5.26	2.41
28	14.47	18.11	2.15	7.76	2.44	5.16	2.30
29	14.62	18.20	2.05	7.67	2.33	5.06	2.20
30	14.77	18.30	1.96	7.57	2.23	4.97	2.10
31	14.91	18.39	1.87	7.48	2.13	4.88	2.00
32	15.05	18.48	1.78	7.40	2.03	4.79	1.91
33	15.19	18.57	1.69	7.31	1.94	4.71	1.82
34	15.31	18.66	1.61	7.23	1.85	4.63	1.73
35	15.44	18.74	1.53	7.15	1.76	4.55	1.65
36	15.56	18.82	1.45	7.08	1.67	4.47	1.56
37	15.68	18.90	1.37	7.00	1.59	4.40	1.48
38	15.80	18.98	1.30	6.93	1.51	4.32	1.41
39	15.91	19.06	1.22	6.86	1.44	4.25	1.33
40	16.02	19.14	1.15	6.79	1.36	4.18	1.26
41	16.13	19.21	1.09	6.72	1.29	4.12	1.19
42	16.23	19.28	1.02	6.66	1.22	4.05	1.12
43	16.33	19.35	.96	6.60	1.15	3.99	1.05
44	16.43	19.42	.89	6.53	1.08	3.93	.99
45	16.53	19.49	.83	6.47	1.02	3.87	.93

Appendix: Radar Detection Tables

```
PROBABILITY OF DETECTION= .9000
CONSTANT FALSE ALARM PROBABILITY
LOG10(PFA)=- 7
```

K	N DB	THRESH DB	NON-FLUC DB	SWER 1 DB	SWER 2 DB	SWER 3 DB	SWER 4 DB
1	0.00	12.07	13.74	21.82	21.82	17.95	17.95
2	3.01	12.81	11.17	19.32	15.44	15.43	13.45
3	4.77	13.36	9.73	17.92	12.71	14.02	11.33
4	6.02	13.80	8.73	16.95	11.05	13.05	9.98
5	6.99	14.18	7.97	16.22	9.89	12.32	9.00
6	7.78	14.51	7.36	15.63	9.00	11.72	8.24
7	8.45	14.80	6.86	15.14	8.30	11.23	7.62
8	9.03	15.07	6.42	14.72	7.71	10.81	7.11
9	9.54	15.31	6.04	14.36	7.20	10.45	6.66
10	10.00	15.54	5.71	14.04	6.77	10.12	6.27
11	10.41	15.75	5.41	13.75	6.39	9.83	5.93
12	10.79	15.95	5.14	13.49	6.04	9.57	5.62
13	11.14	16.13	4.89	13.25	5.74	9.33	5.33
14	11.46	16.31	4.66	13.03	5.46	9.11	5.08
15	11.76	16.47	4.45	12.83	5.20	8.90	4.84
16	12.04	16.63	4.25	12.64	4.96	8.71	4.62
17	12.30	16.78	4.07	12.46	4.74	8.53	4.42
18	12.55	16.93	3.90	12.30	4.54	8.37	4.23
19	12.79	17.06	3.74	12.14	4.35	8.21	4.06
20	13.01	17.20	3.58	11.99	4.17	8.06	3.89
21	13.22	17.33	3.44	11.85	4.00	7.92	3.73
22	13.42	17.45	3.30	11.72	3.84	7.79	3.58
23	13.62	17.57	3.17	11.59	3.69	7.67	3.44
24	13.80	17.68	3.05	11.47	3.54	7.55	3.31
25	13.98	17.79	2.93	11.36	3.41	7.43	3.18
26	14.15	17.90	2.81	11.25	3.28	7.32	3.05
27	14.31	18.00	2.70	11.14	3.15	7.21	2.94
28	14.47	18.11	2.60	11.04	3.03	7.11	2.82
29	14.62	18.20	2.50	10.94	2.92	7.01	2.72
30	14.77	18.30	2.40	10.85	2.81	6.92	2.61
31	14.91	18.39	2.31	10.76	2.70	6.83	2.51
32	15.05	18.48	2.21	10.67	2.60	6.74	2.42
33	15.19	18.57	2.13	10.59	2.50	6.66	2.32
34	15.31	18.66	2.04	10.51	2.41	6.57	2.23
35	15.44	18.74	1.96	10.43	2.32	6.49	2.15
36	15.56	18.82	1.88	10.35	2.23	6.42	2.06
37	15.68	18.90	1.80	10.28	2.14	6.34	1.98
38	15.8	18.98	1.73	10.21	2.06	6.27	1.90
39	15.91	19.06	1.65	10.14	1.98	6.20	1.82
40	16.02	19.14	1.58	10.07	1.90	6.13	1.75
41	16.13	19.21	1.51	10.00	1.83	6.06	1.68
42	16.23	19.28	1.45	9.94	1.75	6.00	1.60
43	16.33	19.35	1.38	9.87	1.68	5.93	1.54
44	16.43	19.42	1.32	9.81	1.61	5.87	1.47
45	16.53	19.49	1.26	9.75	1.54	5.81	1.40

PROBABILITY OF DETECTION= .9500
CONSTANT FALSE ALARM PROBABILITY
LOG10(PFA)=- 7

N	N DB	THRESH DB	NON-FLUC DB	SWER 1 DB	SWER 2 DB	SWER 3 DB	SWER 4 DB
1	0.00	12.07	14.19	24.95	24.95	19.74	19.74
2	3.01	12.81	11.61	22.46	17.23	17.23	14.60
3	4.77	13.36	10.15	21.05	14.07	15.81	12.25
4	6.02	13.80	9.14	20.09	12.19	14.84	10.77
5	6.99	14.18	8.38	19.36	10.89	14.10	9.72
6	7.78	14.51	7.76	18.77	9.91	13.51	8.91
7	8.45	14.80	7.25	18.28	9.13	13.02	8.26
8	9.03	15.07	6.81	17.86	8.49	12.60	7.71
9	9.54	15.31	6.43	17.50	7.95	12.23	7.24
10	10.00	15.54	6.09	17.18	7.48	11.90	6.83
11	10.41	15.75	5.79	16.89	7.07	11.61	6.46
12	10.79	15.95	5.51	16.62	6.70	11.35	6.14
13	11.14	16.13	5.26	16.39	6.37	11.11	5.85
14	11.46	16.31	5.03	16.17	6.07	10.89	5.58
15	11.76	16.47	4.82	15.96	5.80	10.68	5.33
16	12.04	16.63	4.62	15.77	5.55	10.49	5.11
17	12.30	16.78	4.43	15.60	5.32	10.31	4.90
18	12.55	16.93	4.26	15.43	5.10	10.15	4.70
19	12.79	17.06	4.10	15.28	4.90	9.99	4.52
20	13.01	17.20	3.94	15.13	4.71	9.84	4.34
21	13.22	17.33	3.79	14.99	4.53	9.70	4.18
22	13.42	17.45	3.66	14.86	4.36	9.57	4.03
23	13.62	17.57	3.52	14.73	4.21	9.44	3.88
24	13.80	17.68	3.40	14.61	4.05	9.32	3.74
25	13.98	17.79	3.28	14.50	3.91	9.21	3.61
26	14.15	17.90	3.16	14.39	3.77	9.10	3.48
27	14.31	18.00	3.05	14.29	3.64	8.99	3.36
28	14.47	18.11	2.95	14.18	3.52	8.89	3.24
29	14.62	18.20	2.84	14.09	3.40	8.79	3.13
30	14.77	18.30	2.74	13.99	3.28	8.70	3.02
31	14.91	18.39	2.65	13.90	3.17	8.60	2.92
32	15.05	18.48	2.56	13.82	3.07	8.52	2.82
33	15.19	18.57	2.47	13.73	2.97	8.43	2.73
34	15.31	18.66	2.38	13.65	2.87	8.35	2.63
35	15.44	18.74	2.30	13.56	2.77	8.27	2.54
36	15.56	18.82	2.22	13.49	2.68	8.19	2.46
37	15.68	18.90	2.14	13.41	2.59	8.12	2.37
38	15.80	18.98	2.07	13.34	2.51	8.05	2.29
39	15.91	19.06	1.99	13.27	2.42	7.97	2.21
40	16.02	19.14	1.92	13.20	2.34	7.91	2.14
41	16.13	19.21	1.85	13.13	2.26	7.84	2.06
42	16.23	19.28	1.78	13.07	2.18	7.77	1.99
43	16.33	19.35	1.72	13.01	2.11	7.71	1.92
44	16.43	19.42	1.65	12.95	2.04	7.65	1.85
45	16.53	19.49	1.59	12.88	1.97	7.59	1.78

Appendix: Radar Detection Tables

```
PROBABILITY OF DETECTION= .9900
CONSTANT FALSE ALARM PROBABILITY
LOG10(PFA)=- 7
```

N	N DB	THRESH DB	NON-FLUC DB	SWER 1 DB	SWER 2 DB	SWER 3 DB	SWER 4 DB
1	0.00	12.07	14.97	32.05	32.05	23.56	23.56
2	3.01	12.81	12.36	29.54	21.06	21.06	16.92
3	4.77	13.36	10.89	28.13	16.89	19.65	14.05
4	6.02	13.80	9.87	27.16	14.49	18.67	12.32
5	6.99	14.18	9.09	26.43	12.89	17.93	11.10
6	7.78	14.51	8.46	25.84	11.71	17.34	10.19
7	8.45	14.80	7.94	25.37	10.78	16.87	9.45
8	9.03	15.07	7.49	24.96	10.02	16.44	8.84
9	9.54	15.31	7.10	24.59	9.38	16.07	8.32
10	10.00	15.54	6.76	24.27	8.85	15.75	7.87
11	10.41	15.75	6.45	23.98	8.37	15.46	7.47
12	10.79	15.95	6.17	23.72	7.95	15.19	7.12
13	11.14	16.13	5.92	23.48	7.58	14.95	6.79
14	11.46	16.31	5.68	23.26	7.24	14.73	6.50
15	11.76	16.47	5.47	23.05	6.93	14.52	6.24
16	12.04	16.63	5.26	22.87	6.65	14.33	5.99
17	12.30	16.78	5.07	22.66	6.40	14.16	5.77
18	12.55	16.93	4.89	22.50	6.16	13.99	5.56
19	12.79	17.06	4.73	22.34	5.94	13.83	5.36
20	13.01	17.20	4.57	22.20	5.72	13.69	5.17
21	13.22	17.33	4.42	22.06	5.52	13.52	5.00
22	13.42	17.45	4.28	21.92	5.34	13.39	4.84
23	13.62	17.57	4.14	21.80	5.17	13.26	4.68
24	13.80	17.68	4.02	21.68	5.00	13.14	4.53
25	13.98	17.79	3.89	21.56	4.85	13.02	4.39
26	14.15	17.90	3.78	21.45	4.70	12.91	4.26
27	14.31	18.00	3.66	21.35	4.55	12.81	4.13
28	14.47	18.11	3.56	21.25	4.41	12.71	4.01
29	14.62	18.20	3.45	21.15	4.28	12.61	3.89
30	14.77	18.30	3.35	21.06	4.16	12.51	3.78
31	14.91	18.39	3.25	20.97	4.04	12.42	3.67
32	15.05	18.48	3.16	20.88	3.93	12.33	3.56
33	15.19	18.57	3.07	20.80	3.82	12.25	3.46
34	15.31	18.66	2.98	20.71	3.71	12.17	3.36
35	15.44	18.74	2.90	20.64	3.61	12.09	3.27
36	15.56	18.82	2.82	20.56	3.51	12.01	3.18
37	15.68	18.90	2.74	20.48	3.41	11.93	3.09
38	15.8.	18.98	2.66	20.41	3.32	11.86	3.00
39	15.91	19.06	2.58	20.34	3.23	11.79	2.92
40	16.02	19.14	2.51	20.27	3.14	11.72	2.84
41	16.13	19.21	2.44	20.21	3.06	11.65	2.76
42	16.23	19.28	2.37	20.14	2.97	11.59	2.68
43	16.33	19.35	2.30	20.08	2.89	11.52	2.61
44	16.43	19.42	2.24	20.02	2.82	11.47	2.54
45	16.53	19.49	2.17	19.96	2.74	11.41	2.47

PROBABILITY OF DETECTION= .5000
CONSTANT FALSE ALARM PROBABILITY
LOG10(PFA)=- 8

	N DB	THRESH DB	NON-FLUC DB	SWER 1 DB	SWER 2 DB	SWER 3 DB	SWER 4 DB
1	0.00	12.05	12.53	14.08	14.08	13.26	13.26
2	3.01	13.33	10.01	11.55	10.73	10.73	10.35
3	4.77	13.83	8.59	10.14	9.05	9.31	8.81
4	6.02	14.25	7.61	9.16	7.95	8.34	7.77
5	6.99	14.60	6.87	8.42	7.14	7.59	6.99
6	7.78	14.91	6.28	7.83	6.49	7.00	6.38
7	8.45	15.19	5.78	7.33	5.97	6.51	5.87
8	9.03	15.44	5.36	6.91	5.52	6.08	5.43
9	9.54	15.67	4.99	6.54	5.13	5.72	5.06
10	10.00	15.89	4.67	6.22	4.79	5.39	4.72
11	10.41	16.09	4.37	5.92	4.49	5.10	4.43
12	10.79	16.28	4.11	5.66	4.21	4.83	4.16
13	11.14	16.45	3.87	5.42	3.96	4.59	3.91
14	11.46	16.62	3.65	5.20	3.73	4.37	3.69
15	11.76	16.78	3.44	4.99	3.52	4.17	3.48
16	12.04	16.93	3.25	4.80	3.32	3.98	3.29
17	12.30	17.08	3.07	4.62	3.14	3.80	3.11
18	12.55	17.22	2.91	4.46	2.97	3.63	2.94
19	12.79	17.35	2.75	4.30	2.81	3.48	2.78
20	13.01	17.48	2.60	4.15	2.66	3.33	2.63
21	13.22	17.60	2.46	4.01	2.52	3.19	2.49
22	13.42	17.72	2.33	3.88	2.38	3.05	2.35
23	13.62	17.83	2.20	3.75	2.25	2.93	2.22
24	13.80	17.94	2.08	3.63	2.13	2.81	2.10
25	13.98	18.05	1.97	3.52	2.01	2.69	1.99
26	14.15	18.16	1.86	3.41	1.90	2.58	1.87
27	14.31	18.26	1.75	3.30	1.79	2.47	1.77
28	14.47	18.35	1.65	3.20	1.69	2.37	1.67
29	14.62	18.45	1.55	3.10	1.59	2.27	1.57
30	14.77	18.54	1.46	3.01	1.49	2.18	1.47
31	14.91	18.63	1.36	2.92	1.40	2.09	1.38
32	15.05	18.72	1.28	2.83	1.31	2.00	1.29
33	15.19	18.81	1.19	2.74	1.22	1.92	1.21
34	15.31	18.89	1.11	2.66	1.14	1.83	1.12
35	15.44	18.97	1.03	2.58	1.06	1.76	1.04
36	15.56	19.05	.95	2.50	.98	1.68	.97
37	15.68	19.13	.88	2.43	.91	1.60	.89
38	15.80	19.21	.80	2.36	.83	1.53	.82
39	15.91	19.28	.73	2.29	.76	1.46	.75
40	16.02	19.35	.67	2.22	.69	1.39	.68
41	16.13	19.43	.60	2.15	.62	1.32	.61
42	16.23	19.50	.53	2.08	.56	1.26	.54
43	16.33	19.56	.47	2.02	.49	1.20	.48
44	16.43	19.63	.41	1.96	.43	1.13	.42
45	16.53	19.70	.35	1.90	.37	1.07	.36

Appendix: Radar Detection Tables

```
PROBABILITY OF DETECTION= .6000
CONSTANT FALSE ALARM PROBABILITY
LOG10(PFA)=- 8
```

N	N DB	THRESH DB	NON-FLUC DB	SWER 1 DB	SWER 2 DB	SWER 3 DB	SWER 4 DB
1	0.00	12.65	12.90	15.45	15.45	14.19	14.19
2	3.01	13.33	10.35	12.92	11.66	11.66	11.02
3	4.77	13.83	8.92	11.50	9.81	10.24	9.38
4	6.02	14.25	7.94	10.52	8.62	9.26	8.29
5	6.99	14.60	7.19	9.78	7.75	8.51	7.48
6	7.78	14.91	6.59	9.18	7.06	7.92	6.83
7	8.45	15.19	6.10	8.69	6.50	7.42	6.30
8	9.03	15.44	5.67	8.27	6.03	7.00	5.85
9	9.54	15.67	5.30	7.90	5.62	6.63	5.46
10	10.00	15.89	4.97	7.57	5.26	6.30	5.12
11	10.41	16.09	4.67	7.28	4.94	6.01	4.81
12	10.79	16.28	4.41	7.02	4.65	5.75	4.53
13	11.14	16.45	4.17	6.77	4.39	5.50	4.28
14	11.46	16.62	3.94	6.55	4.15	5.28	4.05
15	11.76	16.78	3.73	6.35	3.93	5.08	3.84
16	12.04	16.93	3.54	6.16	3.73	4.89	3.64
17	12.30	17.08	3.36	5.98	3.54	4.71	3.45
18	12.55	17.22	3.19	5.81	3.36	4.54	3.28
19	12.79	17.35	3.04	5.65	3.20	4.38	3.12
20	13.01	17.48	2.89	5.51	3.04	4.23	2.96
21	13.22	17.60	2.75	5.37	2.89	4.09	2.82
22	13.42	17.72	2.61	5.23	2.75	3.96	2.68
23	13.62	17.83	2.48	5.11	2.62	3.83	2.55
24	13.80	17.94	2.36	4.98	2.49	3.71	2.43
25	13.98	18.05	2.24	4.87	2.37	3.60	2.31
26	14.15	18.16	2.13	4.76	2.25	3.48	2.19
27	14.31	18.26	2.03	4.65	2.14	3.38	2.08
28	14.47	18.35	1.92	4.55	2.03	3.28	1.98
29	14.62	18.45	1.82	4.45	1.93	3.18	1.88
30	14.77	18.54	1.73	4.36	1.83	3.08	1.78
31	14.91	18.63	1.64	4.27	1.74	2.99	1.69
32	15.05	18.72	1.55	4.18	1.65	2.90	1.60
33	15.19	18.81	1.46	4.09	1.56	2.82	1.51
34	15.31	18.89	1.38	4.01	1.47	2.74	1.43
35	15.44	18.97	1.30	3.93	1.39	2.66	1.34
36	15.56	19.05	1.22	3.85	1.31	2.58	1.27
37	15.68	19.13	1.15	3.78	1.23	2.50	1.19
38	15.80	19.21	1.07	3.71	1.16	2.43	1.11
39	15.91	19.28	1.00	3.64	1.08	2.36	1.04
40	16.02	19.35	.93	3.57	1.01	2.29	.97
41	16.13	19.43	.86	3.50	.94	2.22	.90
42	16.23	19.50	.80	3.43	.87	2.16	.84
43	16.33	19.56	.74	3.37	.81	2.10	.77
44	16.43	19.63	.67	3.31	.74	2.03	.71
45	16.53	19.70	.61	3.25	.68	1.97	.65

PROBABILITY OF DETECTION= .7000
CONSTANT FALSE ALARM PROBABILITY
LOG10(PFA)=- 8

N	N DB	THRESH DB	NON-FLUC DB	SWER 1 DB	SWER 2 DB	SWER 3 DB	SWER 4 DB
1	0.00	12.65	13.27	17.05	17.05	15.24	15.24
2	3.01	13.33	10.71	14.51	12.70	12.70	11.77
3	4.77	13.83	9.27	13.09	10.66	11.28	10.01
4	6.02	14.25	8.28	12.11	9.35	10.29	8.85
5	6.99	14.60	7.53	11.37	8.41	9.55	7.99
6	7.78	14.91	6.92	10.77	7.68	8.95	7.32
7	8.45	15.19	6.42	10.28	7.08	8.46	6.77
8	9.03	15.44	5.99	9.85	6.57	8.03	6.30
9	9.54	15.67	5.61	9.49	6.14	7.66	5.89
10	10.00	15.89	5.28	9.16	5.76	7.33	5.53
11	10.41	16.09	4.99	8.87	5.43	7.04	5.22
12	10.79	16.28	4.72	8.60	5.13	6.77	4.93
13	11.14	16.45	4.47	8.36	4.85	6.53	4.67
14	11.46	16.62	4.25	8.14	4.60	6.31	4.43
15	11.76	16.78	4.04	7.93	4.37	6.10	4.21
16	12.04	16.93	3.84	7.74	4.16	5.91	4.01
17	12.30	17.08	3.66	7.56	3.96	5.73	3.82
18	12.55	17.22	3.49	7.40	3.78	5.57	3.64
19	12.79	17.35	3.33	7.24	3.60	5.41	3.47
20	13.01	17.48	3.18	7.09	3.44	5.26	3.32
21	13.22	17.60	3.04	6.95	3.29	5.12	3.17
22	13.42	17.72	2.90	6.82	3.14	4.98	3.03
23	13.62	17.83	2.77	6.69	3.00	4.86	2.89
24	13.80	17.94	2.65	6.57	2.87	4.74	2.76
25	13.98	18.05	2.53	6.45	2.75	4.62	2.64
26	14.15	18.16	2.42	6.34	2.63	4.51	2.53
27	14.31	18.26	2.31	6.23	2.51	4.40	2.41
28	14.47	18.35	2.21	6.13	2.40	4.30	2.31
29	14.62	18.45	2.11	6.03	2.29	4.20	2.20
30	14.77	18.54	2.01	5.94	2.19	4.11	2.10
31	14.91	18.63	1.92	5.85	2.10	4.01	2.01
32	15.05	18.72	1.83	5.76	2.00	3.93	1.92
33	15.19	18.81	1.74	5.67	1.91	3.84	1.83
34	15.31	18.89	1.66	5.59	1.82	3.76	1.74
35	15.44	18.97	1.58	5.51	1.74	3.68	1.66
36	15.56	19.05	1.50	5.43	1.65	3.60	1.58
37	15.68	19.13	1.42	5.36	1.57	3.52	1.50
38	15.80	19.21	1.35	5.29	1.50	3.45	1.42
39	15.91	19.28	1.28	5.22	1.42	3.38	1.35
40	16.02	19.35	1.21	5.15	1.35	3.31	1.28
41	16.13	19.43	1.14	5.08	1.28	3.24	1.21
42	16.23	19.50	1.07	5.01	1.21	3.18	1.14
43	16.33	19.56	1.01	4.95	1.14	3.11	1.07
44	16.43	19.63	.95	4.89	1.07	3.05	1.01
45	16.53	19.70	.88	4.83	1.01	2.99	.95

Appendix: Radar Detection Tables

```
PROBABILITY OF DETECTION= .8000
CONSTANT FALSE ALARM PROBABILITY
LOG10(PFA)=- 8
```

N	N DB	THRESH DB	NON-FLUC DB	SWER 1 DB	SWER 2 DB	SWER 3 DB	SWER 4 DB
1	0.00	12.65	13.68	19.11	19.11	16.54	16.54
2	3.01	13.33	11.11	16.58	14.00	14.00	12.60
3	4.77	13.83	9.66	15.16	11.69	12.57	10.75
4	6.02	14.25	8.66	14.18	10.24	11.59	9.51
5	6.99	14.60	7.90	13.43	9.21	10.84	8.60
6	7.78	14.91	7.29	12.84	8.41	10.24	7.89
7	8.45	15.19	6.78	12.34	7.76	9.75	7.30
8	9.03	15.44	6.35	11.92	7.22	9.32	6.81
9	9.54	15.67	5.97	11.55	6.76	8.95	6.39
10	10.00	15.89	5.63	11.22	6.36	8.62	6.02
11	10.41	16.09	5.33	10.93	6.00	8.33	5.69
12	10.79	16.28	5.06	10.66	5.68	8.06	5.39
13	11.14	16.45	4.81	10.42	5.39	7.82	5.12
14	11.46	16.62	4.59	10.20	5.13	7.59	4.87
15	11.76	16.78	4.38	9.99	4.89	7.39	4.64
16	12.04	16.93	4.18	9.80	4.66	7.20	4.43
17	12.30	17.08	4.00	9.62	4.45	7.02	4.24
18	12.55	17.22	3.82	9.45	4.26	6.85	4.05
19	12.79	17.35	3.66	9.30	4.08	6.69	3.88
20	13.01	17.48	3.51	9.15	3.91	6.54	3.72
21	13.22	17.60	3.37	9.01	3.75	6.40	3.56
22	13.42	17.72	3.23	8.87	3.60	6.26	3.42
23	13.62	17.83	3.10	8.74	3.45	6.14	3.28
24	13.80	17.94	2.97	8.62	3.31	6.02	3.15
25	13.98	18.05	2.86	8.51	3.18	5.90	3.02
26	14.15	18.16	2.74	8.40	3.06	5.79	2.90
27	14.31	18.26	2.63	8.29	2.94	5.68	2.79
28	14.47	18.35	2.53	8.19	2.82	5.58	2.68
29	14.62	18.45	2.43	8.09	2.71	5.48	2.58
30	14.77	18.54	2.33	7.99	2.61	5.38	2.47
31	14.91	18.63	2.24	7.90	2.51	5.29	2.38
32	15.05	18.72	2.15	7.81	2.41	5.20	2.28
33	15.19	18.81	2.06	7.73	2.31	5.12	2.19
34	15.31	18.89	1.97	7.65	2.22	5.04	2.10
35	15.44	18.97	1.89	7.57	2.13	4.96	2.02
36	15.56	19.05	1.81	7.49	2.05	4.88	1.93
37	15.68	19.13	1.73	7.41	1.97	4.80	1.85
38	15.8	19.21	1.66	7.34	1.89	4.73	1.78
39	15.91	19.28	1.59	7.27	1.81	4.66	1.70
40	16.02	19.35	1.52	7.20	1.73	4.59	1.63
41	16.13	19.43	1.45	7.13	1.66	4.52	1.56
42	16.23	19.50	1.38	7.07	1.59	4.46	1.49
43	16.33	19.56	1.32	7.00	1.52	4.39	1.42
44	16.43	19.63	1.25	6.94	1.45	4.33	1.35
45	16.53	19.70	1.19	6.88	1.38	4.27	1.29

PROBABILITY OF DETECTION= .9000
CONSTANT FALSE ALARM PROBABILITY
LOG10(PFA)=- 8

N	N DB	THRESH DB	NON-FLUC DB	SWER 1 DB	SWER 2 DB	SWER 3 DB	SWER 4 DB
1	0.00	12.65	14.23	22.40	22.40	18.51	18.51
2	3.01	13.33	11.63	19.87	15.97	15.97	13.96
3	4.77	13.83	10.17	18.44	13.21	14.53	11.81
4	6.02	14.25	9.16	17.46	11.54	13.55	10.44
5	6.99	14.60	8.39	16.71	10.36	12.80	9.45
6	7.78	14.91	7.77	16.12	9.46	12.20	8.68
7	8.45	15.19	7.26	15.62	8.74	11.70	8.05
8	9.03	15.44	6.82	15.20	8.14	11.27	7.53
9	9.54	15.67	6.44	14.83	7.63	10.90	7.08
10	10.00	15.89	6.10	14.50	7.19	10.57	6.68
11	10.41	16.09	5.79	14.21	6.80	10.28	6.33
12	10.79	16.28	5.52	13.94	6.45	10.01	6.01
13	11.14	16.45	5.27	13.70	6.14	9.77	5.73
14	11.46	16.62	5.04	13.48	5.86	9.54	5.47
15	11.76	16.78	4.82	13.27	5.60	9.34	5.23
16	12.04	16.93	4.62	13.08	5.36	9.14	5.01
17	12.30	17.08	4.44	12.90	5.13	8.96	4.80
18	12.55	17.22	4.26	12.73	4.93	8.80	4.61
19	12.79	17.35	4.10	12.57	4.73	8.64	4.43
20	13.01	17.48	3.95	12.42	4.55	8.49	4.26
21	13.22	17.60	3.80	12.28	4.38	8.35	4.10
22	13.42	17.72	3.66	12.15	4.22	8.21	3.95
23	13.62	17.83	3.53	12.02	4.06	8.08	3.81
24	13.80	17.94	3.40	11.90	3.92	7.96	3.67
25	13.98	18.05	3.28	11.78	3.78	7.84	3.54
26	14.15	18.16	3.17	11.67	3.65	7.73	3.42
27	14.31	18.26	3.05	11.57	3.52	7.62	3.30
28	14.47	18.35	2.95	11.46	3.40	7.52	3.18
29	14.62	18.45	2.85	11.36	3.29	7.42	3.08
30	14.77	18.54	2.75	11.27	3.17	7.33	2.97
31	14.91	18.63	2.65	11.18	3.07	7.24	2.87
32	15.05	18.72	2.56	11.09	2.97	7.15	2.77
33	15.19	18.81	2.47	11.00	2.87	7.06	2.68
34	15.31	18.89	2.39	10.92	2.77	6.98	2.59
35	15.44	18.97	2.30	10.84	2.68	6.90	2.50
36	15.56	19.05	2.22	10.76	2.59	6.82	2.41
37	15.68	19.13	2.15	10.69	2.50	6.74	2.33
38	15.80	19.21	2.07	10.62	2.42	6.67	2.25
39	15.91	19.28	2.00	10.54	2.34	6.60	2.17
40	16.02	19.35	1.92	10.48	2.26	6.53	2.10
41	16.13	19.43	1.85	10.41	2.18	6.46	2.02
42	16.23	19.50	1.79	10.34	2.10	6.40	1.95
43	16.33	19.56	1.72	10.28	2.03	6.33	1.88
44	16.43	19.63	1.66	10.22	1.96	6.27	1.81
45	16.53	19.70	1.59	10.16	1.89	6.21	1.75

Appendix: Radar Detection Tables

```
PROBABILITY OF DETECTION= .9500
CONSTANT FALSE ALARM PROBABILITY
LOG10(PFA)=- 8
```

N	N DB	THRESH DB	NON-FLUC DB	SWER 1 DB	SWER 2 DB	SWER 3 DB	SWER 4 DB
1	0.00	12.65	14.65	25.54	25.54	20.31	20.31
2	3.01	13.33	12.04	23.00	17.75	17.75	15.10
3	4.77	13.83	10.57	21.58	14.56	16.32	12.72
4	6.02	14.25	9.55	20.59	12.60	15.33	11.23
5	6.99	14.60	8.77	19.86	11.35	14.58	10.16
6	7.78	14.91	8.15	19.26	10.35	13.98	9.34
7	8.45	15.19	7.63	18.76	9.57	13.48	8.67
8	9.03	15.44	7.19	18.34	8.92	13.05	8.11
9	9.54	15.67	6.80	17.97	8.37	12.68	7.64
10	10.00	15.89	6.46	17.64	7.89	12.35	7.22
11	10.41	16.09	6.15	17.35	7.47	12.06	6.85
12	10.79	16.28	5.88	17.08	7.10	11.79	6.53
13	11.14	16.45	5.62	16.84	6.77	11.54	6.23
14	11.46	16.62	5.39	16.61	6.46	11.32	5.96
15	11.76	16.78	5.17	16.41	6.19	11.11	5.71
16	12.04	16.93	4.97	16.22	5.93	10.92	5.48
17	12.30	17.08	4.78	16.04	5.70	10.74	5.26
18	12.55	17.22	4.61	15.87	5.48	10.57	5.07
19	12.79	17.35	4.44	15.71	5.27	10.41	4.88
20	13.01	17.48	4.29	15.56	5.08	10.26	4.70
21	13.22	17.60	4.14	15.42	4.90	10.12	4.54
22	13.42	17.72	4.00	15.29	4.73	9.99	4.38
23	13.62	17.83	3.86	15.16	4.57	9.86	4.23
24	13.80	17.94	3.74	15.04	4.42	9.74	4.09
25	13.98	18.05	3.62	14.92	4.27	9.62	3.96
26	14.15	18.16	3.50	14.81	4.13	9.51	3.83
27	14.31	18.26	3.39	14.71	4.00	9.40	3.71
28	14.47	18.35	3.28	14.60	3.87	9.30	3.59
29	14.62	18.45	3.18	14.51	3.75	9.20	3.48
30	14.77	18.54	3.08	14.41	3.64	9.10	3.37
31	14.91	18.63	2.98	14.32	3.53	9.01	3.26
32	15.05	18.72	2.89	14.23	3.42	8.92	3.16
33	15.19	18.81	2.80	14.14	3.31	8.83	3.07
34	15.31	18.89	2.71	14.06	3.21	8.75	2.97
35	15.44	18.97	2.63	13.98	3.12	8.67	2.88
36	15.56	19.05	2.55	13.90	3.02	8.59	2.80
37	15.68	19.13	2.47	13.82	2.93	8.52	2.71
38	15.80	19.21	2.39	13.75	2.85	8.44	2.63
39	15.91	19.28	2.32	13.68	2.76	8.37	2.55
40	16.02	19.35	2.24	13.61	2.68	8.30	2.47
41	16.13	19.43	2.17	13.54	2.60	8.23	2.40
42	16.23	19.50	2.11	13.48	2.52	8.17	2.32
43	16.33	19.56	2.04	13.41	2.45	8.10	2.25
44	16.43	19.63	1.97	13.35	2.37	8.04	2.18
45	16.53	19.70	1.91	13.29	2.30	7.98	2.11

PROBABILITY OF DETECTION= .9900
CONSTANT FALSE ALARM PROBABILITY
LOG10(PFA)=- 8

N	N DB	THRESH DB	NON-FLUC DB	SWER 1 DB	SWER 2 DB	SWER 3 DB	SWER 4 DB
1	0.00	12.65	15.39	32.66	32.66	24.12	24.12
2	3.01	13.33	12.76	30.08	21.58	21.58	17.41
3	4.77	13.83	11.27	28.65	17.36	20.15	14.51
4	6.02	14.25	10.24	27.67	14.96	19.16	12.76
5	6.99	14.60	9.45	26.92	13.34	18.41	11.52
6	7.78	14.91	8.82	26.33	12.14	17.80	10.59
7	8.45	15.19	8.29	25.83	11.20	17.30	9.85
8	9.03	15.44	7.84	25.40	10.43	16.87	9.22
9	9.54	15.67	7.45	25.03	9.78	16.50	8.70
10	10.00	15.89	7.10	24.70	9.24	16.17	8.24
11	10.41	16.09	6.79	24.41	8.76	15.90	7.84
12	10.79	16.28	6.51	24.14	8.33	15.63	7.48
13	11.14	16.45	6.25	23.90	7.96	15.39	7.15
14	11.46	16.62	6.01	23.68	7.62	15.16	6.86
15	11.76	16.78	5.79	23.47	7.30	14.95	6.59
16	12.04	16.93	5.58	23.28	7.02	14.76	6.34
17	12.30	17.08	5.39	23.10	6.76	14.56	6.11
18	12.55	17.22	5.21	22.93	6.51	14.39	5.90
19	12.79	17.35	5.05	22.78	6.29	14.23	5.70
20	13.01	17.48	4.89	22.63	6.07	14.08	5.51
21	13.22	17.60	4.74	22.49	5.88	13.94	5.34
22	13.42	17.72	4.59	22.35	5.69	13.80	5.17
23	13.62	17.83	4.46	22.23	5.51	13.67	5.01
24	13.80	17.94	4.33	22.10	5.34	13.55	4.86
25	13.98	18.05	4.20	21.99	5.18	13.43	4.72
26	14.15	18.16	4.09	21.88	5.03	13.32	4.58
27	14.31	18.26	3.97	21.77	4.89	13.21	4.45
28	14.47	18.35	3.86	21.67	4.75	13.11	4.33
29	14.62	18.45	3.76	21.57	4.62	13.01	4.21
30	14.77	18.54	3.66	21.47	4.49	12.91	4.10
31	14.91	18.63	3.56	21.38	4.37	12.82	3.99
32	15.05	18.72	3.46	21.30	4.25	12.73	3.88
33	15.19	18.81	3.37	21.21	4.14	12.65	3.78
34	15.31	18.89	3.29	21.13	4.03	12.56	3.68
35	15.44	18.97	3.20	21.05	3.93	12.48	3.58
36	15.56	19.05	3.12	20.97	3.83	12.41	3.49
37	15.68	19.13	3.04	20.89	3.74	12.33	3.40
38	15.80	19.21	2.96	20.82	3.64	12.26	3.31
39	15.91	19.28	2.88	20.75	3.55	12.18	3.23
40	16.02	19.35	2.81	20.68	3.46	12.11	3.15
41	16.13	19.43	2.74	20.61	3.38	12.05	3.07
42	16.23	19.50	2.67	20.55	3.29	11.98	2.99
43	16.33	19.56	2.60	20.49	3.21	11.92	2.92
44	16.43	19.63	2.54	20.42	3.13	11.85	2.85
45	16.53	19.70	2.47	20.36	3.06	11.80	2.78

Appendix: Radar Detection Tables

PROBABILITY OF DETECTION= .5000
CONSTANT FALSE ALARM PROBABILITY
LOG10(PFA)=- 9

N	N DB	THRESH DB	NON-FLUC DB	SWER 1 DB	SWER 2 DB	SWER 3 DB	SWER 4 DB
1	0.00	13.16	13.06	14.61	14.61	13.78	13.78
2	3.01	13.79	10.50	12.05	11.23	11.23	10.84
3	4.77	14.26	9.06	10.61	9.53	9.79	9.28
4	6.02	14.65	8.07	9.62	8.41	8.80	8.23
5	6.99	14.98	7.32	8.87	7.59	8.05	7.44
6	7.78	15.27	6.71	8.27	6.94	7.44	6.82
7	8.45	15.54	6.21	7.77	6.40	6.94	6.30
8	9.03	15.78	5.78	7.34	5.95	6.51	5.86
9	9.54	16.00	5.41	6.97	5.55	6.14	5.48
10	10.00	16.21	5.08	6.64	5.21	5.81	5.14
11	10.41	16.40	4.78	6.34	4.90	5.51	4.84
12	10.79	16.58	4.52	6.07	4.62	5.25	4.57
13	11.14	16.75	4.27	5.83	4.37	5.00	4.32
14	11.46	16.91	4.05	5.60	4.14	4.78	4.09
15	11.76	17.06	3.84	5.40	3.92	4.57	3.88
16	12.04	17.21	3.65	5.20	3.72	4.38	3.68
17	12.30	17.35	3.47	5.02	3.54	4.20	3.50
18	12.55	17.48	3.30	4.85	3.36	4.03	3.33
19	12.79	17.61	3.14	4.70	3.20	3.87	3.17
20	13.01	17.73	2.99	4.55	3.05	3.72	3.02
21	13.22	17.85	2.85	4.40	2.90	3.58	2.87
22	13.42	17.97	2.71	4.27	2.76	3.44	2.74
23	13.62	18.08	2.58	4.14	2.63	3.31	2.61
24	13.80	18.18	2.46	4.02	2.51	3.19	2.48
25	13.98	18.29	2.34	3.90	2.39	3.07	2.37
26	14.15	18.39	2.23	3.79	2.28	2.96	2.25
27	14.31	18.49	2.13	3.68	2.17	2.86	2.14
28	14.47	18.58	2.02	3.58	2.06	2.75	2.04
29	14.62	18.67	1.92	3.48	1.96	2.65	1.94
30	14.77	18.76	1.83	3.38	1.86	2.56	1.84
31	14.91	18.85	1.74	3.29	1.77	2.47	1.75
32	15.05	18.94	1.65	3.20	1.68	2.38	1.66
33	15.19	19.02	1.56	3.12	1.59	2.29	1.58
34	15.31	19.10	1.48	3.03	1.51	2.21	1.49
35	15.44	19.18	1.40	2.95	1.43	2.13	1.41
36	15.56	19.26	1.32	2.88	1.35	2.05	1.33
37	15.68	19.34	1.24	2.80	1.27	1.97	1.26
38	15.80	19.41	1.17	2.73	1.20	1.90	1.18
39	15.91	19.48	1.10	2.65	1.13	1.83	1.11
40	16.02	19.55	1.03	2.59	1.05	1.76	1.04
41	16.13	19.62	.96	2.52	.99	1.69	.97
42	16.23	19.69	.90	2.45	.92	1.63	.91
43	16.33	19.76	.83	2.39	.85	1.56	.84
44	16.43	19.83	.77	2.32	.79	1.50	.78
45	16.53	19.89	.71	2.26	.73	1.44	.72

PROBABILITY OF DETECTION= .6000
CONSTANT FALSE ALARM PROBABILITY
LOG10(PFA)=- 9

K	N DB	THRESH DB	NON-FLUC DB	SWER 1 DB	SWER 2 DB	SWER 3 DB	SWER 4 DB
1	0.00	13.16	13.40	15.97	15.97	14.71	14.71
2	3.01	13.79	10.83	13.41	12.15	12.15	11.51
3	4.77	14.26	9.38	11.97	10.28	10.71	9.85
4	6.02	14.65	8.38	10.98	9.07	9.71	8.74
5	6.99	14.98	7.62	10.23	8.19	8.96	7.92
6	7.78	15.27	7.02	9.63	7.49	8.36	7.26
7	8.45	15.54	6.51	9.12	6.92	7.85	6.72
8	9.03	15.78	6.08	8.69	6.44	7.42	6.27
9	9.54	16.00	5.70	8.32	6.03	7.05	5.87
10	10.00	16.21	5.37	7.99	5.67	6.72	5.52
11	10.41	16.40	5.07	7.69	5.34	6.42	5.21
12	10.79	16.58	4.80	7.42	5.05	6.15	4.93
13	11.14	16.75	4.55	7.18	4.79	5.91	4.67
14	11.46	16.91	4.33	6.96	4.55	5.68	4.44
15	11.76	17.06	4.12	6.75	4.32	5.47	4.22
16	12.04	17.21	3.92	6.55	4.12	5.28	4.02
17	12.30	17.35	3.74	6.37	3.92	5.10	3.83
18	12.55	17.48	3.57	6.20	3.74	4.93	3.66
19	12.79	17.61	3.41	6.05	3.58	4.77	3.49
20	13.01	17.73	3.26	5.90	3.42	4.62	3.34
21	13.22	17.85	3.12	5.75	3.27	4.48	3.19
22	13.42	17.97	2.98	5.62	3.12	4.34	3.05
23	13.62	18.08	2.85	5.49	2.99	4.21	2.92
24	13.80	18.18	2.73	5.37	2.86	4.09	2.79
25	13.98	18.29	2.61	5.25	2.74	3.97	2.67
26	14.15	18.39	2.50	5.14	2.62	3.86	2.56
27	14.31	18.49	2.39	5.03	2.51	3.75	2.45
28	14.47	18.58	2.28	4.93	2.40	3.65	2.34
29	14.62	18.67	2.18	4.83	2.29	3.55	2.24
30	14.77	18.76	2.09	4.73	2.20	3.46	2.14
31	14.91	18.85	2.00	4.64	2.10	3.36	2.05
32	15.05	18.94	1.91	4.55	2.01	3.27	1.96
33	15.19	19.02	1.82	4.46	1.92	3.19	1.87
34	15.31	19.10	1.73	4.38	1.83	3.11	1.78
35	15.44	19.18	1.65	4.30	1.75	3.02	1.70
36	15.56	19.26	1.58	4.22	1.67	2.95	1.62
37	15.68	19.34	1.50	4.15	1.59	2.87	1.54
38	15.80	19.41	1.42	4.07	1.51	2.80	1.47
39	15.91	19.48	1.35	4.00	1.44	2.72	1.39
40	16.02	19.55	1.28	3.93	1.36	2.66	1.32
41	16.13	19.62	1.21	3.86	1.29	2.59	1.25
42	16.23	19.69	1.15	3.80	1.23	2.52	1.19
43	16.33	19.76	1.08	3.73	1.16	2.46	1.12
44	16.43	19.83	1.02	3.67	1.09	2.39	1.06
45	16.53	19.89	.96	3.61	1.03	2.33	.99

Appendix: Radar Detection Tables

```
PROBABILITY OF DETECTION= .7000
CONSTANT FALSE ALARM PROBABILITY
LOG10(PFA)=- 9
```

N	N DB	THRESH DB	NON-FLUC DB	SWER 1 DB	SWER 2 DB	SWER 3 DB	SWER 4 DB
1	0.00	13.16	13.75	17.57	17.57	15.75	15.75
2	3.01	13.79	11.16	15.00	13.18	13.18	12.24
3	4.77	14.26	9.71	13.56	11.12	11.74	10.46
4	6.02	14.65	8.70	12.57	9.80	10.74	9.29
5	6.99	14.98	7.94	11.82	8.84	9.99	8.42
6	7.78	15.27	7.33	11.21	8.10	9.38	7.74
7	8.45	15.54	6.82	10.71	7.49	8.88	7.18
8	9.03	15.78	6.38	10.28	6.98	8.45	6.70
9	9.54	16.00	6.00	9.91	6.54	8.08	6.29
10	10.00	16.21	5.67	9.57	6.16	7.74	5.93
11	10.41	16.40	5.37	9.28	5.82	7.45	5.61
12	10.79	16.58	5.09	9.01	5.52	7.18	5.31
13	11.14	16.75	4.84	8.76	5.24	6.93	5.05
14	11.46	16.91	4.62	8.54	4.99	6.71	4.81
15	11.76	17.06	4.41	8.33	4.75	6.50	4.59
16	12.04	17.21	4.21	8.14	4.54	6.30	4.38
17	12.30	17.35	4.03	7.96	4.34	6.12	4.19
18	12.55	17.48	3.85	7.79	4.15	5.95	4.01
19	12.79	17.61	3.69	7.63	3.97	5.79	3.84
20	13.01	17.73	3.54	7.48	3.81	5.64	3.68
21	13.22	17.85	3.39	7.33	3.65	5.50	3.53
22	13.42	17.97	3.26	7.20	3.50	5.36	3.39
23	13.62	18.08	3.13	7.07	3.36	5.23	3.25
24	13.80	18.18	3.00	6.95	3.23	5.11	3.12
25	13.98	18.29	2.88	6.83	3.10	4.99	3.00
26	14.15	18.39	2.77	6.72	2.98	4.88	2.88
27	14.31	18.49	2.66	6.61	2.87	4.77	2.77
28	14.47	18.58	2.55	6.51	2.75	4.67	2.66
29	14.62	18.67	2.45	6.41	2.65	4.57	2.55
30	14.77	18.76	2.36	6.31	2.54	4.47	2.45
31	14.91	18.85	2.26	6.22	2.45	4.38	2.36
32	15.05	18.94	2.17	6.13	2.35	4.29	2.26
33	15.19	19.02	2.09	6.04	2.26	4.21	2.17
34	15.31	19.10	2.00	5.96	2.17	4.12	2.09
35	15.44	19.18	1.92	5.88	2.08	4.04	2.00
36	15.56	19.26	1.84	5.80	2.00	3.96	1.92
37	15.68	19.34	1.76	5.73	1.92	3.89	1.84
38	15.80	19.41	1.69	5.65	1.84	3.81	1.76
39	15.91	19.48	1.61	5.58	1.76	3.74	1.69
40	16.02	19.55	1.54	5.51	1.69	3.67	1.62
41	16.13	19.62	1.47	5.44	1.62	3.60	1.55
42	16.23	19.69	1.41	5.38	1.55	3.54	1.48
43	16.33	19.76	1.34	5.31	1.48	3.47	1.41
44	16.43	19.83	1.28	5.25	1.41	3.41	1.35
45	16.53	19.89	1.22	5.19	1.35	3.35	1.28

PROBABILITY OF DETECTION= .8000
CONSTANT FALSE ALARM PROBABILITY
LOG10(PFA)=- 9

N	N DB	THRESH DB	NON-FLUC DB	SWER 1 DB	SWER 2 DB	SWER 3 DB	SWER 4 DB
1	0.00	13.16	14.14	19.63	19.63	17.05	17.05
2	3.01	13.79	11.54	17.07	14.48	14.48	13.13
3	4.77	14.26	10.08	15.62	12.14	13.03	11.20
4	6.02	14.65	9.06	14.63	10.68	12.03	9.94
5	6.99	14.98	8.29	13.88	9.63	11.28	9.02
6	7.78	15.27	7.68	13.27	8.82	10.67	8.29
7	8.45	15.54	7.16	12.77	8.17	10.17	7.70
8	9.03	15.78	6.72	12.34	7.62	9.73	7.20
9	9.54	16.00	6.34	11.96	7.15	9.36	6.77
10	10.00	16.21	6.00	11.63	6.75	9.03	6.40
11	10.41	16.40	5.70	11.33	6.38	8.73	6.06
12	10.79	16.58	5.42	11.07	6.00	8.46	5.76
13	11.14	16.75	5.17	10.82	5.77	8.21	5.49
14	11.46	16.91	4.94	10.59	5.50	7.99	5.24
15	11.76	17.06	4.73	10.39	5.25	7.78	5.01
16	12.04	17.21	4.53	10.19	5.03	7.58	4.79
17	12.30	17.35	4.34	10.01	4.82	7.40	4.59
18	12.55	17.48	4.17	9.84	4.62	7.23	4.41
19	12.79	17.61	4.01	9.68	4.44	7.07	4.23
20	13.01	17.73	3.85	9.53	4.26	6.92	4.07
21	13.22	17.85	3.71	9.39	4.10	6.78	3.91
22	13.42	17.97	3.57	9.25	3.95	6.64	3.77
23	13.62	18.08	3.44	9.12	3.80	6.51	3.63
24	13.80	18.18	3.31	9.00	3.66	6.39	3.49
25	13.98	18.29	3.19	8.88	3.53	6.27	3.37
26	14.15	18.39	3.08	8.77	3.40	6.16	3.25
27	14.31	18.49	2.97	8.66	3.28	6.05	3.13
28	14.47	18.58	2.86	8.56	3.17	5.95	3.02
29	14.62	18.67	2.76	8.46	3.05	5.85	2.91
30	14.77	18.76	2.66	8.37	2.95	5.75	2.81
31	14.91	18.85	2.56	8.27	2.85	5.66	2.71
32	15.05	18.94	2.47	8.18	2.75	5.57	2.61
33	15.19	19.02	2.38	8.10	2.65	5.48	2.52
34	15.31	19.10	2.30	8.01	2.56	5.40	2.43
35	15.44	19.18	2.22	7.93	2.47	5.32	2.35
36	15.56	19.26	2.14	7.85	2.38	5.24	2.26
37	15.68	19.34	2.06	7.78	2.30	5.16	2.18
38	15.80	19.41	1.98	7.71	2.22	5.09	2.10
39	15.91	19.48	1.91	7.63	2.14	5.02	2.03
40	16.02	19.55	1.84	7.56	2.06	4.95	1.95
41	16.13	19.62	1.77	7.50	1.99	4.88	1.88
42	16.23	19.69	1.70	7.43	1.92	4.81	1.81
43	16.33	19.76	1.63	7.36	1.85	4.75	1.74
44	16.43	19.83	1.57	7.30	1.78	4.68	1.68
45	16.53	19.89	1.51	7.24	1.71	4.62	1.61

Appendix: Radar Detection Tables

PROBABILITY OF DETECTION= .9000
CONSTANT FALSE ALARM PROBABILITY
LOG10(PFA)=- 9

N	N DB	THRESH DB	NON-FLUC DB	SWER 1 DB	SWER 2 DB	SWER 3 DB	SWER 4 DB
1	0.00	13.16	14.06	22.91	22.91	19.01	19.01
2	3.01	13.79	12.04	20.35	16.44	16.44	14.42
3	4.77	14.26	10.56	18.90	13.66	14.99	12.24
4	6.02	14.65	9.54	17.91	11.96	13.99	10.80
5	6.99	14.98	8.76	17.15	10.77	13.23	9.85
6	7.78	15.27	8.14	16.55	9.86	12.62	9.07
7	8.45	15.54	7.62	16.05	9.13	12.12	8.44
8	9.03	15.78	7.17	15.62	8.53	11.69	7.90
9	9.54	16.00	6.79	15.24	8.01	11.31	7.44
10	10.00	16.21	6.44	14.91	7.57	10.97	7.04
11	10.41	16.40	6.14	14.61	7.17	10.68	6.69
12	10.79	16.58	5.86	14.35	6.82	10.40	6.37
13	11.14	16.75	5.60	14.10	6.50	10.16	6.08
14	11.46	16.91	5.37	13.87	6.22	9.93	5.82
15	11.76	17.06	5.15	13.67	5.95	9.72	5.58
16	12.04	17.21	4.95	13.47	5.71	9.53	5.35
17	12.30	17.35	4.77	13.29	5.48	9.35	5.15
18	12.55	17.48	4.59	13.12	5.27	9.17	4.95
19	12.79	17.61	4.43	12.96	5.08	9.01	4.77
20	13.01	17.73	4.27	12.81	4.89	8.86	4.60
21	13.22	17.85	4.12	12.66	4.72	8.72	4.44
22	13.42	17.97	3.98	12.53	4.56	8.58	4.28
23	13.62	18.08	3.85	12.40	4.40	8.45	4.14
24	13.80	18.18	3.72	12.28	4.25	8.33	4.00
25	13.98	18.29	3.60	12.16	4.11	8.21	3.87
26	14.15	18.39	3.48	12.05	3.98	8.10	3.74
27	14.31	18.49	3.37	11.94	3.85	7.99	3.62
28	14.47	18.58	3.26	11.83	3.73	7.89	3.51
29	14.62	18.67	3.16	11.74	3.61	7.79	3.40
30	14.77	18.76	3.06	11.64	3.50	7.69	3.29
31	14.91	18.85	2.96	11.55	3.39	7.60	3.19
32	15.05	18.94	2.87	11.46	3.29	7.51	3.09
33	15.19	19.02	2.78	11.37	3.19	7.42	2.99
34	15.31	19.10	2.69	11.29	3.09	7.34	2.90
35	15.44	19.18	2.61	11.21	3.00	7.25	2.81
36	15.56	19.26	2.53	11.13	2.91	7.18	2.73
37	15.68	19.34	2.45	11.05	2.82	7.10	2.64
38	15.80	19.41	2.37	10.98	2.73	7.02	2.56
39	15.91	19.48	2.30	10.91	2.65	6.95	2.48
40	16.02	19.55	2.23	10.84	2.57	6.88	2.41
41	16.13	19.62	2.16	10.77	2.49	6.81	2.33
42	16.23	19.69	2.09	10.70	2.42	6.75	2.26
43	16.33	19.76	2.02	10.64	2.34	6.68	2.19
44	16.43	19.83	1.96	10.57	2.27	6.62	2.12
45	16.53	19.89	1.89	10.51	2.20	6.56	2.05

PROBABILITY OF DETECTION= .9500
CONSTANT FALSE ALARM PROBABILITY
LOG10(PFA)=- 9

N	N DB	THRESH DB	NON-FLUC DB	SWER 1 DB	SWER 2 DB	SWER 3 DB	SWER 4 DB
1	0.00	13.16	15.07	26.05	26.05	20.80	20.80
2	3.01	13.79	12.43	23.49	18.22	18.22	15.55
3	4.77	14.26	10.94	22.04	15.00	16.77	13.14
4	6.02	14.65	9.91	21.05	13.08	15.77	11.63
5	6.99	14.98	9.13	20.29	11.75	15.01	10.55
6	7.78	15.27	8.50	19.69	10.75	14.40	9.72
7	8.45	15.54	7.98	19.19	9.95	13.89	9.04
8	9.03	15.78	7.53	18.76	9.30	13.46	8.48
9	9.54	16.00	7.14	18.38	8.74	13.08	8.00
10	10.00	16.21	6.79	18.05	8.26	12.75	7.58
11	10.41	16.40	6.48	17.75	7.83	12.45	7.21
12	10.79	16.58	6.20	17.48	7.46	12.18	6.87
13	11.14	16.75	5.94	17.24	7.12	11.93	6.57
14	11.46	16.91	5.71	17.01	6.81	11.71	6.30
15	11.76	17.06	5.49	16.80	6.53	11.50	6.04
16	12.04	17.21	5.29	16.61	6.27	11.30	5.81
17	12.30	17.35	5.10	16.43	6.04	11.12	5.60
18	12.55	17.48	4.92	16.26	5.81	10.95	5.39
19	12.79	17.61	4.75	16.10	5.61	10.79	5.20
20	13.01	17.73	4.60	15.95	5.41	10.64	5.03
21	13.22	17.85	4.45	15.80	5.23	10.49	4.86
22	13.42	17.97	4.30	15.67	5.06	10.36	4.70
23	13.62	18.08	4.17	15.54	4.89	10.23	4.55
24	13.80	18.18	4.04	15.42	4.74	10.10	4.41
25	13.98	18.29	3.92	15.30	4.59	9.98	4.27
26	14.15	18.39	3.80	15.19	4.45	9.87	4.14
27	14.31	18.49	3.69	15.08	4.32	9.76	4.02
28	14.47	18.58	3.58	14.98	4.19	9.66	3.90
29	14.62	18.67	3.48	14.88	4.07	9.56	3.79
30	14.77	18.76	3.37	14.78	3.95	9.46	3.68
31	14.91	18.85	3.28	14.69	3.84	9.37	3.57
32	15.05	18.94	3.18	14.60	3.73	9.28	3.47
33	15.19	19.02	3.09	14.51	3.63	9.19	3.37
34	15.31	19.10	3.01	14.43	3.52	9.11	3.28
35	15.44	19.18	2.92	14.35	3.43	9.03	3.19
36	15.56	19.26	2.84	14.27	3.33	8.95	3.10
37	15.68	19.34	2.76	14.19	3.24	8.87	3.01
38	15.80	19.41	2.68	14.11	3.15	8.80	2.93
39	15.91	19.48	2.61	14.04	3.07	8.72	2.85
40	16.02	19.55	2.53	13.97	2.98	8.65	2.77
41	16.13	19.62	2.46	13.90	2.90	8.59	2.69
42	16.23	19.69	2.39	13.83	2.83	8.52	2.62
43	16.33	19.76	2.33	13.77	2.75	8.45	2.55
44	16.43	19.83	2.26	13.71	2.68	8.39	2.48
45	16.53	19.89	2.20	13.65	2.60	8.33	2.41

Appendix: Radar Detection Tables

PROBABILITY OF DETECTION= .9900
CONSTANT FALSE ALARM PROBABILITY
LOG10(PFA)=- 9

N	N DB	THRESH DB	NON-FLUC DB	SWER 1 DB	SWER 2 DB	SWER 3 DB	SWER 4 DB
1	0.00	13.16	15.78	33.15	33.15	24.62	24.62
2	3.01	13.79	13.12	30.57	22.05	22.05	17.85
3	4.77	14.26	11.61	29.12	17.79	20.59	14.92
4	6.02	14.65	10.57	28.12	15.37	19.59	13.15
5	6.99	14.98	9.78	27.36	13.74	18.83	11.90
6	7.78	15.27	9.14	26.76	12.53	18.22	10.96
7	8.45	15.54	8.61	26.25	11.58	17.71	10.21
8	9.03	15.78	8.16	25.82	10.80	17.28	9.58
9	9.54	16.00	7.76	25.45	10.14	16.90	9.04
10	10.00	16.21	7.41	25.11	9.60	16.59	8.58
11	10.41	16.40	7.09	24.82	9.11	16.29	8.17
12	10.79	16.58	6.81	24.55	8.68	16.02	7.80
13	11.14	16.75	6.55	24.30	8.30	15.77	7.47
14	11.46	16.91	6.31	24.07	7.95	15.54	7.18
15	11.76	17.06	6.08	23.87	7.64	15.33	6.91
16	12.04	17.21	5.88	23.67	7.35	15.14	6.66
17	12.30	17.35	5.69	23.49	7.08	14.95	6.43
18	12.55	17.48	5.51	23.32	6.83	14.78	6.21
19	12.79	17.61	5.33	23.16	6.61	14.62	6.01
20	13.01	17.73	5.17	23.01	6.39	14.47	5.82
21	13.22	17.85	5.02	22.87	6.19	14.33	5.64
22	13.42	17.97	4.88	22.73	6.00	14.19	5.47
23	13.62	18.08	4.74	22.60	5.82	14.04	5.31
24	13.80	18.18	4.61	22.48	5.65	13.91	5.16
25	13.98	18.29	4.49	22.36	5.49	13.80	5.01
26	14.15	18.39	4.37	22.25	5.33	13.68	4.88
27	14.31	18.49	4.25	22.14	5.19	13.57	4.75
28	14.47	18.58	4.14	22.04	5.05	13.47	4.62
29	14.62	18.67	4.03	21.94	4.92	13.37	4.50
30	14.77	18.76	3.94	21.84	4.79	13.27	4.38
31	14.91	18.85	3.84	21.75	4.67	13.18	4.27
32	15.05	18.94	3.74	21.66	4.55	13.09	4.17
33	15.19	19.02	3.65	21.57	4.44	13.00	4.06
34	15.31	19.10	3.56	21.49	4.33	12.92	3.96
35	15.44	19.18	3.47	21.41	4.23	12.84	3.87
36	15.56	19.26	3.39	21.33	4.13	12.76	3.77
37	15.68	19.34	3.31	21.26	4.03	12.68	3.68
38	15.80	19.41	3.23	21.18	3.93	12.61	3.60
39	15.91	19.48	3.15	21.11	3.84	12.53	3.51
40	16.02	19.55	3.08	21.04	3.75	12.46	3.43
41	16.13	19.62	3.01	20.97	3.66	12.39	3.35
42	16.23	19.69	2.94	20.91	3.58	12.33	3.27
43	16.33	19.76	2.87	20.84	3.50	12.26	3.20
44	16.43	19.83	2.80	20.78	3.42	12.20	3.12
45	16.53	19.89	2.73	20.72	3.34	12.14	3.05

PROBABILITY OF DETECTION= .5000
CONSTANT FALSE ALARM PROBABILITY
LOG10(PFA)=-10

N	N DB	THRESH DB	NON-FLUC DB	SWER 1 DB	SWER 2 DB	SWER 3 DB	SWER 4 DB
1	0.00	13.62	13.53	15.08	15.08	14.26	14.26
2	3.01	14.21	10.94	12.50	11.67	11.67	11.29
3	4.77	14.65	9.48	11.04	9.96	10.22	9.71
4	6.02	15.01	8.48	10.04	8.83	9.21	8.64
5	6.99	15.33	7.72	9.28	7.99	8.45	7.85
6	7.78	15.60	7.11	8.67	7.33	7.84	7.21
7	8.45	15.86	6.60	8.16	6.79	7.33	6.69
8	9.03	16.09	6.17	7.73	6.33	6.90	6.24
9	9.54	16.30	5.79	7.35	5.93	6.52	5.86
10	10.00	16.50	5.45	7.01	5.58	6.19	5.51
11	10.41	16.68	5.15	6.71	5.27	5.89	5.21
12	10.79	16.85	4.88	6.44	4.99	5.62	4.93
13	11.14	17.02	4.64	6.20	4.73	5.37	4.68
14	11.46	17.17	4.41	5.97	4.50	5.14	4.45
15	11.76	17.32	4.20	5.76	4.28	4.93	4.24
16	12.04	17.46	4.00	5.56	4.08	4.74	4.04
17	12.30	17.60	3.82	5.38	3.89	4.55	3.85
18	12.55	17.73	3.65	5.21	3.72	4.38	3.68
19	12.79	17.85	3.49	5.05	3.55	4.22	3.52
20	13.01	17.97	3.34	4.90	3.40	4.07	3.36
21	13.22	18.08	3.19	4.75	3.25	3.93	3.22
22	13.42	18.20	3.06	4.62	3.11	3.79	3.08
23	13.62	18.30	2.93	4.49	2.98	3.66	2.95
24	13.80	18.41	2.80	4.36	2.85	3.54	2.83
25	13.98	18.51	2.68	4.24	2.73	3.42	2.71
26	14.15	18.61	2.57	4.13	2.61	3.30	2.59
27	14.31	18.70	2.46	4.02	2.50	3.20	2.48
28	14.47	18.79	2.36	3.92	2.40	3.09	2.38
29	14.62	18.88	2.26	3.82	2.30	2.99	2.28
30	14.77	18.97	2.16	3.72	2.20	2.89	2.18
31	14.91	19.06	2.07	3.63	2.10	2.80	2.09
32	15.05	19.14	1.98	3.54	2.01	2.71	1.99
33	15.19	19.22	1.89	3.45	1.93	2.62	1.91
34	15.31	19.30	1.81	3.37	1.84	2.54	1.82
35	15.44	19.38	1.73	3.29	1.76	2.46	1.74
36	15.56	19.45	1.65	3.21	1.68	2.38	1.66
37	15.68	19.53	1.57	3.13	1.60	2.30	1.58
38	15.80	19.60	1.50	3.06	1.52	2.23	1.51
39	15.91	19.67	1.42	2.98	1.45	2.16	1.44
40	16.02	19.74	1.35	2.91	1.38	2.09	1.37
41	16.13	19.81	1.29	2.85	1.31	2.02	1.30
42	16.23	19.87	1.22	2.78	1.24	1.95	1.23
43	16.33	19.94	1.15	2.71	1.18	1.89	1.17
44	16.43	20.00	1.09	2.65	1.11	1.82	1.10
45	16.53	20.07	1.03	2.59	1.05	1.76	1.04

Appendix: Radar Detection Tables

```
PROBABILITY OF DETECTION= .6000
CONSTANT FALSE ALARM PROBABILITY
LOG10(PFA)=-10
```

N	N DB	THRESH DB	NON-FLUC DB	SWER 1 DB	SWER 2 DB	SWER 3 DB	SWER 4 DB
1	0.00	13.62	13.85	16.44	16.44	15.17	15.17
2	3.01	14.21	11.25	13.85	12.58	12.58	11.95
3	4.77	14.65	9.79	12.40	10.70	11.13	10.26
4	6.02	15.01	8.78	11.39	9.48	10.12	9.14
5	6.99	15.33	8.01	10.63	8.58	9.36	8.31
6	7.78	15.60	7.40	10.02	7.88	8.75	7.65
7	8.45	15.86	6.88	9.51	7.31	8.24	7.10
8	9.03	16.09	6.45	9.08	6.82	7.80	6.64
9	9.54	16.30	6.07	8.70	6.40	7.42	6.24
10	10.00	16.50	5.73	8.36	6.03	7.09	5.89
11	10.41	16.68	5.43	8.06	5.71	6.79	5.57
12	10.79	16.85	5.15	7.79	5.41	6.52	5.29
13	11.14	17.02	4.90	7.54	5.14	6.27	5.03
14	11.46	17.17	4.68	7.32	4.90	6.04	4.79
15	11.76	17.32	4.46	7.11	4.67	5.83	4.57
16	12.04	17.46	4.27	6.91	4.46	5.64	4.37
17	12.30	17.60	4.08	6.73	4.27	5.45	4.18
18	12.55	17.73	3.91	6.56	4.09	5.28	4.00
19	12.79	17.85	3.75	6.40	3.92	5.12	3.83
20	13.01	17.97	3.59	6.24	3.76	4.97	3.68
21	13.22	18.08	3.45	6.10	3.60	4.82	3.53
22	13.42	18.20	3.31	5.96	3.46	4.69	3.39
23	13.62	18.30	3.18	5.83	3.32	4.56	3.25
24	13.80	18.41	3.06	5.71	3.19	4.43	3.13
25	13.98	18.51	2.94	5.59	3.07	4.31	3.10
26	14.15	18.61	2.82	5.48	2.95	4.20	2.89
27	14.31	18.70	2.71	5.37	2.84	4.09	2.78
28	14.47	18.79	2.61	5.26	2.73	3.99	2.67
29	14.62	18.88	2.51	5.16	2.62	3.89	2.57
30	14.77	18.97	2.41	5.07	2.52	3.79	2.47
31	14.91	19.06	2.32	4.97	2.42	3.70	2.37
32	15.05	19.14	2.23	4.88	2.33	3.61	2.28
33	15.19	19.22	2.14	4.80	2.24	3.52	2.19
34	15.31	19.30	2.05	4.71	2.15	3.44	2.10
35	15.44	19.38	1.97	4.63	2.07	3.35	2.02
36	15.56	19.45	1.89	4.55	1.98	3.27	1.94
37	15.68	19.53	1.81	4.48	1.90	3.20	1.86
38	15.80	19.60	1.74	4.40	1.83	3.12	1.78
39	15.91	19.67	1.67	4.33	1.75	3.05	1.71
40	16.02	19.74	1.60	4.26	1.68	2.98	1.64
41	16.13	19.81	1.53	4.19	1.61	2.91	1.57
42	16.23	19.87	1.46	4.12	1.54	2.85	1.50
43	16.33	19.94	1.39	4.06	1.47	2.78	1.43
44	16.43	20.00	1.33	4.00	1.41	2.72	1.37
45	16.53	20.07	1.27	3.93	1.34	2.65	1.31

PROBABILITY OF DETECTION= .7000
CONSTANT FALSE ALARM PROBABILITY
LOG10(PFA)=-10

N	N DB	THRESH DB	NON-FLUC DB	SWER 1 DB	SWER 2 DB	SWER 3 DB	SWER 4 DB
1	0.00	13.62	14.18	18.03	18.03	16.21	16.21
2	3.01	14.21	11.57	15.44	13.62	13.62	12.67
3	4.77	14.65	10.10	13.98	11.53	12.16	10.87
4	6.02	15.01	9.08	12.98	10.20	11.15	9.68
5	6.99	15.33	8.31	12.22	9.23	10.39	8.81
6	7.78	15.60	7.69	11.60	8.48	9.77	8.11
7	8.45	15.86	7.18	11.10	7.87	9.26	7.54
8	9.03	16.09	6.74	10.66	7.35	8.83	7.06
9	9.54	16.30	6.35	10.28	6.91	8.45	6.65
10	10.00	16.50	6.01	9.95	6.52	8.11	6.28
11	10.41	16.68	5.71	9.65	6.18	7.81	5.96
12	10.79	16.85	5.43	9.37	5.87	7.54	5.66
13	11.14	17.02	5.18	9.13	5.59	7.29	5.39
14	11.46	17.17	4.95	8.90	5.33	7.06	5.15
15	11.76	17.32	4.74	8.69	5.09	6.85	4.92
16	12.04	17.46	4.54	8.49	4.88	6.65	4.72
17	12.30	17.60	4.35	8.31	4.67	6.47	4.52
18	12.55	17.73	4.18	8.14	4.48	6.30	4.34
19	12.79	17.85	4.02	7.98	4.31	6.14	4.17
20	13.01	17.97	3.86	7.82	4.14	5.99	4.01
21	13.22	18.08	3.72	7.68	3.98	5.84	3.85
22	13.42	18.20	3.58	7.54	3.83	5.70	3.71
23	13.62	18.30	3.44	7.41	3.69	5.57	3.57
24	13.80	18.41	3.32	7.29	3.55	5.45	3.44
25	13.98	18.51	3.20	7.17	3.43	5.33	3.32
26	14.15	18.61	3.08	7.06	3.30	5.22	3.20
27	14.31	18.70	2.97	6.95	3.19	5.11	3.08
28	14.47	18.79	2.87	6.84	3.07	5.00	2.97
29	14.62	18.88	2.77	6.74	2.96	4.90	2.87
30	14.77	18.97	2.67	6.65	2.86	4.81	2.77
31	14.91	19.06	2.57	6.55	2.76	4.71	2.67
32	15.05	19.14	2.48	6.46	2.66	4.62	2.58
33	15.19	19.22	2.39	6.38	2.57	4.53	2.48
34	15.31	19.30	2.31	6.29	2.48	4.45	2.40
35	15.44	19.38	2.22	6.21	2.39	4.37	2.31
36	15.56	19.45	2.14	6.13	2.31	4.29	2.23
37	15.68	19.53	2.07	6.05	2.23	4.21	2.15
38	15.80	19.60	1.99	5.98	2.15	4.14	2.07
39	15.91	19.67	1.92	5.91	2.07	4.06	2.00
40	16.02	19.74	1.85	5.84	2.00	3.99	1.92
41	16.13	19.81	1.78	5.77	1.92	3.93	1.85
42	16.23	19.87	1.71	5.70	1.85	3.86	1.78
43	16.33	19.94	1.64	5.64	1.78	3.79	1.71
44	16.43	20.00	1.58	5.57	1.72	3.73	1.65
45	16.53	20.07	1.52	5.51	1.65	3.67	1.58

Appendix: Radar Detection Tables

PROBABILITY OF DETECTION= .8000
CONSTANT FALSE ALARM PROBABILITY
LOG10(PFA)=-10

N	N DB	THRESH DB	NON-FLUC DB	SWER 1 DB	SWER 2 DB	SWER 3 DB	SWER 4 DB
1	0.00	13.62	14.56	20.10	20.10	17.50	17.50
2	3.01	14.21	11.93	17.50	14.91	14.91	13.55
3	4.77	14.65	10.45	16.04	12.55	13.44	11.60
4	6.02	15.01	9.43	15.04	11.07	12.43	10.32
5	6.99	15.33	8.65	14.27	10.01	11.67	9.39
6	7.78	15.60	8.03	13.66	9.20	11.06	8.66
7	8.45	15.86	7.51	13.15	8.54	10.54	8.06
8	9.03	16.09	7.06	12.72	7.98	10.11	7.56
9	9.54	16.30	6.68	12.34	7.51	9.73	7.12
10	10.00	16.50	6.33	12.00	7.09	9.39	6.74
11	10.41	16.68	6.03	11.70	6.73	9.09	6.40
12	10.79	16.85	5.75	11.43	6.40	8.82	6.10
13	11.14	17.02	5.50	11.18	6.10	8.57	5.82
14	11.46	17.17	5.26	10.95	5.83	8.34	5.57
15	11.76	17.32	5.05	10.74	5.59	8.13	5.33
16	12.04	17.46	4.85	10.55	5.36	7.93	5.12
17	12.30	17.60	4.66	10.36	5.14	7.75	4.92
18	12.55	17.73	4.48	10.19	4.95	7.57	4.73
19	12.79	17.85	4.32	10.03	4.76	7.41	4.55
20	13.01	17.97	4.16	9.88	4.59	7.26	4.38
21	13.22	18.08	4.02	9.73	4.42	7.12	4.23
22	13.42	18.20	3.88	9.60	4.26	6.98	4.08
23	13.62	18.30	3.74	9.47	4.12	6.85	3.94
24	13.80	18.41	3.61	9.34	3.98	6.72	3.80
25	13.98	18.51	3.49	9.22	3.84	6.60	3.68
26	14.15	18.61	3.38	9.11	3.71	6.49	3.55
27	14.31	18.70	3.27	9.00	3.59	6.38	3.44
28	14.47	18.79	3.16	8.90	3.47	6.28	3.32
29	14.62	18.88	3.06	8.79	3.36	6.17	3.22
30	14.77	18.97	2.96	8.70	3.25	6.08	3.11
31	14.91	19.06	2.86	8.60	3.15	5.98	3.01
32	15.05	19.14	2.77	8.51	3.05	5.89	2.92
33	15.19	19.22	2.68	8.43	2.95	5.81	2.82
34	15.31	19.30	2.59	8.34	2.86	5.72	2.73
35	15.44	19.38	2.51	9.26	2.77	5.64	2.65
36	15.56	19.45	2.43	8.18	2.68	5.56	2.56
37	15.68	19.53	2.35	8.11	2.60	5.48	2.48
38	15.80	19.60	2.27	8.03	2.52	5.41	2.40
39	15.91	19.67	2.20	7.96	2.44	5.34	2.32
40	16.02	19.74	2.13	7.89	2.30	5.27	2.25
41	16.13	19.81	2.06	7.82	2.28	5.20	2.17
42	16.23	19.87	1.99	7.75	2.21	5.13	2.10
43	16.33	19.94	1.92	7.69	2.14	5.06	2.04
44	16.43	20.00	1.86	7.62	2.07	5.00	1.97
45	16.53	20.07	1.79	7.56	2.00	4.94	1.90

PROBABILITY OF DETECTION= .9000
CONSTANT FALSE ALARM PROBABILITY
LOG10(PFA)=-10

N	N DB	THRESH DB	NON-FLUC DB	SWER 1 DB	SWER 2 DB	SWER 3 DB	SWER 4 DB
1	0.00	13.62	15.05	23.37	23.37	19.46	19.46
2	3.01	14.21	12.41	20.78	16.86	16.86	14.83
3	4.77	14.65	10.92	19.32	14.06	15.39	12.63
4	6.02	15.01	9.88	18.31	12.35	14.38	11.23
5	6.99	15.33	9.10	17.55	11.14	13.62	10.21
6	7.78	15.60	8.47	16.94	10.23	13.00	9.42
7	8.45	15.86	7.94	16.43	9.49	12.49	8.78
8	9.03	16.09	7.50	15.99	8.88	12.05	8.24
9	9.54	16.30	7.11	15.62	8.36	11.67	7.78
10	10.00	16.50	6.76	15.28	7.91	11.33	7.38
11	10.41	16.68	6.45	14.98	7.51	11.03	7.02
12	10.79	16.85	6.17	14.71	7.15	10.76	6.69
13	11.14	17.02	5.91	14.46	6.83	10.51	6.40
14	11.46	17.17	5.67	14.23	6.54	10.28	6.14
15	11.76	17.32	5.46	14.02	6.27	10.07	5.89
16	12.04	17.46	5.25	13.82	6.03	9.87	5.66
17	12.30	17.60	5.06	13.64	5.80	9.69	5.45
18	12.55	17.73	4.89	13.47	5.59	9.51	5.26
19	12.79	17.85	4.72	13.31	5.39	9.35	5.07
20	13.01	17.97	4.56	13.15	5.20	9.20	4.90
21	13.22	18.08	4.41	13.01	5.03	9.05	4.74
22	13.42	18.20	4.27	12.87	4.86	8.92	4.58
23	13.62	18.30	4.14	12.74	4.71	8.79	4.44
24	13.80	18.41	4.01	12.62	4.56	8.66	4.30
25	13.98	18.51	3.88	12.50	4.42	8.54	4.16
26	14.15	18.61	3.77	12.39	4.28	8.43	4.04
27	14.31	18.70	3.65	12.28	4.15	8.32	3.91
28	14.47	18.79	3.55	12.17	4.03	8.21	3.80
29	14.62	18.88	3.44	12.07	3.91	8.11	3.69
30	14.77	18.97	3.34	11.97	3.80	8.01	3.58
31	14.91	19.06	3.24	11.88	3.69	7.92	3.48
32	15.05	19.14	3.15	11.79	3.58	7.83	3.38
33	15.19	19.22	3.06	11.70	3.48	7.74	3.28
34	15.31	19.30	2.97	11.61	3.38	7.66	3.19
35	15.44	19.38	2.89	11.53	3.29	7.58	3.10
36	15.56	19.45	2.81	11.45	3.20	7.50	3.01
37	15.68	19.53	2.73	11.38	3.11	7.42	2.92
38	15.80	19.60	2.65	11.30	3.02	7.35	2.84
39	15.91	19.67	2.58	11.23	2.94	7.27	2.76
40	16.02	19.74	2.50	11.16	2.86	7.20	2.69
41	16.13	19.81	2.43	11.09	2.78	7.13	2.61
42	16.23	19.87	2.36	11.02	2.70	7.06	2.54
43	16.33	19.94	2.29	10.96	2.63	7.00	2.47
44	16.43	20.00	2.23	10.89	2.56	6.93	2.40
45	16.53	20.07	2.16	10.83	2.48	6.87	2.33

Appendix: Radar Detection Tables

PROBABILITY OF DETECTION= .9500
CONSTANT FALSE ALARM PROBABILITY
LOG10(PFA)=-10

N	N DB	THRESH DB	NON-FLUC DB	SWER 1 DB	SWER 2 DB	SWER 3 DB	SWER 4 DB
1	0.00	13.62	15.44	26.51	26.51	21.24	21.24
2	3.01	14.21	12.78	23.92	18.64	18.64	15.96
3	4.77	14.65	11.28	22.46	15.40	17.17	13.53
4	6.02	15.01	10.24	21.45	13.47	16.16	12.00
5	6.99	15.33	9.45	20.69	12.12	15.39	10.91
6	7.78	15.60	8.82	20.08	11.11	14.78	10.06
7	8.45	15.86	8.29	19.57	10.30	14.27	9.38
8	9.03	16.09	7.84	19.13	9.64	13.83	8.81
9	9.54	16.30	7.44	18.75	9.08	13.45	8.32
10	10.00	16.50	7.09	18.42	8.59	13.11	7.90
11	10.41	16.68	6.78	18.12	8.16	12.81	7.52
12	10.79	16.85	6.50	17.84	7.78	12.53	7.19
13	11.14	17.02	6.24	17.59	7.44	12.28	6.88
14	11.46	17.17	6.00	17.37	7.13	12.05	6.60
15	11.76	17.32	5.78	17.15	6.85	11.84	6.35
16	12.04	17.46	5.57	16.96	6.58	11.64	6.11
17	12.30	17.60	5.38	16.77	6.34	11.46	5.89
18	12.55	17.73	5.20	16.60	6.12	11.29	5.69
19	12.79	17.85	5.04	16.44	5.91	11.12	5.50
20	13.01	17.97	4.88	16.29	5.71	10.97	5.32
21	13.22	18.08	4.73	16.14	5.53	10.83	5.15
22	13.42	18.20	4.58	16.01	5.35	10.69	4.99
23	13.62	18.30	4.45	15.88	5.19	10.56	4.84
24	13.81	18.41	4.32	15.75	5.03	10.43	4.70
25	13.98	18.51	4.19	15.63	4.88	10.31	4.56
26	14.15	18.61	4.07	15.52	4.74	10.20	4.43
27	14.31	18.70	3.96	15.41	4.61	10.09	4.30
28	14.47	18.79	3.85	15.31	4.48	9.98	4.18
29	14.62	18.88	3.75	15.21	4.36	9.88	4.07
30	14.77	18.97	3.64	15.11	4.24	9.78	3.96
31	14.91	19.06	3.55	15.02	4.12	9.69	3.85
32	15.05	19.14	3.45	14.93	4.01	9.60	3.75
33	15.19	19.22	3.36	14.84	3.91	9.51	3.65
34	15.31	19.30	3.27	14.75	3.81	9.43	3.55
35	15.44	19.38	3.19	14.67	3.71	9.34	3.46
36	15.56	19.45	3.10	14.59	3.61	9.27	3.37
37	15.68	19.53	3.02	14.52	3.52	9.19	3.28
38	15.80	19.60	2.95	14.44	3.43	9.11	3.20
39	15.91	19.67	2.87	14.37	3.35	9.04	3.12
40	16.02	19.74	2.80	14.30	3.26	8.97	3.04
41	16.13	19.81	2.73	14.23	3.18	8.90	2.96
42	16.23	19.87	2.66	14.16	3.10	8.83	2.89
43	16.33	19.94	2.59	14.10	3.02	8.77	2.81
44	16.43	20.00	2.52	14.03	2.95	8.70	2.74
45	16.53	20.07	2.46	13.97	2.88	8.64	2.67

PROBABILITY OF DETECTION= .9900
CONSTANT FALSE ALARM PROBABILITY
LOG10(PFA)=-10

N	N DB	THRESH DB	NON-FLUC DB	SWER 1 DB	SWER 2 DB	SWER 3 DB	SWER 4 DB
1	0.00	13.62	16.12	33.60	33.60	25.06	25.06
2	3.01	14.21	13.45	31.00	22.46	22.46	18.25
3	4.77	14.65	11.93	29.53	18.18	21.00	15.30
4	6.02	15.01	10.88	28.52	15.74	19.98	13.50
5	6.99	15.33	10.08	27.76	14.09	19.21	12.25
6	7.78	15.60	9.43	27.14	12.88	18.59	11.29
7	8.45	15.86	8.90	26.63	11.92	18.08	10.52
8	9.03	16.09	8.44	26.20	11.13	17.67	9.90
9	9.54	16.30	8.04	25.82	10.48	17.28	9.35
10	10.00	16.50	7.69	25.48	9.92	16.94	8.88
11	10.41	16.68	7.37	25.20	9.42	16.64	8.47
12	10.79	16.85	7.08	24.93	8.99	16.36	8.10
13	11.14	17.02	6.82	24.68	8.60	16.11	7.78
14	11.46	17.17	6.58	24.45	8.26	15.88	7.47
15	11.76	17.32	6.35	24.24	7.94	15.67	7.20
16	12.04	17.46	6.14	24.05	7.65	15.47	6.95
17	12.30	17.60	5.95	23.86	7.38	15.29	6.71
18	12.55	17.73	5.77	23.69	7.13	15.12	6.49
19	12.79	17.85	5.60	23.53	6.89	14.95	6.28
20	13.01	17.97	5.43	23.38	6.68	14.80	6.09
21	13.22	18.08	5.28	23.23	6.47	14.66	5.91
22	13.42	18.20	5.14	23.10	6.28	14.52	5.74
23	13.62	18.30	5.00	22.97	6.10	14.39	5.58
24	13.80	18.41	4.87	22.82	5.93	14.26	5.43
25	13.98	18.51	4.74	22.70	5.77	14.14	5.28
26	14.15	18.61	4.62	22.58	5.62	14.03	5.15
27	14.31	18.70	4.50	22.47	5.47	13.92	5.01
28	14.47	18.79	4.39	22.37	5.33	13.81	4.89
29	14.62	18.88	4.29	22.27	5.19	13.71	4.76
30	14.77	18.97	4.19	22.17	5.06	13.61	4.65
31	14.91	19.06	4.09	22.08	4.94	13.52	4.53
32	15.05	19.14	3.99	21.99	4.82	13.43	4.43
33	15.19	19.22	3.90	21.90	4.71	13.34	4.32
34	15.31	19.30	3.81	21.82	4.60	13.26	4.22
35	15.44	19.38	3.72	21.73	4.49	13.18	4.12
36	15.56	19.45	3.63	21.66	4.39	13.07	4.03
37	15.68	19.53	3.55	21.58	4.29	13.00	3.94
38	15.80	19.60	3.47	21.50	4.19	12.92	3.85
39	15.91	19.67	3.39	21.43	4.10	12.85	3.77
40	16.02	19.74	3.32	21.36	4.01	12.78	3.68
41	16.13	19.81	3.25	21.29	3.92	12.71	3.60
42	16.23	19.87	3.18	21.23	3.84	12.64	3.52
43	16.33	19.94	3.11	21.16	3.76	12.57	3.45
44	16.43	20.00	3.04	21.10	3.68	12.51	3.37
45	16.53	20.07	2.98	21.03	3.60	12.45	3.30

Index

Accuracy, 35–38, 39–40
Airborne radar, 103, 126
Aircraft, 13, 69
Ambiguity, 83–85, 88–89, 103
Angular resolution, 48
Antenna: auxiliary, 113; beamwidth, 35; and clutter, 53, 69–70, 80, 103, 106; efficiency, 4; and jamming, 112, 113; and loss, 19, 33, 106, 109; and motion, 69–70; pulse bursts, 10; and pulse-Doppler, 83, 92; and range, 3; receive gain, 3, 5, 19, 80, 103, 106, 109; and temperature, 14, 15; transmit gain, 3, 4–5
Ares, M., 89, 99–100
Array antennas, 4–5, 33, 92
Aspect angles, 123, 125–126, 128
Atmospheric conditions, 6; and attenuation, 30, 103, 106, 109; and loss, 30, 112

Backscatter, 74, 79, 125
Bandwidth, 24, 26, 40, 123
Barrage jamming, 111–112
Barlow, E.J., 58
Barton, D.K., 23, 30, 37, 38, 40, 41, 62, 75, 90
Beam shape loss, 20–22
Beamwidth, 70; and accuracy, 35–36; and antenna, 35; and loss, 19–20, 21–22
Blake, L.V., 12, 30
Blind speeds, 45–46, 47, 50–51, 70, 83
Boltzman's constant, 3, 15, 111
Bradley, G.A., 20
Brennan, L.E., 14

Cancelers, 43, 45, 46–47, 58; multiple, 49–50, 53; and phase instabilities, 65–66; and pulse jitter, 67–68; and second-time-around clutter, 71–73; and spectral density, 105
Cathode ray tube, (CRT) display, 9, 23
Chaff, 58–62, 75

Closing velocity, 13, 14
Clutter: and antenna, 53, 69–70, 80, 103, 106; and attenuation, 51–52; ground, 58, 75–76, 103, 116; improvement factor, 53, 73–74, 81–82; magnitude, 74–82; and MTI operation, 43–82, 103–110; precipitation, 30, 56–58, 74, 75; and pulse-Doppler, 82–102; and range, 76, 79, 81, 103; sea, 56, 74–75, 124; and sidelobes, 80–81; and velocity, 63, 82, 83, 86, 103; and wind, 56–58, 74
Clutter-locking technique, 48–49, 53, 58
Coding, 116
Coherent Memory Filter, 95
Cohn, G.I., 95
Collapsing loss, 23–24
Comb filters, 27
Computers, 92, 103–110; and radar cross-section, 123–124
Cone-cylindrical targets, 124, 125
Conical-scan radar, 37, 38
Constant false alarm rate (CFAR) action, 95, 134–135
Cruise missiles, 124
CW radar, 82

Deception repeaters, 117
Decorrelation, 9
Delay line cancelers, 43, 45, 95. *See also* Cancelers
Detection, probability of, 6, 7, 12, 22; and pulse-Doppler operation, 95; and radar cross-section models, 123–125, 130; and signal-to-noise ratio, 133–138
Dolph-Chebyshev weighting, 88
Doppler frequency, 49, 55–58; and MTI, 45, 47–48, 53–56. *See also* Pulse-Doppler

Electronic countermeasures, 111–117
Error averaging, 38. *See also* Accuracy

281

False alarm, probability of, 6, 7, 12; and bandwidth, 24; and detection probability, 133–138; and pulse-Doppler, 95; and pulse number, 95
False alarm rate, 134–135, 137
Fan-beam radar, 6, 20, 32, 35–36
Fehlner, L.F., 133–134
Filtering: and accuracy, 37; and clutter, 43; comb, 95, 97–98; feedback, 95; IF, 90, 91–92; losses, 11, 24–26, 27–30, 98, 106, 108–109; matched, 24, 25, 38, 86, 88, 89; and pulse-Doppler operation, 82, 84, 86, 88, 90–95, 97; rectangular, 91, 105–106
Flesher, G.T., 95
FM coding, 116
Fourier transform, 105
Frequency diversity, 113
Frequency hopping, 113

Galejs, J., 95, 97–98
Gaussian beam shape, 36
Grisetti, R.S., 58
Ground clutter, 58, 75–76, 103; and pulse-width discrimination, 116

Hamming amplitude weighting, 88

Input-noise temperature, 3, 14–17, 111–112
Integration, 9, 10–12, 14, 134, 138; and pulse-Doppler, 82, 90–95, 97; time, 92
Intrapulse coding, 116

Jamming. *See* Electronic countermeasures

Kirkpatrick, G.M., 58
Krason, H., 75–76

Lawson, J.L., 12, 30–31
Log-periodic arrays, 5
Losses, 14, 15, 16–33; and antenna, 19, 33, 106, 109; and atmosphere, 30, 112; and beamwidth, 19–20, 21–22; and cathode ray tube display, 23; and filtering, 11, 24–26, 27–30, 98, 116, 108–109; and pulse-Doppler, 97–98; and range gating, 11, 23, 26–27, 98, 106, 108–109; and receiver mismatch, 24–26, 30; and scanning, 19–20

Mallett, J.D., 14
Manasse, R., 41
Marcum, J.I., 7, 23, 24, 91, 133–136
Microwave component, 18, 19
Microwave receive network, 14, 15, 16
Missiles, 13, 30
Monopulse radar, 37–38
Morris, J.D., 137
Motion of target, 13, 43–44, 65
MTI, 43; and amplitude processing, 46–47, 50, 69; and antenna, 69–70; clutter-locked, 48–49, 53, 58; coherent, 43–45; digital, 51; gain, 108–109; improvement factor, 73–74; limitations on, 51–73; and multiple-canceler systems, 49–50; and phase-processing, 43–46, 47, 49, 50, 65–66, 67–69; and pulse-Doppler, 105–106; and pulse repetition frequency, 45, 50–51, 70, 103–105; and range gating, 103–110; vector processing, 47–48; and velocity, 103
Multiple-pulse waveforms, 4

Nathanson, F.E., 43, 82
Newton-Raphson iteration, 137
Noise capture, 11
Nonrectangular pulses, 4
North, D.O., 97

Observations, number of, 6, 13–14
Off-broadside loss, 33, 112
Operator: integration, 12; loss, 30–33

Parameter discrimination, 113–117
Pattern loss. *See* Beam shape loss
Pencil beam radar, 20, 22, 36–38
Phase: averaging, 49; blind, 47, 50; instabilities, 100; and MTI

Index

operation, 43–46, 47, 49, 50, 65–66, 67–69; and pulse-Doppler operation, 97, 100
Polarization, 123, 125, 126
PPI shape, 35
Precipitation: clutter, 57–58, 74, 75; loss, 30, 112
Propagation factor, 3, 6
Pulse bursts, 10, 82
Pulse compression, 4, 116
Pulse-Doppler, 43, 82–102; and ambiguity, 83–85, and amplitude, 88–89, 100; and antenna, 83, 92; and detection probability, 133–138; efficiency, 97–98; and energy, 95–98; and interpulse period, 83–86, 95, 99–100; limitations, 100–102; and phase, 97, 100; and pulse train, 83–88, 91–95; and range, 82, 106; and range gating, 84, 92, 98, 103–110; and receivers, 90–95
Pulse jamming, 111–117
Pulse jitter, 67–69
Pulse repetition frequency (PRF), 12, 42; and jamming, 113–114; and MTI operation, 45, 50–51, 70, 103–105; and pulse-Doppler, 83; staggered, 70; and visibility, 110
Pulse trains, 42, 82; and pulse-Doppler, 83–88, 91–95; staggered, 86–87, 88; uniform, 83–86

Quadrature processing, 48, 92, 103

Radar cross-sections, 107; averaging, 126–130; models, 123–130
Radiation Laboratory Series, 12
Rain, 30, 75
Randig, G., 75–76
Range: and antenna, 3; and clutter, 76, 79, 81, 103; equations, 3–33; measurement accuracy, 39–40; and pulse-Doppler, 82, 106; and resolution, 40–41; and signal-to-noise ratio, 3, 6, 11, 14, 133
Range gating: and accuracy, 37; losses, 11, 23 26–27, 98, 106, 108–109;

and MTI, 103–110; and pulse-Doppler operation, 84, 92, 98, 103–110; straddling, 26–27, 106
Receiver, 106, 108, 113
Receiver-mismatch loss, 24–26, 30
Reflector antennas, 4
Reilly, J.P., 43
Resnick, J.B., 85, 86
Resolution, 38, 40–41, 42
Rotary joints, 4

Santa, M.M., 58
Satellite tracking, 30, 124
S/C ratio. *See* Signal-to clutter ratio
Scanning loss, 19–20
Scanning radar, 7–9, 10
Scintillation effects, 5, 6, 7–9
Sea clutter, 56, 74–75, 124
Search radar, 90
Second-time-around clutter, 70–73
Self-screening jammer, 112
Sequential-lobing radar, 37, 38
Sidebands, 100
Sidelobe, 80–91, 85; blanking, 113; and jammer, 112, 113
Signal-to-clutter ratio, 98–100, 103, 106, 108–109
Signal-to-noise (S/N) ratio; and accuracy, 35–36, 37, 38, 39, 41; "average," 124; and detection probability, 133–138; and integration, 14, 97–98; and loss, 22, 106–107; and motion of target, 13; and range, 3, 6, 11, 14, 133
Skolnick, M.I., 30–31, 39, 50, 75
Snow, 30
Spatial discrimination, 113
Squint angle, 37–38
Stack factor, 21–22
Stationary targets, 13
Step-scanning radars, 73
Swerling, P., 7–9, 35–36, 91, 133–134

Taylor amplitude weighting, 88
Thermal noise, 39–40
Time jitter, 100
Tracking radar, 82, 90

Transmitter: instabilities, 103, 106, 108; noise, 97, 108

Uhlenbeck, G.E., 12, 30–31

Velocity, 63, 82 83, 86; closing, 13, 14; and MTI operation, 103
Video display, 9, 23, 24, 32

Wainstein, L.A., 48
Waveforms, 3, 4, 6; and pulse-Doppler operation, 84–85, 86, 88–89, 100, 105–106; spectral density, 103–106
Wideband noise. *See* Barrage jamming
Wind and clutter, 56–58, 74

Yagi arrays, 5

Zubakov, V.D., 48

About the Author

Christian G. Bachman, D.Sc., P.E., is the founder and president of KRYBOK Associates, a technical management consultancy located in Wilmington, Massachusetts. He has over twenty-five years' experience in engineering and management of programs in electronic-systems technology, and is the author of *Laser Radar Systems and Techniques* and *Radar Targets*.

He received the B.S.E.E. from Oregon State University, the M.S.E.E. from the State University of New York, and the D.Sc. in engineering from California Western University. He has completed additional graduate work at the University of California at Los Angeles, Cornell University, and the University of Michigan. He is a graduate of the National Defense University's Industrial College of the Armed Forces, and a registered professional engineer in Massachusetts.